IMBALANCE OF POWER

IMBALANCE OF POWER

An Analysis of
Shifting U.S.-Soviet Military Strengths
by
JOHN M. COLLINS

And

Net Assessment Appraisal
by
ANTHONY H. CORDESMAN

San Rafael, California
London, England

PRESIDIO PRESS
1114 Irwin Street
San Rafael, California
94901

Library of Congress Cataloging in Publication Data

Collins, John M
Imbalance of power.

Includes bibliographical references.
1. United States—Foreign relations—Russia. 2. Russia—Foreign relations—United States. 3. United States—Armed forces. 4. Russia—Armed forces. 5. North Atlantic Treaty Organization. I. Cordesman, Anthony H., joint author. II. Title.
E183.8.R9C57 355.03'3073 77-91933
ISBN 0-89141-059-7

Printed in the United States of America

To Harry
for many reasons

PUBLISHER'S NOTE

This book is divided into two parts which run concurrently. The basic study compares United States and Soviet military capabilities and was written by John M. Collins, strategic analyst of the Research Staff of the Library of Congress.

The second part, by Anthony H. Cordesman, former assistant to the Deputy Secretary of Defense and Secretary of the Defense Intelligence Board, consists of a preface and several net assessment appraisals which are interspersed throughout the Collins study. These appraisals appear in a different type face from the basic study and are further identified by a band on each outside margin.

CONTENTS

FOREWORD

by Senator Howard H. Baker, Jr.

In ancient Greece, the tyrant of Syracuse is reputed to have forced Damocles, a member of his court, to eat a banquet under a sword suspended by a single thread. He did this to demonstrate the precariousness of even a king's fate. This story is probably a myth, but there is nothing mythical about the sword suspended over the free world. There are now Russian submarines patrolling off our coasts whose missiles could strike our cities with negligible warning. Russian ICBMs could strike every American city many times over in less than an hour. More than half of our population could be dead in less time than it took to transmit a declaration of war in World War II.

We have learned to live under this sword with confidence and hope. Unlike Athens in the Peloponnesian War, our freedoms have grown under the stress of the thirty years of the nuclear era. We have real allies, not captive clients or dependents. We have learned we can defend ourselves, but show restraint. We have learned that other nations can be both free and neutral. We have learned that we can pay the cost of national defense—and still improve the equality and quality of our society. We have learned we can defend without ever ceasing the search for arms control and peace.

We sometimes forget, in the agony of meeting each new challenge this era brings, that we have faced more challenges successfully since 1945 than any other society in history. We have made many mistakes, and we will make many more, but as a nation we have grown under pressure. No one can look back on the last generation and doubt our courage, or our progress.

But this growing freedom and maturity has a price. We are constantly dependent on our military strength to minimize the risk that nuclear weapons will be used against the United States. We are dependent on the military strength not only for our lives, but for that added margin of security that allows us to pursue broad social goals, to permit wide ranging dissent and differences of view, to defend our allies, to give other nations the choice of independence and neutrality, and to pursue arms control without weakness. A secure military balance is the fundamental price we must pay to pursue still other, even greater objectives.

We have tended to lose sight of this reality in recent years:

- A long period of strategic nuclear superiority has led many of us to take our fundamental security for granted. But we have no monopoly on technological innovation, tactics of the modern battlefield, or strategic planning. We have gradually accepted Soviet strategic nuclear parity, focusing on hopes of detente and a new limit on strategic arms. As a society, we have failed to focus on the consequences of future competition with a Soviet Union which is willing to commit vast resources to increasing its superior strategic power and, to many observers, achieve strategic nuclear superiority. We have let hope for the future blind us to the fact that we still face decades of military competition before there is any probability that an armed world of hostile nations can be united into a secure peace.
- The military balance grows increasingly complex, more difficult to analyze and understand as each new development surfaces. Even with the full resources of Congress, it is difficult to cut through the morass of technical terms and special interest pleading, to determine how new weapon systems (ours, and the Soviets') will affect our present and future security. Congress is only now coming to realize that it must shape defense policy through a comprehensive assessment of the balance, and that it cannot focus on technical features, price tags, or individual programs without examining how these affect our overall military capabilities.

The two studies that make up this book probably provide the most authoritative, complete assessment of the U.S./Soviet balance that the American citizen and Congress have had available in unclassified form. They describe not only the current balance, but the broad trends in the balance that will shape our future national security.

John Collins' study provides a unique service in making available to the public the information necessary for an informed and open debate on the defense needs of this nation and its allies.

Anthony Cordesman has added the new techniques of net assessment to provide a different view of that balance focusing on the different ways that military strength can be counted and military capabilities can be measured. He has looked beyond force numbers to differences in strategy, tactics, and doctrine.

Neither author would claim that his work is comprehensive or without error. John Collins' study, for instance, was never intended to stand on its own, but was commissioned by Congress as a companion piece to his earlier works on Soviet and U.S. national defense issues. Both authors have emphasized the uncertainties in their judgments and analysis. However, I feel that both can take pride in providing a new beginning for strategic studies.

I commend this book because I believe that all Americans should become more concerned with strategic studies and with their military security. They cannot, like Damocles, rely on indifference and hope, and let fate shape their future. They should understand the military balance, and John Collins and Anthony Cordesman have given them a sophisticated, intellectual framework with which to shape that understanding.

We forget how much of every tax dollar we pay is dedicated to the cost of wars—past, present, and future—particularly, if we count the cost of veterans' benefits and interest on our national debt (which derives largely from the cost of military conflict). We forget that four out of every ten male citizens over eighteen years of age have either worn or are wearing the uniform of our country.

Thus, it is important that every American pay close attention to the trends in U.S. and Soviet military strength shown in the Collins-Cordesman work. John Collins warns us of a present danger, but I feel the most critical question lies in whether we will maintain in the future the determination and strength necessary to deter the Soviet Union, and to insure a peaceful world.

My concern is that we renew our commitment to the future, and have no illusions about the nature or character of the Soviet Union's unprecedented military buildup. We must pursue a safe and meaningful SALT II agreement; we must pursue safe and meaningful mutual force reductions between NATO and the Warsaw Pact; and we must not over-react to the militarism of a totalitarian Soviet society that lacks adequate political and economic mechanisms for converting its growing economic wealth into consumer goods and an improved life for its citizens. But, we must also not let Soviet power grow to the point where our deterrents weaken, where our resolve and capability are uncertain, and where we face a rising threat of war.

It is my hope that without over-reacting to military issues or obsessive concern with our security, America and the western world will retain our recognition of the threat posed by the rapid changes toward an increasingly unfavorable military balance. If we plan prudently, we can meet the challenges of strategic nuclear parity, the growing strength of the Warsaw Pact, and the problems of the military in the post-Viet Nam era; but to do so, we need to understand the dynamics which John Collins and Anthony Cordesman present here.

November 21, 1977

PREFACE

by ANTHONY H. CORDESMAN

There have been few attempts to compare the overall military balance between the United States and the Soviet Union, since it is a complex task, and one requiring massive research. No matter how dedicated and successful the research effort, no comparison can treat every major factor, describe every major issue and uncertainty, and apply every relevant methodology. A full assessment of the U.S.-Soviet military balance is as impossible to achieve as it is important to attempt.

It is also not surprising that many of the attempts made to compare U.S. and Soviet forces have been designed to support given political or policy views. Few analysts or sponsors are likely to take on such a challenge without a clear motive or purpose.

THE MILITARY BALANCE, published annually by the International Institute for Strategic Studies (I.I.S.S.), is the only such attempt which has established a wide-spread reputation for objectivity. Most others have been little more than pamphlets or tracts, and many have been published to advocate conservative or liberal positions on national security policy.

John M. Collins's study of contemporary trends in the U.S. and Soviet military balance goes far beyond the work done by the I.I.S.S. in its annual publication. It provides both detail and depth which the I.I.S.S. cannot provide in its limited treatment of U.S. and Soviet forces in its assessment of all the world's military forces.

The result is a pioneering attempt to provide an unclassified analysis of major factors which affect the balance without advocating a particular political or policy view. His study draws directly on the Department of Defense and the U.S. intelligence community for most of his data, rather than on secondary sources or his own opinions. He arrays data on the balance which have never before been presented in one report, and ties together many major factors which shape U.S. and Soviet military capabilities.

As Collins notes, no pioneering effort can be comprehensive, or free of institutional and unintentional bias. Collins notes in his introduction, for instance, that he must rely on unclassified data and this limits what he can discuss. He must rely on intelligence on Soviet and Communist forces which suffers from significant institutional bias. He must be selective and omit comparisons which might lead to different views and conclusions. And, he must use simple quantitative comparisons as his methodology and omit the use of advanced analytic techniques which might often provide more useful insights into the balance.

His analysis of force trends is also not an attempt at net assessment, but rather an attempt to catalogue key measures of current military forces. Inevitably, this means that the study must ignore important differences in the strategic position of the United States and the Soviet Union, because it was never intended to assess the broad cultural, ideological, economic, and technological factors which ultimately shape Soviet and U.S. military capability, or the historical factors and trends which have shaped each side's present force posture. Collins chooses to provide depth on the composition of current forces rather than a broad historical perspective.

There are many areas where more detail could be provided, where other experts have different views, or even where other portrayals of the balance seem equally valid. The net assessment appraisals which I have added to his are my attempt to broaden the scope of the analysis in this book, and to explain alternative views or data when these seem particularly relevant.

All studies of the balance must be written, however, in the face of serious problems which limit any attempt at net assessment. The greatest weaknesses in the study result from basic problems in the state of the "art" which no analyst can currently overcome.

A. THE PROBLEM OF INTELLIGENCE

No study of the balance can ultimately be better than the data that are released by, or "leaked" from, United States intelligence agencies. Moreover, at least until the recent intelligence reorganization by the Carter Administration, only the Defense Intelligence Agency (DIA) had the charter and staff required for comprehensive counts of Soviet force strength. The Central Intelligence Agency, the military service intelligence staffs, and other elements of the U.S. intelligence community made major contributions to analyzing Soviet military forces, as did analysts outside the intelligence community and in other nations, but DIA has been the key link in shaping all free world estimates of Soviet force strength.

Unfortunately, this often leads to an exaggeration of Soviet capabilities, or more importantly, to an exaggeration of Western knowledge of Soviet capabilities. DIA tends to exaggerate other aspects of threat capability; this tendency is encouraged by official users of defense intelligence. DIA tends to credit the Soviet Union with capability when it does not know, and has a long tradition of providing answers whether it has sufficient data or not. It also has a tendency to "mirror image" Soviet capabilities against those of U.S. forces or technology when it lacks actual intelligence, without indicating that such "mirror imaging" is the actual source of its estimates. And, these tendencies are compounded by other problems which affect the validity of intelligence estimates:

1. Both military and civilian bureaucracies need high estimates of the threat to justify force levels, new weapons, and defense research. With some exceptions, most users of intelligence want high estimates of the threat.
2. Intelligence officers are compartmented specialists. They often lack practical experience with the real world problems in the threat forces they describe. They lack the background and training to judge what might go wrong with threat forces and plans.
3. Few intelligence officers have extensive training in measuring military effectiveness. They are not familiar with test and evaluation techniques, historical research on weapons or force effectiveness, or operations research. They usually are prevented from comparing U.S. and foreign systems by informal pressures from the Joint Chiefs, the service staffs, or civilian decision-makers.

4. Intelligence officers are rarely required to compare U.S., Allied, and threat forces directly. In general, they generate data using different standards, measurement methods, assumptions, and definitions from United States forces data. These differences often lead to estimates which disguise biases in favor of threat forces. Such biases include exaggerated estimates of threat sortie rates, kill probabilities, rates of fire, readiness, circular errors of probability, system reliability, mobilization and build-up rates, and munitions stocks.
5. DIA evolved from service intelligence branches with a tradition that intelligence counted the strength of the threat and estimated its location, but did not judge its comparative tactical and military effectiveness. This was partly the result of pressures by the more prestigious plans and operations branches of the military services and the Joint Staff to cause the intelligence branches to stay away from estimates reflecting on U.S. capabilities. Accordingly, in spite of recent major efforts at reform, intelligence still tends to concentrate too much on enemy order-of-battle and technical performance of threat equipment, and to pay too little attention to threat training, build-up capability, tactics, operations and maintenance and similar "soft" factors.
6. In contrast, many intelligence officers have personal experience with our allies. They see them (warts and all) and often with more than a touch of American parochialism. Many intelligence users also have no incentive to seek high estimates of Allied capability. The justification for U.S. programs is as much the lack of Allied capabilities as the presence of threat capabilities. This leads to an inverse tendency of U.S. intelligence to underestimate Allied capabilities.
7. Estimates of threat capabilities are increasingly dependent on estimates of technology and weapons systems performance. Many aspects of weapons performance are, however, not even theoretically visible or detectable through intelligence sources. For example, it is extremely difficult to estimate factors like reliability, mean time between failures (MTBF), and equipment availability rates even for U.S. systems until they are proven in war. Few weapons have ever approached their estimated or theoretical technical performance capability in actual combat, yet experts continue to act as if the "next" system would behave without problems.
8. Users have demanded and received intrinsically impossible estimates of threat capabilities which go far into the future, or into unknowable areas of speculation. The Office of Defense Research and Engineering, for example, has forced DIA to make predictions of Soviet capability that go so far into the future where it is unlikely the Soviets have such plans. Since the only data available are U.S. plans or capabilities, DIA is forced to "mirror

image." It is not surprising that the intelligence officers forced to do such work have tended to make guesses which maximize threat capabilities.

9. These tendencies are compounded when intelligence estimates threat capabilities for future years. These involve the greatest areas of uncertainty and are most subject to the tendency to assume high capability in the absence of concrete knowledge. This is why estimates of trends in Soviet forces tend to be so bleak. The enemy we know is invariably preferable to the enemy we will know.

It is unclear that the intelligence community can ultimately be blamed for its biases. The problems that limit the quality of most intelligence estimates are ultimately made by users, and only a few top level users have sought objectivity rather than estimates to meet their policies and needs. Bureaucratic and career pressures that compartment and limit many aspects of U.S. intelligence grow out of a long tradition that intelligence is a servant and not a partner.

Any comment that intelligence tends to exaggerate threat capabilities when it lacks data must also be kept in careful perspective. U.S. intelligence often performs superbly in measuring basic aspects of Soviet forces such as the strength of combat units and major weapons, and in detecting the existence of new Soviet weapons, deployments, and technology. A "bias" or "tendency" does not mean that the basic numbers in the Collins study or these net assessment appraisals are wrong, but when the answer is uncertain, or an issue raised, the estimate sometimes gives the threat high capability. Moreover, as Albert Wholstetter has documented in "Legends of the Strategic Arms Race," USSI Report 75-1, United States Strategic Institute, Washington, D.C., 1975, intelligence can also underestimate critical aspects of the threat. It has done so consistently in estimating future Soviet strategic forces since the famous "missile gap" of the early 1960s.

B. THE PROBLEM OF METHODOLOGY

As his introduction clearly describes, Collins's basic methodology in this study is to compare total current U.S. and Soviet force strengths and similar measures of NATO and Warsaw Pact forces in light of recent trends, and they are easy to understand. They do not require a sophisticated knowledge of tactics, war gaming, or measures of military effectiveness, and are explicit since major weapons and combat units are easy to define and count. They do not involve extensive judgments by the analysts, and the reader can understand and trace most of the analysis. But Collins has added his best judgment to produce a product that differs diametrically in many respects from "official policy."

At the same time, as Collins notes in many places in his study, quasi-"static" force comparisons have major limitations which the user should clearly understand:

1. They cannot accurately reflect differences in training, readiness, morale, and many other critical aspects of military capability.
2. They do not reflect many of the qualitative differences between the equipment compared. This can disguise major differences in performance capability which are significantly more important than equipment numbers.
3. They are not fully explicit. Almost all aspects of force strength can be counted using very different categories and definitions. A count that includes all artillery, for example, disguises critical differences in range, mobility, and crew protection.
4. If the count of weapons or unit strength is not modified by some measure of effectiveness (MOE), it does not indicate the capability of what is compared in war. If the count is modified by such a measure such as firepower score or kill probability, it then becomes judgmental and ceases to be explicit.
5. Most static force comparisons are made of similar types of equipment. Yet, antitank weapons do not fight antitank weapons, bombers do not fight bombers, and ballistic missile submarines do not fight ballistic missile submarines.
6. Comparisons of total national force strengths are often unrealistic in the sense they involve forces which can never engage each other in wars. Wars will inevitably be fought by only a portion of available forces.
7. War is a dynamic and complex process. Units are constantly lost in combat, they maneuver, they reinforce, or they alter in force strength and weapons mix. Even the most sophisticated static force comparison is a "snapshot." It artificially freezes the balance in a given moment, when the real balance shifts over time. No matter how well the analyst chooses his comparison, it is the dynamic process of war which may actually determine comparative military capability.
8. No comparison can count everything. Almost inevitably, the broader the comparison, the more that must be omitted.

Unfortunately, analysts who cannot publish classified information are forced to rely on "static measures" of the balance for most of their analysis. Yet, even those who spend their lives trying to develop such MOEs, with full access to classified information, differ sharply on the value of given measures, and stress the inadequacy or uncertainty of most of the measures available.

Further, the only methodology which can measure the dynamic process of war and its effect on the balance is war gaming and simulation. Yet, for all the theoretical value of computerized and manual war games, they are not feasible tools for analysts who lack the resources of government or government-sponsored research groups, and even the games and simulations available within government do not yet provide a convincing way of modeling or simulating much of the dynamic interaction of forces in war. Even the most complex computerized war game is still an endless series of compromises with reality. Without exception, models of large scale combat or combined arms must grossly abstract or ignore critical factors shaping the balance of the forces compared.

This limitation in war gaming and simulation, or "dynamic" analysis, is all too familiar to experts, but its importance is easy to illustrate to nonexperts as well. Armored warfare has dominated combat in Europe since the German attack on Poland. However, no war game to date can begin to adequately simulate large scale armored maneuvers even if air forces are not played. Most of the advanced war games used in the Pentagon cannot realistically simulate a large scale armored breakthrough, simultaneous land and air warfare, or the major differences in tactics and force structure between individual NATO and Warsaw Pact forces.

The Collins study does not attempt to speculate broadly on the future, but it does spell out significant trends (and thus improves greatly over the static analyses available heretofore). As the net assessment appraisals in this book clearly show, avoiding such speculation is an important way of ensuring that the data presented represent accurate information, and of ensuring that speculation does not go beyond the proper limits of analysis.

At the same time, however, he does not discuss many of the current debates about the future balance which now influence strategic studies, and which are the sources of major differences of opinion regarding United States and Soviet military capabilities. Experts are now debating the following major issues in discussing the implications of current force trends for the future balance:

1. The "Maturing" Soviet Threat

Many experts now feel that the Soviet Union will reach qualitative military parity, or "maturity" with U.S. forces at some point in the early 1980s. They are not concerned with particular trends in the present balance, but rather with the overall implications if the West should lose its past advantage in military technology, individual unit quality, and production capability.

These experts point to evidence that the Soviet Union is spending far more on research and development than the U.S., that its current military equipment production base is larger, and that it is producing more officers and civilians with military-oriented training in sciences and engineering. Further, such experts feel that Soviet force structures, logistics, tactics, training, and maintenance have evolved to the point where Soviet ground forces are generally as "mature" as U.S. ground forces. They also feel that this maturity will steadily tilt the balance in favor of the Soviets, because they will spend more on defense. They will thus steadily upgrade and expand their forces more rapidly than the U.S. Soviet strategic forces are currently roughly equivalent to U.S. forces; Soviet tactical air forces may equal those of the U.S. in the mid-1980s; and Soviet naval forces may overcome severe training, morale, and organizational problems by some time in the late 1980s.

Other experts feel that the Soviet Union cannot catch up with the West this quickly, and argue that current trends still indicate that serious limitations will continue to exist in the quality of Soviet forces and Soviet military technology. The evidence on both sides of this debate is presented in the net assessment appraisals later in this book, and is highly controversial. Yet, there is concensus on the following points:

- If the Soviet Union continues to put more real resources into improving the quality of its forces, it will equal the U.S. in force quality at some time no later than the late 1980s.
- There are no intrinsic ideological or economic problems in the Soviet system that will prevent it reaching the same level of military effectiveness as U.S. forces.

- There are diminishing returns in any continued U.S. attempt to maintain its lead in military capability. As the gross gaps in weapons and force quality narrow, it begins to cost the U.S. a disproportionate amount to maintain or restore a given qualitative lead even where this is technically possible.

The reasons for this consensus go further than analysis of comparative defense expenditures. They are based on the rapid expansion of the Soviet economy, and certain basic resource advantages the U.S.S.R. enjoys over the United States. The following charts compare these aspects of U.S. and Soviet society. They disguise major uncertainties and disagreements among experts, but they do represent the best intelligence available at the time they were written.

Chart One compares the gross national product of the U.S. and U.S.S.R. It indicates a clear U.S. lead in overall economic power using the traditional measure of the value of all goods and services in the economy.

CHART ONE

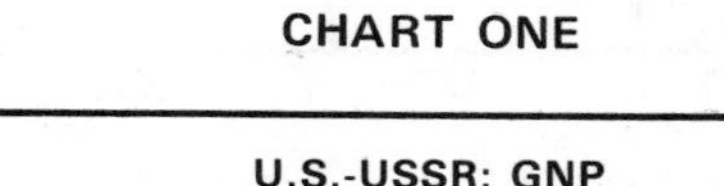

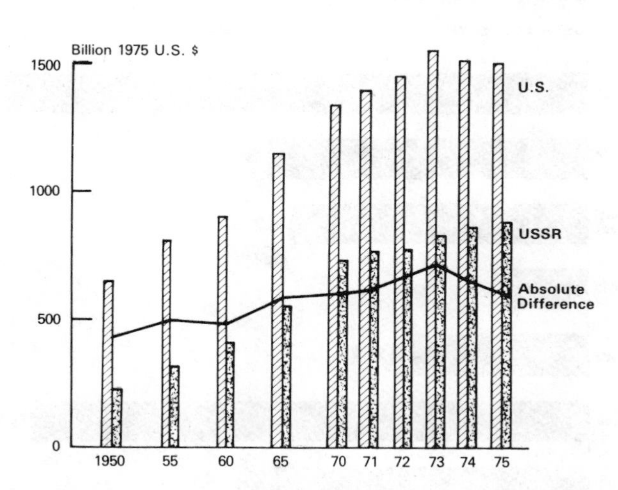

Source: "Allocation of Resources in the Soviet Union and China," Subcommittee on Priorities and Economy in Government, Joint Economic Committee, 94th Congress, 24 May and 15 June, 1976, p. 5

CHART TWO

Comparative Consumer and Service Orientation of the U.s. and Soviet Economies

A. U.S.-USSR: New Fixed Investment

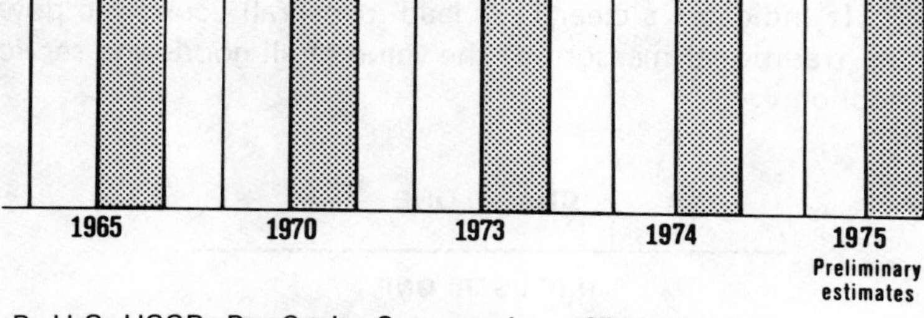

B. U.S.-USSR: Per Capita Consumption, 1974

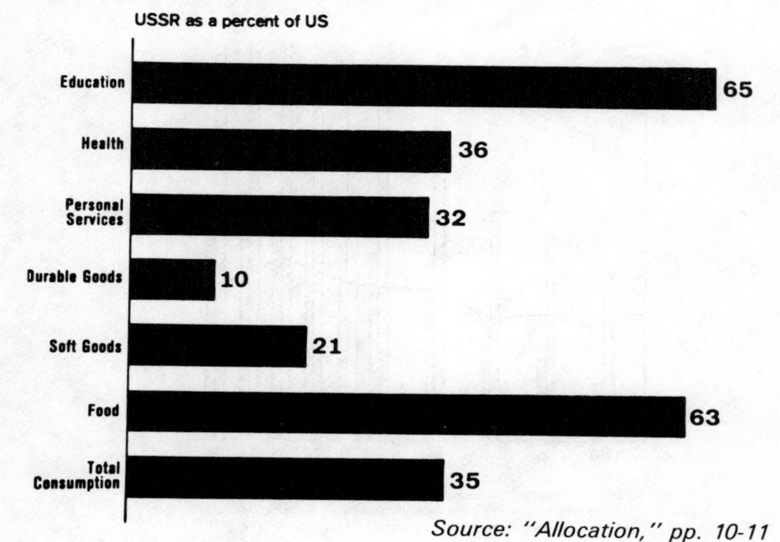

Source: "Allocation," pp. 10-11

The problem is that GNP is a poor measure of a nation's capability to support a defense effort even when comparable and reliable data are available. The U.S. is a consumer economy, and the U.S.S.R. is not. The U.S. is oriented towards consumer services which do not increase basic industrial capacity; the U.S.S.R. is just the reverse. This is illustrated in Chart Two.

It is for this reason that the U.S.S.R., with such a smaller GNP, can exceed the U.S. in industrial growth, in crude steel production, and in the key measures of defense research, development, and production capability discussed later in this book. These overall trends in industrial capacity are illustrated in Chart Three.

CHART THREE

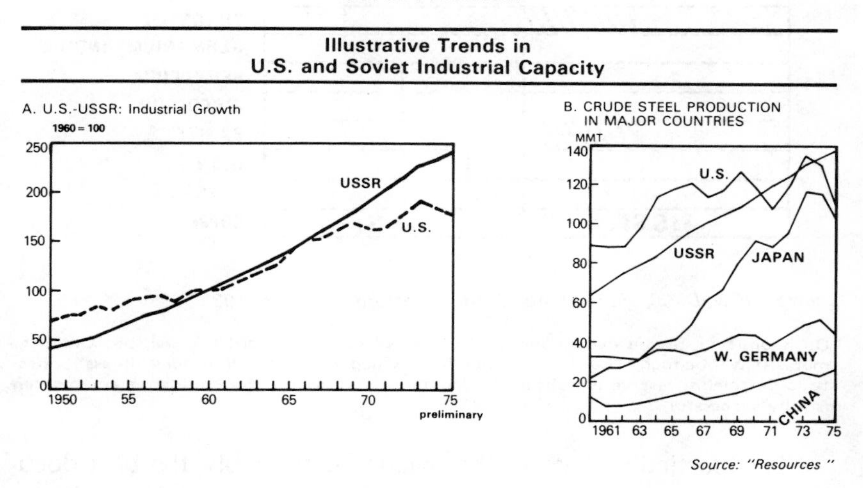

Moreover, while the U.S.S.R. is far more vulnerable than the U.S. to breakdowns of consumer supply and agricultural production, it is immune to dependence on imports of many critical strategic materials, and can allocate these to the defense sector. The relative dependence of each national or critical import is shown in Chart Four.

CHART FOUR

USSR/US DEPENDENCE ON IMPORTS OF STRATEGIC MATERIAL

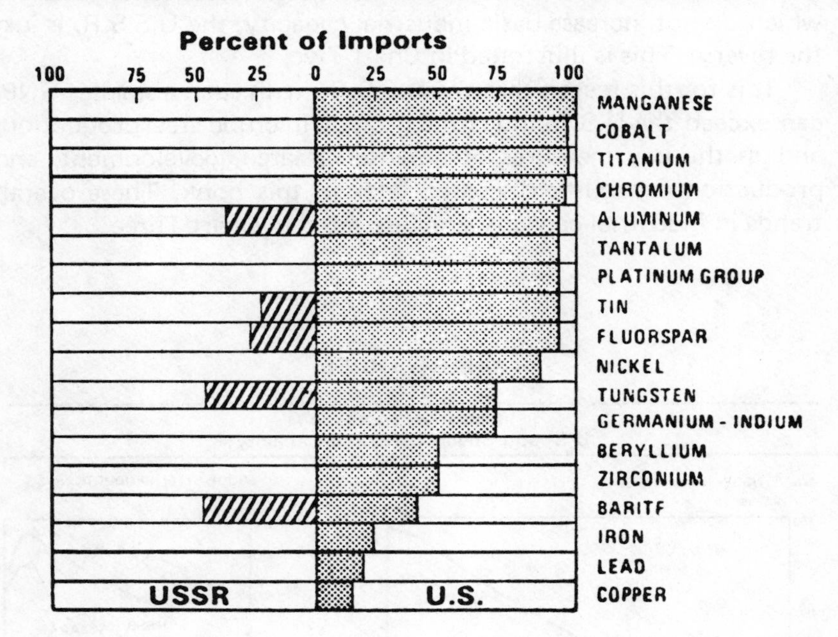

Source: DOD, DJCS, United States Military Posture, FY 78, p. 103

*Oil is omitted. Soviet dependence on future oil imports is controversial, but CIA estimates it will be high. The U.S. will be critically dependent on oil imports unless its strategic petroleum reserve is built up to one billion barrels and the President's energy plan is fully successful.

This particular shift in the balance is probably the best documented, and least controversial, shift projected in U.S. and Soviet military capabilities. Most of the debate over Soviet "maturity" is over "when," rather than "if." However, experts in strategic studies have made three very different interpretations of the result of "maturity":

- The Soviet Union is achieving such "maturity" through incredible sacrifice. It is doing so either to so dominate the military balance as to make it capable of forcing its views and policies upon the world or to wage and win a war with the U.S.

- The Soviet Union has evolved an economic and political structure which concentrates surplus national resources on increasing capital industry and technology, and puts the output into military forces. Unlike the United States, where most increase in wealth goes to the consumer sector, economic growth in the Soviet Union is channeled into the military. For political and bureaucratic reasons, the Soviet Union cannot make major shifts in total resources into the consumer sector. As the U.S.S.R. becomes a "wealthy nation," it creates a militaristic machine with no ultimate purpose, which it simply cannot turn off or bring into proper proportion.
- The Soviet Union has evolved from a state which has constantly been invaded. It has struggled for years to equal the West and has strong feelings of technical and cultural inferiority for which it must now compensate. Its ideology makes it further fear the West, and it is reacting to a past U.S. lead in strategic forces, fears of Germany, and a growing Chinese threat. The U.S.S.R. will never cease improving its military strength until it equals or exceeds the West and China in all aspects of military capability. Its reactions are defensive and xenophobic, not aggressive. Parity will not threaten the U.S., but finally give the U.S.S.R. the security it needs to agree to arms control and a stable balance.

Although there are passionate advocates of each position, the truth is that no expert can know which interpretation is right. All three interpretations probably are partly correct, in that the current Soviet leaders have not fixed on any one objective but react to a shifting mix of all three motivations.

But, this argument has another side. The Soviets must try to interpret the future capabilities of the United States, whose force plans undergo sudden political shifts. U.S. political rhetoric is confusing, and includes many "threatening" statements. At least part of the motive for the "maturity" of Soviet military capabilities may be Soviet inability to understand or predict the course of U.S. strategy and military capabilities.

2. *The Instability of U.S. Force Planning*

Much of the concern that U.S. experts express over the buildup of Soviet forces reflects their concern with problems in U.S. force planning. Put simply, some experts feel the balance is shifting toward the U.S.S.R. because the U.S. government cannot pursue a consistent strategy for improving its forces, and is not using its existing resources effectively. Recent examples of this "United States threat to the United States" include:

- development and cancellation of the B-1 bomber;
- uncertainties and instability in the sizing of the SSBN and SSN force;
- uncertainty over the upgrading and/or replacement of the Minuteman missile;
- uncertainty over the steps to be taken in reducing the vulnerability of strategic forces, and in improving the U.S. command and control system;
- lack of follow-on improvement of tactical nuclear forces other than improvements in nuclear rounds and weapons and the upgrading of the Pershing missile;
- inability to produce and deploy a new tank for more than ten years;
- inability to design, produce and deploy a supporting family of armored fighting vehicles during more than a decade;
- lack of a consistent plan to overcome the major qualitative defects in the artillery mix;
- continuing inability to reach a firm decision on future advanced attack helicopter strength;
- continuing delays in the improvement of anti-armor rounds for tanks, in providing an anti-armor capability for artillery, and in improving anti-armor mine capability;
- lack of clear plans for dealina with chemical and biological warfare;
- lack ot progress in giving the Army anti-air weaponry superior or equal to threat torces, and in deploying replacements for the Nike, Hawk, Vulcan and Chapparral missiles;
- inability to exploit a lead in anti-tank guided missile technology in sufficient numbers, and with suitable crew protection and night or poor weather combat capability;
- lack of Army equipment designed for combat in built-up or urban areas;
- lack of stable plans for future carrier forces and continuing uncertainty as to the configuration and mission of the surface fleet;
- continuing inability to deploy ship-to-ship and surface-to-surface missiles in the fleet in suitable numbers, with suitable capability, and using some clear strategy;
- on-going debate over the nuclear power for the surface fleet and uncertainty as to the "high"-"low" mix of the future fleet;
- major fluctuations in plans to upgrade the air defenses of the fleet;
- lack of any stable plan for developing new surface escorts;
- lack of ability to decide on the modernization of U.S. amphibious assault forces, and continuing uncertainty over future sea lift capability;
- inability to give Marine Corps forces adequate contingency capability against modern armor;
- continuing instability in the overall mix of air defense, air attack, and other tactical air forces to be deployed in the NATO theater;

- inability to rationalize and standardize land, air, or naval tactics and weaponry with NATO and other Allies;
- uncertainty as to the nature of future U.S. strategic mobility;
- uncertainty as to the future forces to be available for combat outside the NATO area, and as to future military commitments;
- growing uncertainty over the future of volunteer and reserve forces.

Any discussion of the reasons for the instability in U.S. force planning would be interminable, and there are arguments for and against virtually every individual decision that has contributed to this instability. The fact remains, however, that the U.S. has seemed less able than the Soviets to translate resources into forces. This probably will continue to be a major factor shaping the future balance.

3. The Other Side of the Future

It is scarcely surprising that most Western analysts of the balance focus on what the Soviet Union might do to radically de-stabilize the future balance, and ignore the Soviet view of the West. The Soviets do, however, write their own assessments of Western forces and, stripped of their turgid rhetoric, these assessments often seem to reflect the same fears and concerns about the West that their Western counterparts show about the U.S.S.R.

As Collins notes throughout his study, the United States has the capacity to quickly make major increases in the size and quality of its strategic forces. No Soviet planner who has read the U.S. debate over the size of U.S. strategic forces should feel any confidence as to what the U.S. will ultimately do. The U.S. can also radically increase its conventional forces, and deploy them in what the Soviets must view as highly unpredictable ways. The American intervention in Vietnam must still puzzle Soviet planners.

Moreover, the "maturity" of Soviet military forces discussed earlier will be achieved, if at all, largely because the West does not choose to compete with the same proportion of its total resources. The West collectively has the manpower, economic capacity, and technical capability to outmatch any conceivable force the Soviets can develop and, unlike the U.S.S.R., the West has no practical reason to fear China in the near future.

The United States and its allies also have an "unpredictable" tendency to change leaders and political objectives. It is impossible to determine how seriously Soviet leaders really take their rhetoric regarding Western "adventurism" or the risk of some "last crisis of capitalism," but it seems likely that behind their rhetoric regarding the "inevitable triumph" of communism lurks a real fear that the West might attack or blunder into war.

The Soviet view of the future balance may, therefore, be as fearful of the West as the West is of developments in the U.S.S.R. Unfortunately, little has been done to determine what the Soviets fear, or how they view their own vulnerabilities. Such assessment is badly needed to put Western fears of the future in proper perspective.

C. THE PROBLEM OF UNINTENTIONAL BIAS:

National security studies almost inevitably are conducted by experts who believe in strong military forces and a strong national defense. They are also written by analysts who would like to make policy, to plan forces, and to direct military strategy. They are conducted by experts who develop their expertise in an environment where a constant competition goes on between national defense and other ways of spending the federal budget.

Only a few defense experts have a different interest in arms control, reducing budgets, or shifting defense resources to other uses. The vast majority of national security experts are conditioned by their work and professional positions to develop an unintentional bias toward larger U.S. forces and toward a pessimistic view of the threat.

As a result, analysts of the balance tend to cease concentrating on the evidence, and shift their focus to attempts to make policy, strategy, budget, and weapons procurement recommendations. They cease to examine the facts, and concentrate on organizing them to prove something. The result is that their resulting assessment of the balance becomes what the Pentagon calls "advocacy analysis": an assessment of the balance which is tailored to "sell," rather than measure each side's capability.

One of the merits of the Collins study is that its biases usually stop at the warning stage, and rarely go on to policy solutions. This is characteristic of the most useful work done to assess the balance.

D. NEXT STEPS IN NET ASSESSMENT:

At the start of this introduction, I referred to this study as a pioneering work. It should be clear that this is not a personal judgment that John M. Collins's assessment of the balance is always objective or comprehensive. At the same time, however, the reader should clearly understand that it approaches the "state of art in assembling basic force numbers," and that many of the problems in the study are the unavoidable result of problems in the institutions from which data must be drawn and in the methodology available for net assessment.

INTRODUCTION

by SENATOR JESSE A. HELMS

(from Senator Helms' address to the Senate, August 5, 1977, requesting consent for the Collins report to be printed in the Congressional Record.)

John M. Collins is Senior Specialist in National Defense with the Congressional Research Service of the Library of Congress. He is well known for his authoritative and definitive analyses of defense problems, particularly the study "United States/Soviet Military Balance, a Frame of Reference for Congress," published last year, and "United States and Soviet City Defense, Considerations for Congress." Both received unanimous favorable comment from every quarter.

Mr. Collins has now prepared a new study, "American and Soviet Armed Services, Strengths Compared, 1970-76." This study is more exhaustive than the previous work on the military balance since it shows the trends for the 6-year period. But more important, for the first time, it assembles in one place a total compendium of unclassified statistics from the best official sources, chosen according to a universal rule of counting.

Those who deal with unclassified statistics are faced with the problem of different sets of numbers which count force levels and equipment available in different ways—some items actually deployed, others in stock or in training use, others without proper distinctions between tactical and strategic deployment, and so forth. Mr. Collins has taken his statistics directly from unclassified DOD computer printouts, and from declassified DIA estimates, and has arranged them in totals of true comparability. The product has been thoroughly reviewed by the defense and intelligence communities, as well as by a number of specialists in the Library of Congress.

Thus for the first time public debate can take place on an unclassified basis with the assurance of reliable and uniform statistics. Those who have access to classified statistics may find some discrepancies, based on different counting methods, the need to protect intelligence sources, or new methods of breaking down statistical components. That is a question, however, that need not inhibit public discussion. Were this the only service that Mr. Collins had performed, the study would still be a major contribution.

It is important to understand also what the Collins study is not. It makes no recommendations, and contains no options. It does not discuss weapons systems which, at the end of 1976, were still under development or assigned only to training squadrons. There is no discussion here of the MX mobile strategic missile system, the B-1 bomber, or the F-18 and F-15 fighters. Such discussions more appropriately belong to Congress, the administration, and the defense community; there will be better discussions as a result of this factual analysis of the trends of comparable strengths between 1970 and 1976.

That debate could well begin on top of the wreckage of some of the myths which Mr. Collins has demolished in this study. One is the myth that levels of defense spending, taken by themselves, are anything more than rough indicators of military effort. On the Soviet side, as Mr. Collins points out, true levels of spending are disguised by spreading military costs into what we would consider the civilian sector of the budget, and there is no accurate method of translating ruble expenditures into dollar expenditures—particularly when we do not know the physical characteristics of Soviet equipment not in our possession, nor their efficiency in producing it.

Far more important, perhaps, is the use to which spending is put. Military postures are presumably shaped to fit a predetermined strategy. The important symmetries are not in spending or in numbers of equipment in various classes, but rather in meeting the actual threat posed by an opponent. What emerges from a study of the trends between 1970 and 1976 is a growing asymmetry between Soviet strategy and U.S. strategy. Debate on military preparedness should not be about budgets, but about strategies.

Mr. Collins himself draws no conclusions, but a study of the data presented reveals, among others, the following asymmetries between the United States and the Soviets:

I. STRATEGIC NUCLEAR ASYMMETRIES

First. Assured destruction is still the heart of our strategic defense concept, but two legs of the Triad are significantly weaker than they were in 1970.

Second. One leg—our undefended fixed-site ICBM's—is weaker because the Soviets are developing a massive counter-silo capability.

Third. The other leg—our strategic bombers—is significantly weaker because of aging equipment and great Soviet strides in air defense across the world: in SAM's, new kinds of interceptor aircraft, and improved warning.

Fourth. Assured destruction itself is significantly degraded because of Soviet strategic defenses, civil defenses that are far ahead of ours—U.S. civil defense being almost nonexistent—and emphasis on ABM research which far exceeds our own.

II. SOVIET GROUND COMBAT POWER ASYMMETRIES

First. Soviet ground combat power dwarfs that of the United States whose forces are so spare that they are stripped of flexibility.

Second. The psychological effect of Soviet ground forces being two or three times that of the United States creates an impression of invulnerability that may or may not be matched at actual combat performance but which may hamper our leadership role.

III. NAVAL OFFENSIVE COMBAT POWER ASYMMETRIES

First. U.S. fleets are in danger of destruction by surprise attacks from Soviet cruise missiles which could fire almost at point blank range. The U.S. Navy has neither any comparable counter-capability nor any defense against such attacks.

Second. U.S. capability for antisubmarine warfare still lacks a major conceptual breakthrough for effectiveness, and, in any case, lacks the sheer numbers which are necessary for successful attrition against the numerically larger Soviet force. These deficiencies constitute a serious threat to U.S. naval and merchant shipping.

IV. NATO DEFENSE CAPABILITY ASYMMETRIES

First. The stark asymmetries in firepower and manpower between NATO and Warsaw Pact forces, including the Soviet forces, is so great that it belies the value of our declared strategy for the protection of Western Europe. Only 5 out of 19 active U.S. divisions—two Army, three Marine—are free to contend with non-NATO contingencies without significantly undercutting U.S. capabilities in Europe.

Second. Soviet IRBM's and MRBM's could destroy NATO's ports, airfields, and supply complexes and control centers at the onset of any war. NATO has no defense against such attack and no comparable countercapability.

Third. Increasing Soviet threats to NATO's tactical air power could cause NATO to lose land battles.

Fourth. A forward defense, up to the perimeter of West Germany, is both a political and military necessity, especially since the withdrawal of France from active participation; but NATO's defenses are dangerously thin when matched against Soviet military threats.

Most of America's national defense problems are self-inflicted wounds caused by short-sighted policies and stale strategic concepts. Until this country begins relating strategy and force posture more effectively, we can bleed the Treasury dry without improving U.S. national security.

Our emphasis should be upon survivable systems, not merely on more aircraft, submarines, missiles, and warheads; we may need more aircraft, submarines, missiles, and warheads, but on the basis of strategies, not numbers. For example, there are still sound reasons for the deployment of the B-1, but merely to match Soviet deployment of the Backfire is not one of them. Our reasons for needing the B-1, and Soviet reasons for deploying the Backfire, are based on different strategic considerations.

The data which has been presented by Mr. Collins makes more imperative public discussion of our capabilities. The purported defense options offered to the President in the Presidential Review Memorandum 10, PRM 10, for example, are said to range from "rough equivalence" down to considerably less than that. The quotations from PRM 10 published in the press dealing with conceding the loss of Germany in our secret NATO strategy are given substance by a review of the stark facts outlined in the Collins study. The time has come not only to get our house in order, but also to get our strategies to protect that house in order.

I also wish to express deep appreciation for the kind cooperation of the able and distinguished chairman of the Committee on Armed Services, Mr. Stennis, and other members of the committee with whom I have discussed this study. I also want to thank Mr. Frank Sullivan, of the committee staff, for his helpful comments and suggestions.

Everybody is entitled to his own views;
everybody is not entitled to his own facts.

James R. Schlesinger
Pacem in Terris IV
December 2, 1975

Background, Purpose and Scope

The Soviet Union, alone among all countries in the world, can seriously challenge the United States across the conflict spectrum. America's armed services must be able to deter or, if necessary, deal decisively with associated problems, while retaining sufficient strength in reserve to safeguard U.S. regional interests against possible threats by lesser powers.[1]

What military balance would best satisfy U.S. national security needs, however, is subject to constant contention.[2] Polarized positions and partisan stands dominate many discussions. Consequent misinformation makes it difficult for Congress to assess the situation accurately and take appropriate action.

[1] Presidential Review Memorandum (PRM) 10, the Carter Administration's first comprehensive survey of U.S. national security requirements, reportedly recognizes five sorts of conflict that conceivably could involve the commitment of U.S. armed forces in support of this country's security interests: strategic nuclear war; combat in Central Europe between NATO and the Warsaw Pact; an "East-West" war elsewhere; altercations in East Asia, typified by a showdown in Korea; and assorted contingencies, such as the conflict in Vietnam. The Soviet Union would be our principal opponent in the first two cases and a possible opponent in cases three and four.

[2] The U.S./Soviet *military* balance is just one component of the U.S./Soviet *strategic* balance, which is just one aspect of the U.S. *global* balance with other powers that determines our total defense demands. Political, economic, geographic, social, psychological, scientific, and technological assets that are central to any strategic balance are considered here only as they directly affect relative strengths of U.S. and Soviet armed services, along with respective allies.

This concise survey, in consonance with the opening quote, seeks to serve a four-fold purpose: furnish facts; outline opinions; sharpen issues; stimulate debate.

The product complements an earlier study by the Congressional Research Service, so that each stands alone, but addresses the subject from different angles.[3] The present report goes into much greater depth.

Care has been taken to silhouette trends, showing which are strong, which are weak, which are shifting, and which are steady. Statistical summaries that originally covered just two years a decade apart (1965, 1975) now span the 1970s. That seven-year spread, combined with system characteristics, indicates qualitative as well as quantitative changes in capabilities on both sides, because it confirms which weapons are phasing in and out on what specific cycles. Extensive sections compare defense budgets, manpower, science/technology, and other topics that the foundation document disregarded or quickly dismissed.

U.S. statistics in the main were drawn from computer printouts produced by the Department of Defense (DOD) and the four military services. Defense Intelligence Agency (DIA) furnished most Soviet figures. Some differ in detail from those in classified publications, but the patterns portrayed are dependable.[4]

Data are displayed unconventionally whenever so doing clarifies comparisons. U.S. strategic nuclear manpower figures, for example, commonly are subsumed in Army, Navy, and Air Force totals. They are segregated here to conform with Soviet Strategic Rocket Forces, which in turn are subdivided to identify manning levels for theater nuclear missiles. Also, division counts are somewhat different than Congress is used to seeing, and so on.

More than 100 knowledgeable officials reviewed the first draft in DOD (especially the Director of Net Assessment and DIA); the

[3] U.S. Congress, Senate, Committee on Armed Services, *United States/Soviet Military Balance: A Frame of Reference for Congress. A study by the Congressional Research Service,* 94th Congress, 2d Session (Washington, D.C.: U.S. Government Printing Office, 1976), 86 pp.

[4] All assessments of the U.S./Soviet military balance depend directly or indirectly on DOD and its affiliates for U.S. force statistics. The intelligence community collects almost all information concerning the Soviet side. Some disagreements derive from dissatisfaction with basic data, but most disputes deal with differences in interpretation.

Statistics for 1975 in this document and the study cited in footnote 3 sometimes differ because better data became available. Those contained herein take precedence.

Joint Staff (especially J-5, Plans and Policy); Army; Navy; Air Force; Marine Corps; and Military Sealift Command. Several specialists at the Library of Congress also contributed. The author, who considered all constructive comments but sought no concurrence, reserved the right to accept or reject. Conflicting opinions were often resolved by citing contrasting viewpoints.

The resultant net assessment, designed to influence U.S. defense policy without promoting particular programs, prescribes no courses of action. It simply affords an analytical tool that Congress, if so inclined, could use to assist its continuing review of U.S. force requirements and connected requests for funds.

[Chapter 1]

Analytical Problems

STANDARD CAUTIONS

Readers should be aware that *any* appraisal of the U.S./Soviet military balance contains many subjective decisions. Defects in the data base confront analysts with severe constraints at the onset. Simple alterations in assumptions can create radically different conclusions, even if the input is constant. The following discourse identifies sample problems and pitfalls.

QUANTITATIVE COMPARISONS

Quantitative analyses are the inescapable starting point for comparing competing forces, but difficulties in compiling compatible figures make the mere matching of raw statistics a complex matter. U.S. intelligence estimates of Soviet strengths, for example, are inexact, because of incomplete collection capabilities and uncertainties introduced by the Kremlin's cover and deception plans. Conclusions in some cases indicate nothing more than an order of magnitude.[1]

[1] Soviet secrecy laws and procedures are outlined in U.S. Congress, Senate, Staff Report prepared for use of the Committee on Aeronautical and Space Sciences: *Soviet Space Programs, 1971-75,* Vol. 2 (Washington, D.C.: U.S. Government Printing Office, August 30, 1976), pp. 87-89.

Just deciding how and what to count can be confusing. Calculations may involve total inventories on both sides, or only those items in operational units. Stockpiles, particularly prepositioned equipment, can be included or ignored. Determining what fits in which category can be equally complicated. U.S. and Soviet definitions of "heavy" bombers and ICBMs differ drastically. Dual-purpose systems, such as air superiority aircraft that double as air defense interceptors, produce similar counting problems.

Large missiles, divisions, ships, and so on in any given class count the same as small ones. Old weapons count the same as new. Service troops count the same as combatants in basic manpower comparisons.

Since there is no accepted conversion formula, no one knows for sure how many bombers compensate for how many ballistic missiles, or what respective weights should be assigned to armor as opposed to anti-tank forces. Translating simple sums into comparative capabilities that connote *quantitative* "superiority," "inferiority," or "essential equivalence" (sometimes called "equal aggregates") is increasingly difficult.

Assessing statistics just to ascertain who is *numerically* "behind" or "ahead" thus is an imprecise art. Schematics, such as Graph 1, in which seventeen subsequent statistical summaries are condensed, are subject to all sorts of odd interpretations, unless tied tightly to analytical texts.

QUALITATIVE COMPARISONS

Quality can compensate for quantity to uncertain extents. It can also confer superior capabilities on forces that are the same size or smaller than those they oppose. Comparing qualitative characteristics of weapons and equipment, however, is a convoluted process, because key indicators are often concealed. Speeds, service ceilings, combat radii, and payload capacities of most Soviet aircraft, for example, are subjects of speculation in the United States.

Backfire bombers are a case in point. Several U.S. aircraft corporations recently estimated effective ranges, using different sets of data from different sources that created different conclusions. Assessments at the lower end of the scale suggest a range limita-

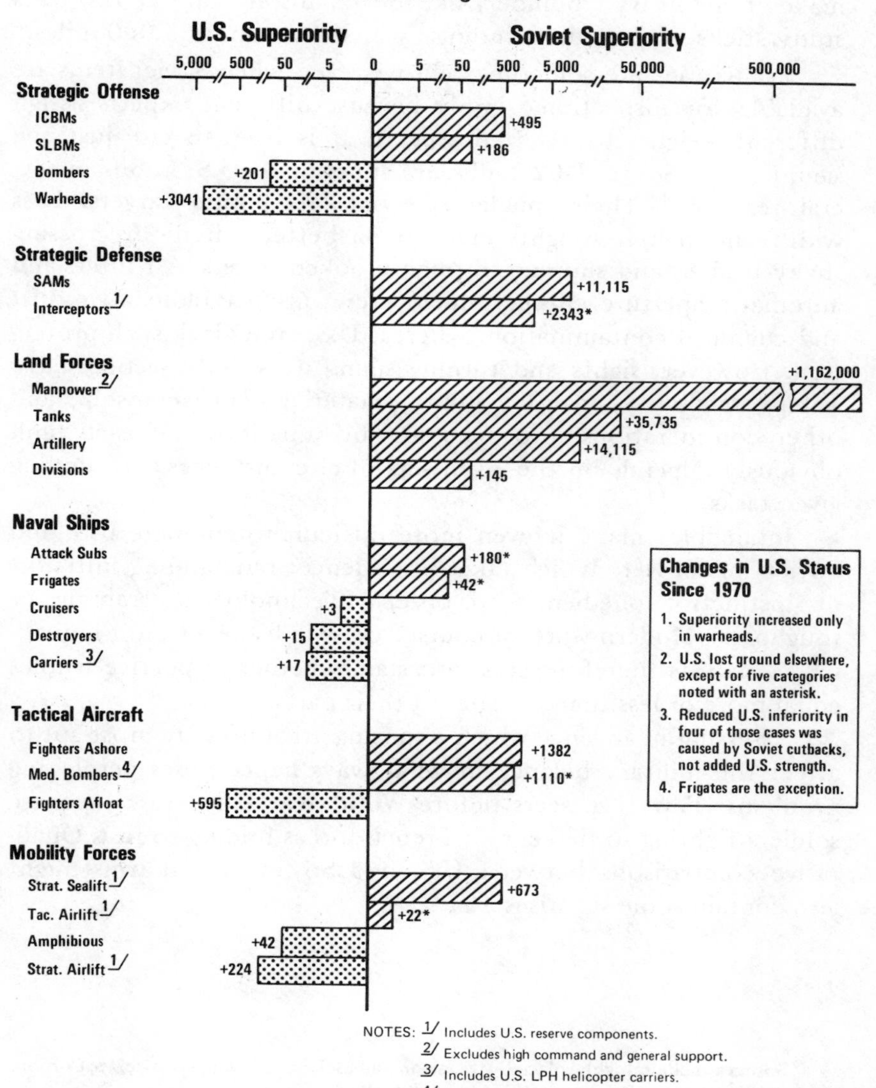

NOTES: 1/ Includes U.S. reserve components.
2/ Excludes high command and general support.
3/ Includes U.S. LPH helicopter carriers.
4/ Includes Navy.
5/ Some Soviet strength estimates are subject to substantial error.

tion of roughly 3,500 nautical miles. If correct, that would classify Backfires in the "medium" bomber category, along with Badgers, Blinders, and FB-111s. Other analyses, however, credit Backfire with nearly twice that range (6,000+ miles), which clearly would make it a "heavy" bomber, like Bears, Bisons, and B-52s. This study sticks with median findings, which were about 5,000 miles.[2]

Evaluations are sometimes elusive, even when Soviet items are available for inspection, simply because different experts assign different weights to assorted aspects. It is true, to cite just one sample, that Soviet T-62 tanks are superior to U.S. M-60s in several respects.[3] Their smaller size presents poorer targets. Less width and lighter weights make them better suited for crossing shaky bridges and slipping through crooked streets. Air filters and automatic aperture closings shield crews against radioactive dust and chemical contamination, whereas U.S. crews lack such protection. However, sights and turning spans are strictly second class. Taking reliability, maintainability, materials, craftsmanship, and other considerations into account, the superiority of each tank obviously depends on the given sets of circumstances surrounding given tasks.

Intangibles make it even more difficult to compare U.S. and Soviet manpower. Which takes precedence: outstanding initiative or instinctive obedience to orders? Technological training or toughness? Modern wars, of course, are won by teams, not individuals. Analysts therefore must ascertain whether respective wholes equal more or less than the sum of their parts.

Assessing in advance which opposing attributes are most apt to affect the military balance in what ways hence poses perplexing problems. Few U.S. seers before World War II foresaw Japanese soldiers fighting so fiercely, or French forces folding so fast. Qualitative comparisons between U.S. and Soviet men-at-arms might also contain some surprises.

[2] Sources include highly placed officials in the McDonnell-Douglas Aircraft Corporation and the Defense Department, February 22, 1977.

[3] *Comparative Characteristics of Main Battle Tanks* (Fort Knox, Kentucky: U.S. Army Armor School, June 1973), pp. 11-1 through 11-4, 14-1 through 14-4.

[Chapter 2]

Building Blocks

PROSPECTUS

Three basic building blocks contribute to military capabilities: money, manpower, and materiel. Those indices, however, are often misapplied in comparing competitors. This short section provides perspective.

DEFENSE BUDGETS

Comparing defense budgets, an interesting intellectual pursuit, is among the most publicized, but least meaningful, means of measuring military power.

Purpose of Comparing Budgets

Budget studies concerning the U.S./Soviet military balance most often emphasize one of two issues, sometimes both:[1] economic "burdens" that armed forces impose on respective societies, with special concern for national abilities to sustain particular pressures over specified periods at the expense of domestic prior-

[1] Andrew W. Marshall, "Comparisons of US and SU Defense Expenditures. A study prepared by Director of Net Assessment, Department of Defense, September 16, 1975." Contained in *Allocation of Resources in the Soviet Union and China, 1975,* p. 155.

ities; and the magnitude of respective military expenditures, with special attention to trends.

Methods of Calculating Soviet Budgets

The annual Soviet State Budget contains a solitary statistic in a single-line entry entitled "Defense." The figure fluctuates slightly to suit political purposes, but has stayed fairly constant at 17-some-odd billion rubles for several years.[2] That scrap of unsubstantiated information reportedly reflects most Soviet outlays for personnel, operations, maintenance, and military construction. Additional costs may be concealed elsewhere, along with research, development, and procurement.[3] For example, the Ministry of Education bears expenses for extensive basic training conducted in civilian schools. Most money for moving military units and materiel comes from the Ministry of Transportation. And the Welfare Ministry administers military retired pay.[4]

U.S. calculations of Soviet defense budgets therefore must be devised, either in dollars or rubles. Both methods begin with the same data base, which includes a detailed account of physical accoutrements and activities that constitute Soviet defense efforts for any given year.[5]

Dollar computations speculate how much it would cost to reproduce Moscow's military establishment in the United States, then contrast consequent estimates with confirmed U.S. expenditures. Such comparisons reveal rough budgetary relationships and trends, but no more.[6]

Ruble computations assist U.S. analysts in assessing actual Soviet expenditures and their impact on that country's economy. Appraisals rely on real Soviet prices to the extent possible, but straightforward statistics are scarce for about one-third of all military items. Costs in such cases are first computed in dollars, then

[2] Central Intelligence Agency, *Estimated Soviet Defense Spending in Rubles, 1970-1975* (May 1976), p. 5; Daniel O. Graham, "The Soviet Military Budget Controversy," *Air Force Magazine* (May 1976), p. 34.

[3] W. T. Lee, "Soviet Defense Expenditures," in *Arms, Men, and Military Budgets: Issues for Fiscal Year 1977,* William Schneider, Jr., and Francis P. Hoeber, eds. (New York: Crane, Russak & Co., 1976), p. 261.

[4] Graham, "The Soviet Military Budget Controversy," p. 34.

[5] CIA, *Estimated Soviet Defense Spending in Rubles,* p. 5.

[6] Central Intelligence Agency, *A Dollar Comparison of Soviet and U.S. Defense Activities, 1965-1975* (February 1976), 8 pp.

converted to rubles, using U.S. intelligence estimates of relative production and efficiency as indices. Determining ruble costs of U.S. forces is such a convoluted process that no reputable comparisons of defense expenditures exist in that coin.[7]

Comparative Economic Burdens

Official U.S. estimates over the past several years have suggested that Soviet outlays for defense equalled a steady 6 to 8 percent of that country's growing Gross National Product (GNP).[8] Recently revised evaluations by CIA, based on better measurement data, now indicate that the share is twice that amount, or 11 to 13 percent,[9] compared with a projected 5.1 percent for the United States in 1977.[10] Other authorities place the Soviet proportion as high as 20 percent.[11]

Increased U.S. estimates of the Soviet defense burden came as no surprise to experienced students of the subject. As the Congressional Budget Office put it, current conclusions "serve principally to resolve a paradox . . . . How could the Soviets squeeze such a large defense establishment out of such a small fraction [6 to 8 percent] of their GNP? It now appears they were not particularly efficient; rather, we simply underestimated how much of their budget went to defense."[12]

One, conclusion, however, seems inescapable. Both sides could sustain current rates of expenditure indefinitely without serious strains. The United States devoted almost 40 percent of its GNP to national defense during World War II, but the civilian standard of living was still higher than in most countries today.[13] The annual

[7] CIA, *Estimated Soviet Defense Spending in Rubles,* p. 5.

[8] Marshall, "Comparisons of US and SU Defense Expenditures," p. 170; and CIA, *Estimated Defense Spending in Rubles,* p. 16.

[9] CIA, *Estimated Soviet Defense Spending in Rubles,* p. 16.

[10] *Special Analyses, Budget of the United States Government, Fiscal Year 1977* (Washington, D.C.: U.S. Government Printing Office, 1976), p. 12.

[11] The Soviet defense budget equals 10 to 20 percent of GNP, according to Marshall in "Comparisons of US and SU Defense Expenditures," p. 164. Graham cites 15 to 20 percent in "The Soviet Military Budget Controversy," pp. 36, 37.

[12] Letter and accompanying staff paper submitted by the Congressional Budget Office to Brock Adams, Chairman of the House Budget Committee, subject: Replies to Chairman Adams' questions in his letter of April 15, 1976. Dated July 21, 1976. Page 3 of cover letter.

[13] Statistics from John B. Braden, *Comparison of U.S. Government Budget Plan Data. . . .Related to Total Budget Outlays and to Gross National Product (GNP), Fiscal Years 1940-1976* (Washington, D.C.: Congressional Research Service, February 7, 1975), p. 7.

h in Soviet GNP seems sufficient to allow increased defense spending and slow improvements in living standards to proceed simultaneously.[14]

Comparative Defense Spending

Estimated total dollar costs of Soviet defense programs in 1976 exceeded U.S. budget authorizations by about 32 percent, according to CIA (40 percent, if pensions are excluded). Soviet expenditures have expanded about 3 percent per year since 1970, whereas U.S. outlays expressed in constant dollars have contracted steadily.[15]

Questionable Confidence in Estimates

Conclusions sketched above are subject to question, since dollar and ruble estimates of Soviet defense budgets both are subject to sizable error. To begin with, it is almost impossible to compile a sound data base. Some counts of Soviet manpower and materiel are incomplete. Others admittedly are incorrect. The costs of Soviet weapons, equipment, and construction depend in large part on their physical and technological characteristics, which in many cases (such as ballistic missiles) are imprecisely known to U.S. analysts.[16]

Assuming that U.S. estimates of Soviet force size and structure were entirely accurate, cost estimates would still be ambiguous "because there is no appropriate or universally agreed set of . . . rules, and thus there is no objective standard by which to measure error." Assumptions and arbitrary judgments consequently abound.[17] Two examples, one related to dollars, the other to rubles, serve as illustrations.

Determining dollar prices for Soviet items not in our possession is a very subjective process. Some weapons, when obtained, prove less sophisticated and cheaper by far than formerly presumed. Others, such as ZSU 23-4 anti-aircraft guns and BMP infantry combat vehicles, turn out to be much "more costly in dollars than their closest US counterparts."[18]

[14] CIA, *Estimated Soviet Defense Spending in Rubles,* p. 17.

[15] CIA, *Dollar Comparison of Soviet and U.S. Defense Activities,* p. 3 as amended by DOD comments on the draft of this study.

[16] Congressional Budget Office, Replies to Chairman Adams' Questions, pp. 1-2.

[17] *Ibid.,* pp. 3-4.

[18] Marshall, "Comparisons of US and SU Defense Expenditures," p. 169.

Manipulations by the Kremlin, which make it impossible for U.S. analysts to know what a ruble is worth, complicate computations.[19] Moreover, prices vary to fit the market. Trucks sold to collective farms may cost 40,000 rubles. The charge to some other State enterprise may be one-fourth that amount. Foreign sales customers may receive identical vehicles for fewer than 4,000 rubles. Our intelligence community is uncertain as to where Soviet armed forces fit on that sliding scale, but it surmises that some parts of the Soviet civil economy subsidize defense spending.[20]

U.S. estimates of Soviet operations/maintenance expenditures are even less exact. Calculations concerning research, development, test, and evaluation (RDT&E)—which contributes to our opponent's future capabilities—are shakier still.[21] Some Soviet budget categories, such as military assistance and civil defense, escape assessment entirely, because evidence is inadequate.[22] Problems are compounded when allies (such as NATO and the Warsaw Pact) are considered.

In sum, the extent to which counting and costing errors overstate or understate Soviet defense budgets is an open debate.[23]

Practical Applications

It may indeed be true that "properly conceived and executed analyses of comparative U.S. and Soviet defense expenditures can provide valuable insights into the status and trends of the two defense efforts."[24] Even so, costs do not measure effectiveness. "Budgets are, in an important sense, little more than a summary of other data. We perceive changes in military capabilities *without* the aid of defense costing calculations" (emphasis added).[25]

One fact, however, is foreboding. More than half of every U.S. defense dollar is devoted to pay and allowances. The Soviets, with far lower pay scales and a controlled economy, can afford to maintain a larger force and modernize at a more rapid rate, because a much greater share of their money can be spent on machines.

[19] The Soviet official rate of exchange for foreign trade purposes is $1.35 per ruble.
[20] Graham, "The Soviet Military Budget Controversy," p. 34.
[21] CIA, *Estimated Soviet Defense Spending in Rubles,* p. 15.
[22] Marshall, "Comparisons of US and SU Defense Expenditures," p. 167.
[23] Congressional Budget Office, Replies to Chairman Adams, pp. 1, 4, 6, 7, 8.
[24] Marshall, "Comparisons of US and SU Defense Expenditures," p. 153.
[25] Congressional Budget Office, Replies to Chairman Adams' cover letter, p. 2.

NET ASSESSMENT APPRAISAL

The trends in defense expenditures that Collins is discussing are summarized in Charts One, Two and Three below:

CHART ONE
U.S. AND SOVIET DEFENSE PROGRAM TRENDS
(U.S. Outlays and Estimated Dollar Costs of Soviet Programs)
(Constant FY 1978 Dollars)

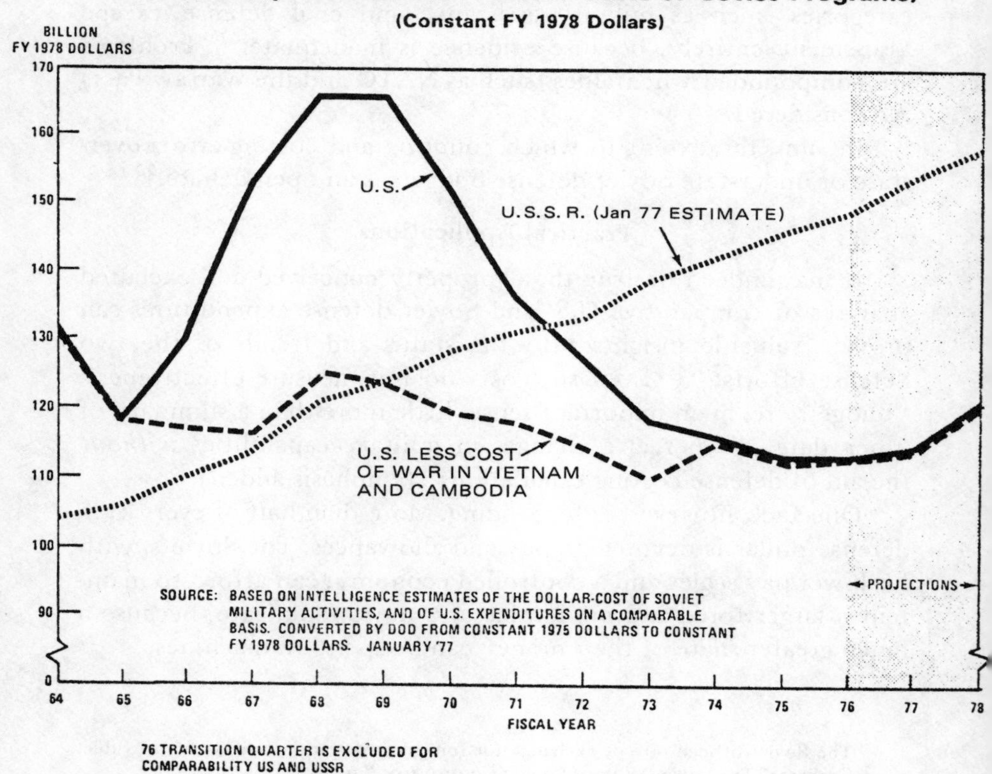

76 TRANSITION QUARTER IS EXCLUDED FOR COMPARABILITY US AND USSR

CHART TWO

ESTIMATED SOVIET DEFENSE EXPENDITURES AND ANNUAL RATES OF GROWTH (IN CONSTANT RUBLES)

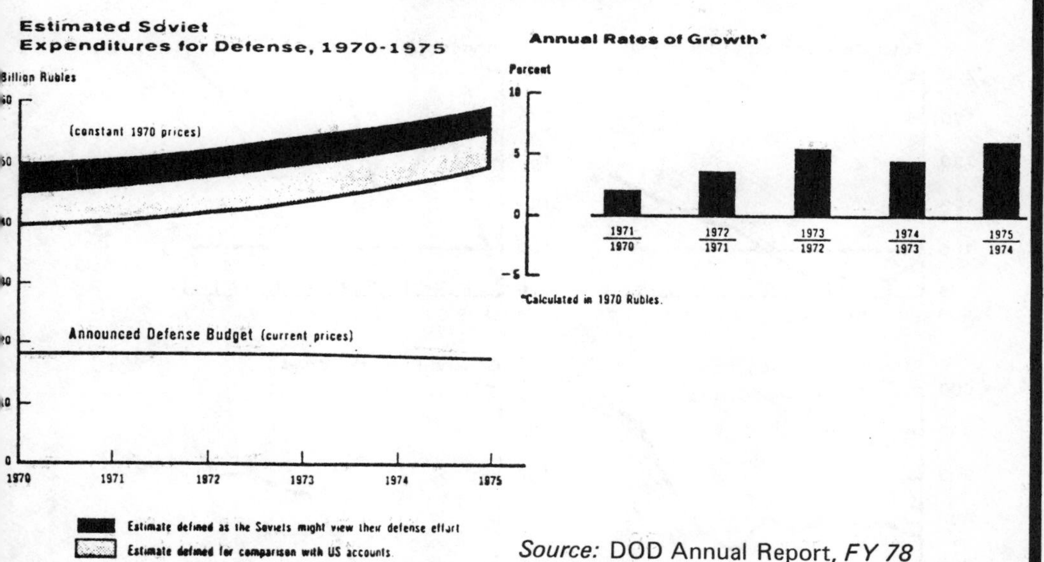

Source: DOD Annual Report, *FY 78*

As John M. Collins notes, estimates of Soviet defense expenditures are highly uncertain. This uncertainty is reflected in the Director of the CIA's comments on the figures:

> As a result of the latest round of National Intelligence Estimates, the intelligence community revised its figures on the production rate of the Backfire bomber and of some major ground force weapons. It also revised estimates of the deployment rates of several strategic missiles and tactical aircraft.
>
> In the spring of last year it completed a major interagency study of Soviet military manpower. That study resulted in an upward revision of the estimated level of active military manpower, and an offsetting decrease in estimated civilian manpower. It also made some changes in the distribution of manpower along the Soviet military services.
>
> It also improved its knowledge of the technical characteristics of Soviet weapons. For example, it has determined the characteristics of the new Soviet strategic missiles more precisely.

CHART THREE

DOLLAR COST OF SOVIET PROGRAMS AS A PERCENTAGE OF U.S. DEFENSE EXPENDITURES*

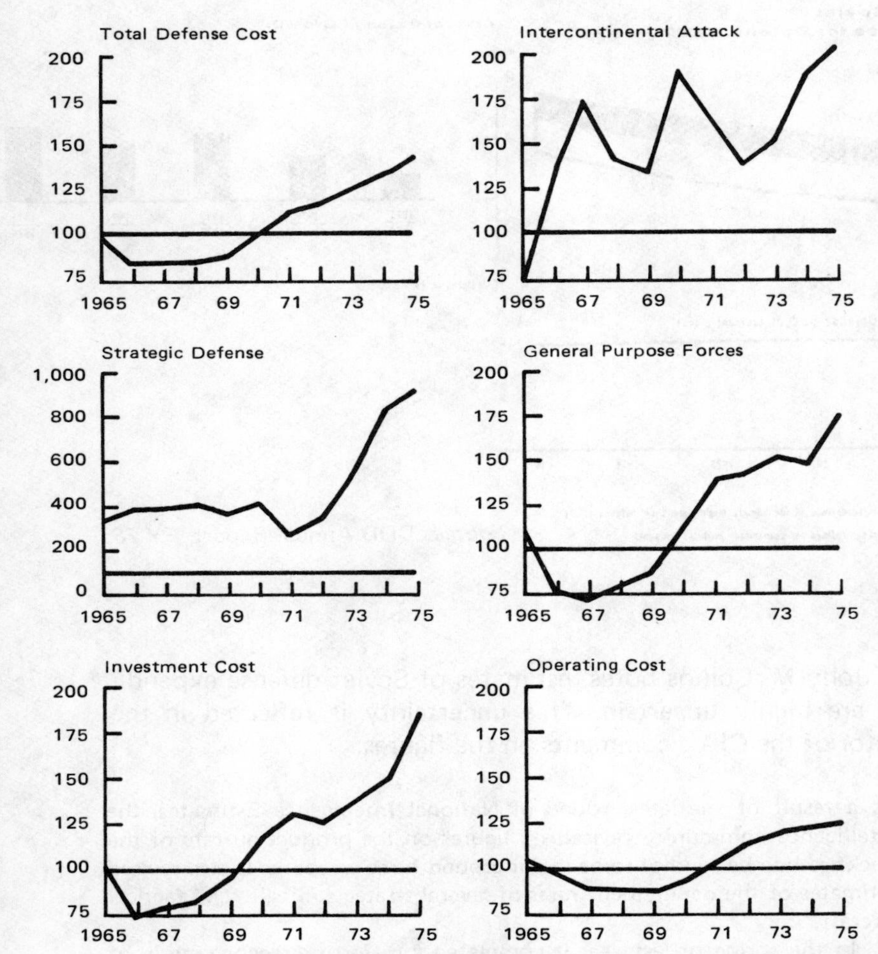

*Values are in 1974 dollars. *Note* that the scale for the "Strategic Defense" graph is significantly different from that of the other graphs in this figure. Department of Defense "total obligational authority" data have been adjusted for comparability with Soviet data.

SOURCE: Donald G. Brennan, "The Soviet Military Build-up and Its Implications for the Negotiations on Strategic Arms Limitations," Orbis, *Volume 2, Number 1, Spring 1977*

As part of a continuing effort to improve costing techniques, it conducted an extensive survey aimed at identifying aspects of Soviet military programs which were not explicitly accounted for in our previous estimates.

This year, it includes, for the first time, explicit estimates for Soviet preinduction military training and utilities for military facilities.

It also made major improvements in the past year in our methodologies for estimating the dollar costs of Soviet weapon systems. Some of the top weapons experts in the U.S. military and in industry helped intelligence.

Source: Testimony of the Hon. George Bush, Director of the CIA," Allocation of Resources in the Soviet Union and China - 1976," Subcommittee on Priorities and Economy in Governments Joint Economic Congress of the United States, 94th Congress, 24 May and 15 June, 1976.

While there is a continuing debate over the validity of such figures, it would be difficult to interpret the meaning of Soviet defense budgets even if perfect intelligence were available. Communist states do not budget on the basis of comparable prices, but by using complex scheduling matrixes which often charge the budget drastically different prices for the same materials on services when they go to different sectors of the economy. Major changes and adjustments in resources are made according to some set of priorities fixed by the planner so that high priority users get the resources they need and other users do not. Since the military sector usually seems to have priority, it may well receive the advantage of special pricing and resource priorities worth billions of dollars to a Western economy. Even the Soviets would have no comparable way of costing such advantages.

This may help illustrate why it is almost impossible to determine what ultimately constitutes a military expenditure in the Soviet Union. Much of the Soviet investment in heavy industry, for example, must be structured to meet military needs. A large portion of the massive Soviet investment in education must also be so structured. In fact, there is virtually no sector of the Soviet economy which is not driven or distorted by the need to support Soviet military forces. This represents a massive cost to the U.S.S.R. in real economic terms, but there is virtually no way to estimate it.

These uncertainties are compounded by the fact that there is no way to relate estimates of Soviet expenditures to efficiency and to convert them into dollar equivalents. Unlike the West, the Soviet system and budget only indirectly relate price to efficiency. For example, large inefficiencies may be concealed within the special prices charged to the military sector, while real Soviet efficiencies through large scale production or research activity are untraceable.

The ultimate result is that the West has no way of making valid comparisons between U.S. and Soviet expenditures. For all the effort that went into generating Charts One and Two, the West is deprived of one of the most important tools for comparing defense efforts by the shape and nature of the Soviet economic system. And, this situation cannot be corrected by trying to cost Soviet forces in Western prices. Even in the United States, major military facilities, equipment, and combat units do not dominate defense expenditures. It is "invisibles" like R&D, manpower, general purpose installations, logistics, operations and maintenance which make up most of the budget. There is no practical way that Soviet expenditures on such "invisibles" can be accurately costed, and attempts to modify comparable U.S. costs for such activities are little more than speculation.

DEFENSE MANPOWER

Manpower levies, like defense budgets, have some bearing on national economies, but are only marginally meaningful as an index for comparing rival military establishments in most respects, as this survey shows.

Quantitative Considerations

THE STATISTICAL BASE

U.S. armed services try to keep careful statistics concerning respective manpower levels. All sorts of activities and administrative actions depend on such lists: pay and allowances, rations, quarters and other construction, clothing and personal equipment, medical support, training facilities, and miscellaneous services are representative. Reserve components and civilians, as well as uniformed regulars, are taken into account.

Conversely, the U.S. intelligence community accorded low priority to Soviet personnel statistics until the recent past, except for combat forces. Other problems were more pressing.[26]

Analysts traditionally scrutinized functional categories. Strategic attack forces, for example, included long-range aviation, plus all ballistic missiles ashore and afloat. Army, navy, and air force general purpose contingents were segregated into special groups. Command and general support (CGS) forces from all services were considered as a separate class. Results, which related manpower statistics to *missions* instead of *organizations,* contained significant oversights, double counting, and other inconsistencies.

A comprehensive reassessment, completed in 1975, reduced such shortcomings by combining functional and organizational approaches.[27] For the first time, intelligence specialists from CIA and DIA addressed discrete, clearly defined entities: Strategic Rocket Forces (SRF); strategic defense forces (the PVO); and integrated ground, air, and naval services that included respective support. Navy statistics, for example, combined strategic nuclear submarines, general purpose forces, and naval infantry (herein called marines). The Ministry of Defense and Main Political Directorate were catalogued separately.[28]

Sharp statistical revisions resulted.

The following review reflects current tabulations. Comparisons with U.S. forces are confined to a single year, because the U.S. intelligence community has never published an agreed adjustment of estimated Soviet personnel strengths for the early part of this decade.[29]

[26] This subsection is predicated on personal conversations between the author and Defense Intelligence Agency (DIA) analysts in September and October, 1976, together with written data furnished by DIA.

[27] Classified details are contained in Defense Intelligence Agency, *Reassessment, Soviet Armed Forces Personnel Strengths (U),* (Washington: July 1975), 8 pp.; and Part II (November 1975), 61 pp.

[28] Organizational and functional approaches complement each other by double-checking intelligence collection and processing capabilities. A third method, which deals with demography, relates raw manpower with requirements by indicating how many men of military age are annually available for induction, how many reservists reach the statutory age of obligatory service each year, and so on. For the latter, see especially Murray Feshbach and Stephen Rapawy, "Soviet Population and Manpower Trends and Policies," in U.S. Congress, Joint Committee Print, *Soviet Economy in a New Perspective, A Compendium of Papers Submitted to the Joint Economic Committee* (Washington, D.C.: U.S. Government Printing Office, October 14, 1976), pp. 113-14, 144-52.

[29] Trend tables and graphs in other documents, including those disseminated by DOD, should be treated cautiously, for reasons described above.

ACTIVE ARMED FORCES

Post-Vietnam cutbacks have *physically subtracted* almost a million men from active U.S. roles since 1970 (from 3,088,000 to 2,095,000).[30] The U.S. *paper recomputation added* almost a million to previously estimated Soviet levels near the end of that period, largely in the command/support category.[31]

Official confidence in Soviet statistics is only about plus or minus 15 percent, but current consequences still show in stark relief on Graph 2 and Figure 1. The Soviet Army alone exceeds the aggregate of all active U.S. forces by almost half a million. Total Soviet active military manpower is more than twice ours (4,437,000 to 2,095,000).

Nearly half a million paramilitary border guards and internal security troops supplement the regular establishment. Many are armed with automatic weapons, aircraft, and armored vehicles.[32] KGB divisions, like the Nazi SS, fought well during World War II, and could today if called on for homeland defense.

Perhaps 70,000 political officers, who parallel the military chain of command at almost every level, are soldiers in every sense.[33] Other forces are not. These include 400,000 support troops committed to construction projects, transportation, and part-time farm labor.[34] Their training is spotty and superficial, and units lack arms. Still, most of them contribute to military capabilities that bear on the balance of power. Railroads run by men in uniform, for example, are the key to internal strategic mobility.[35]

[30] U.S. active military strengths were mainly derived from Manpower Statistics by Defense Planning and Programming Category for each military service and DOD agencies, a series of unpublished working papers prepared for OASD (M&RA), Program Division. Figures for 1970-74 were dated March 29, 1976. Figures for 1975 and 1976 were dated October 21, 1975 and August 26, 1976, respectively. OASD (M&RA) furnished supplemental statistics on U.S. strategic offensive and defensive forces in October, 1976.

[31] Soviet active manpower strengths were derived from diverse official sources.

[32] Border Guards belong to the KGB, internal security troops to the MVD, according to John Erickson, who mentions accoutrements in United States Strategic Institute, Soviet-Warsaw Pact Force Levels, USSI Report 76-2 (Washington, D.C.: 1976), pp. 20-21.

[33] DIA identifies about 12,000 personnel in the Main Political Directorate. The remaining 58,000 are diffused in military units.

[34] Congressman Les Aspin, for example, calls 2,755,000 Soviet armed forces "nonthreatening" to the United States. He credits each side with about 2,000,000 "comparable opposing forces." *Disarmament and International Views* (May 1976), pp. 5-7.

[35] DIA comments on the draft of this study, March 2, 1977.

Graph 2
COMPARATIVE MANPOWER
Statistical Summary
(In Thousands. Note Different Scales)

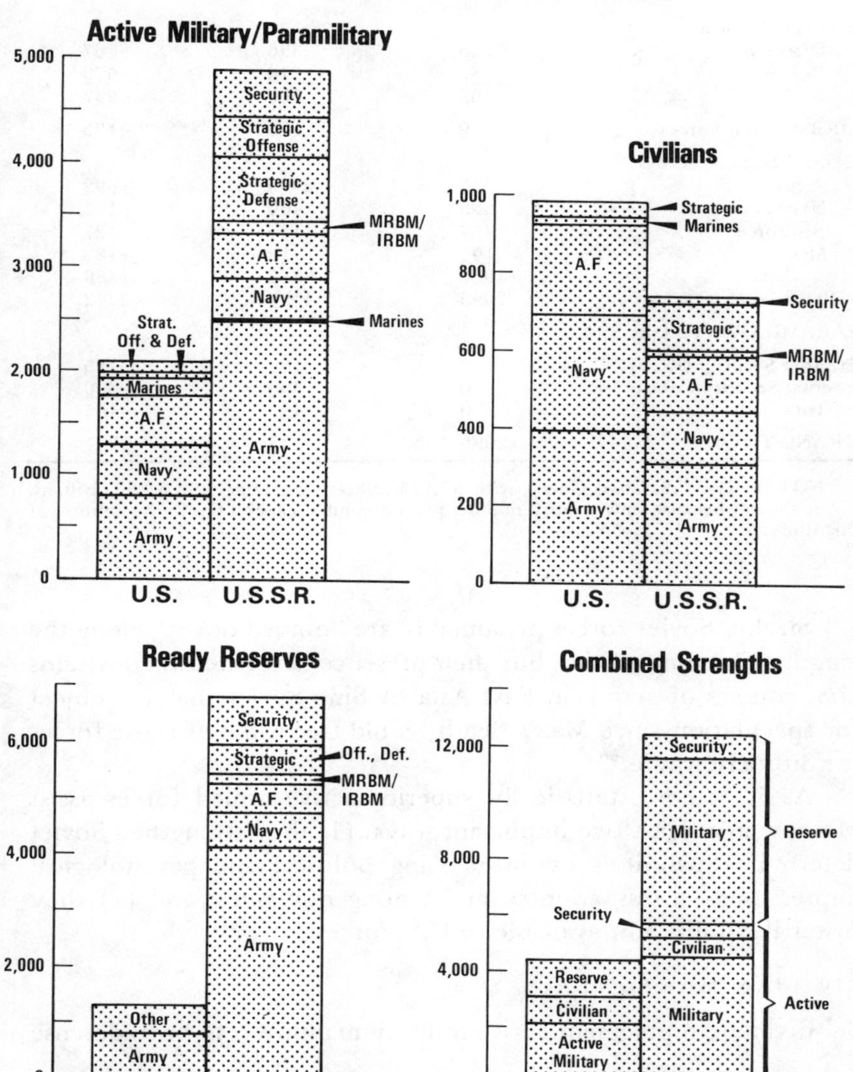

FIGURE 1 U.S. AND SOVIET
Paramilitary Manpower Strengths, FY 1976 (In Thousands)

	United States	*Soviet Union*	*U.S. Standing*
ICES			
Offensive	123	390	-267
Defensive	40	610	-570
Total	163	1000	-837
MRBM/IRBM Forces	0	125	-125
General Purpose Forces			
Army	778	2470	-1692
Navy	505	400	+105
Air Force	457	430	+27
Marines	192	12	+180
Total	1932	3312	-1380
Military Total	2095	4437	-2342
PARAMILITARY FORCES			
Frontier Security (KGB)	0	155	-155
Internal Security (MVD)	0	300	-300
Total	0	455	-455
GRAND TOTAL	2095	4892	-2797

NOTE: The U.S. Coast Guard, with 37,475 military personnel and 6,850 civilians, is not shown, because only a small number possess combat capabilities in the context of this study.

Sizable Soviet forces presumably are "pinned down" along the lengthy Chinese frontier, but their presence nevertheless constrains U.S. courses of action in East Asia. A Sino-Soviet thaw, a subject for speculation since Mao's death, could free some of those forces for duty elsewhere.[36]

As it stands, statistically superior active armed forces assist Moscow in at least two important ways: (1) they strengthen Soviet deterrent capabilities by influencing political and psychological impressions in this country and among our allies; and (2) they foster flexibility not available to U.S. forces.

CIVILIAN MANPOWER

Civilians supplement active military manpower in both defense

[36] U.S. leaders are compelled to consider the eventual possibility of a new Sino-Soviet entente since, to paraphrase Lord Palmerston, there are no such things as permanent friends or permanent enemies, only permanent interests.

establishments to provide continuity and special skills. Once again, U.S. statistics are solid.[37] Those for the Soviets are so spongy that confidence in present estimates approximates plus or minus 25 percent at best, a very high margin of error.[38]

Still, evidence seems to indicate that U.S. civilian strengths exceed the Soviets' somewhat in absolute terms, and are triple proportionately. We employ one civilian for every two military men. Their ratio is one for six or seven (see Figure 2). As a result, overall personnel comparisons that merge active military manpower with civilians reduce this country's quantitative deficit from 2:1 to 5:3 for forces in being. Narrowing the numerical gap, however, by no means indicates that civilian and military strengths are interchangeable. Civilians can substitute for uniformed specialists,

FIGURE 2 U.S. AND SOVIET
Civilian Manpower Strengths, FY 1976 (In Thousands)

	United States	*Soviet Union*	*U.S. Standing*
MILITARY SERVICES			
Strategic Nuclear			
Offensive	22	50	-28
Defensive	15	65	-50
Total	37	115	-78
MRBM/IRBM Forces	0	15	-15
General Purpose Forces			
Army	389	305	+84
Navy	300	135	+165
Air Force	235	145	+90
Marines	19	0	+19
Total	943	585	+358
Military Service Total	980	715	+265
PARAMILITARY FORCES			
Frontier Security	0	8	-8
Internal Security	0	9	-9
Total	0	17	-17
GRAND TOTAL	980	732	+248

[37] U.S. civilian statistics include direct and indirect hire. The latter involves paying foreign governments, who in turn provide civilians to assist U.S. forces overseas. Contract services, which contribute a large but unknown number of man hours, supplement U.S. civilians shown on figures and graphs in this study. See Note 30 for sources of U.S. civilian statistics.

[38] Soviet civilian manpower strengths were derived from diverse official sources.

they are not combat forces in any sense of the word, nor are
readily redeployable in most instances.

READY RESERVES

U.S. Ready Reserve strengths[39] have dropped dramatically since 1970, from 2,661,000 to 1,308,000.[40] Our Marine Corps, which lost 56 percent of assigned personnel, suffered worst, but the Army and Navy were also sliced in half. Decline will continue, because fewer forces annually enter reserve status at a slower rate from a smaller establishment than during days of the draft, when conscripts served two-year terms.

FIGURE 3 U.S. AND SOVIET
Ready Reserve Strengths, FY 1976 (In Thousands)

	United States	*Soviet Union*	*U.S. Standing*
MILITARY SERVICES			
Strategic Nuclear			
Offensive	4	520	-516
Defensive	13	0[1]	+13[1]
Total	17	520	-503
MRBM/IRBM Forces	0	170	-170
General Purpose Forces			
Army	798	4140	-3342
Navy	203	625	-422
Air Force	206	490	-284
Marines	84	0	+84
Total	1291	5255	-3964
Military Total	1308	5945	-4637
PARAMILITARY FORCES	0	855[2]	-855
GRAND TOTAL	1308	6800	-5492

[1] Soviet strategic defensive forces revert to army reserve after discharge.
[2] No breakout between Soviet frontier and internal security forces is available.

[39] U.S. Ready Reserves include personnel who enlist directly into Reserve and National Guard units of the top-priority Selected Reserve, prior service personnel who elect to join Selected Reserve units after stints with the active establishment, and prior service personnel released to Individual Ready Reserves, which are affiliated with no organization, but act as fillers.

[40] U.S. Ready Reserve strengths were derived from Directorate for Information, Operations, and Control, OASD (Compt), *Selected Manpower Statistics,* April 15, 1971, p. 86; April 15, 1972, p. 89; April 15, 1973, p. 91; May 15, 1974, p. 91; May 15, 1975, p. 99; June 76, p. 98. See also *Guard and Reserve Manpower: Strengths and Statistics,* OASD (RA), (June 1976), flyleaf.

Soviet forces released from active service in the last five years are counterparts of the U.S. Ready Reserve for purposes of this study, although they are not precisely comparable, and statistics shown in Figure 3 are questionable.[41] Their regular Air Force, for example, outnumbers the Soviet Navy and has a shorter term of service, yet it has accumulated 165,000 fewer reserves, according to unclassified intelligence estimates.[42] Nevertheless, it is certain that Soviet Army reserves alone would dwarf the combined size of all U.S. reserve components if their estimated numbers were reduced by half. That gap will grow, as U.S. reserves contract.

AGGREGATES ASSESSED

Soviet active military regulars, not counting security forces, exceed the entire U.S. establishment, including civilians and reserves (Figure 4). Soviet military regulars and reserves almost triple the U.S. total (4,383,000 to 11,097,000). Even if U.S.

FIGURE 4 U.S. AND SOVIET
Combined Manpower Statistics Compared, FY 1976 (In Thousands)

	United States	*Soviet Union*	*U.S. Standing*
ACTIVE FORCES			
Military Services			
Military	2095	4437	-2342
Civilian	980	715	+265
Total	3075	5152	-2077
Paramilitary Forces			
Military	0	455	-455
Civilian	0	17	-17
Total	0	472	-472
Active Total	3075	5624	-2549
RESERVE FORCES			
Military	1308	5945	-4637
Paramilitary	0	855	-855
Total	1308	6800	-5492
GRAND TOTAL	4383	12,424	-8041

[41] Soviet forces released from active service during the period 1972-76 total about 5,945,000. U.S. servicemen released during that same time frame total 3,111,000.

[42] U.S. intelligence analysts currently are reviewing estimates of Soviet reserve manpower strengths.

experts have overestimated Soviet active military strengths by 15 percent and all other personnel by 25 percent, the Kremlin still would have almost twice as many people in its military machine as we do (9,764,000 to 4,383,000).

Those statistics, however, convey a false impression, except for special cases. Just as increased costs often fail to increase effectiveness, quantitatively superior personnel strengths frequently fail to create superior capabilities. Threats posed by naval flotillas and fighter squadrons, for example, depend on material, not human, mass. Direct correlations between personnel statistics and power are confined not just to general purpose ground forces, both army and marines, but specifically to "cutting edge" elements that match man against man in mortal combat. (See Chapter 4.)

Qualitative Considerations

Whereas quantitative manpower comparisons are meaningful mainly in the ground combat context, qualitative characteristics affect the U.S./Soviet military balance in many important ways.

Each side has distinctive strengths and weaknesses. Some, like Soviet stamina and U.S. zeal, seem more or less constant. Others are continually changing. Dissidence and drug abuse, which degraded U.S. capabilities in the early 1970s, have largely disappeared.[43] Soviet city dwellers with mechanical skills are supplanting peasants.[44]

This section is confined to a few general considerations. Comments keyed to specific armed services appear elsewhere, when appropriate.

SELECTED U.S. PROBLEMS

U.S. land, sea, and air forces all feature technical skills that demand high-caliber manpower. The quality of non-prior-service accessions, as measured by educational levels and mental capacity,

[43] The Defense Manpower Commission (DMC) "found no evidence that any unit had been affected negatively by socioeconomic changes, either as to performance or mission capability. Generally, commanders [told DMC members] that these are the concerns of Washington, not of the field." Bruce Palmer, Jr., and Curtis W. Tarr, "A Careful Look at Defense Manpower," *Military Review* (September 1976), p. 7.

[44] Frederick C. Turner, "How Tall Are the Russians?" opening remarks in a panel discussion at the annual convention of the Association of the United States Army (Washington, October 12, 1976), p. 4.

reportedly is higher today than in fiscal year 1964, the last year in which we had a peacetime draft.[45] That bright trend, however, may be transitory, because current U.S. unemployment rates create a "buyer's market" for our All-Volunteer Force. Economic recovery could quickly reduce the roster of qualified recruits.[46]

Since 80 percent of all first-term U.S. enlisted men revert to reserve status after three or four active duty years,[47] retaining prime personnel is a pressing problem. The consequent turnover, which causes instability within each Service, complicates training and reduces readiness, especially at echelons where combined arms coordination is essential. The present predilection of careerists to retire after twenty years makes room at the top for younger men, but robs our armed forces of many mature and experienced members.

U.S. Ready Reserves are now better than ever in many respects,[48] but qualitative shortcomings are more sharply pronounced there than in the active Services. Part-time leaders and part-time training impair proficiency least in Air Force airlift and air defense organizations. Most other elements suffer, despite strong command emphasis for the last several years.[49]

SELECTED SOVIET PROBLEMS

The 1967 Law of Universal Service theoretically obligates all 18-year-old Soviet males to serve the State in active armed forces. About 80 percent are conscripted annually, but those committed to "hot spots" constitute the cream of the crop. Culls go to construction gangs and general labor.[50]

Pre-induction preparation starts in grammar schools. Average results are approximately equal to a month of active basic train-

[45] Donald H. Rumsfeld, *Annual Defense Department Report, FY 1977* (Washington, D.C.: Department of Defense, January 27, 1976), p. 289.

[46] *Ibid.*, and Palmer and Tarr, "A Careful Look at Defense Manpower," p. 11.

[47] Rumsfeld, *Annual Defense Department Report, FY 1977*, p. 288.

[48] *Ibid.*, p. 293.

[49] Defense Manpower Commission, Defense Manpower: *The Keystone of National Security, Report to the President and the Congress* (Washington, D.C.: April 1976), p. 2.

[50] Herbert Goldhamer, *The Soviet Soldier: Soviet Military Management at the Troop Level* (New York: Crane, Russak & Co., 1975), pp. 4, 36-37, 69; Turner, "How Tall Are the Russians?" p. 2; William T. Lee, "Military Economics in the USSR," *Air Force Magazine* (March 1976), p. 50; Feshbach and Rapawy, "Soviet Population and Manpower Trends," p. 148.

ing. Before they enter service, about a third of all inductees take additional courses from DOSAAF.[51] Specialists spend six months in a "cram course" before being certified for duty with tactical units.[52]

Nevertheless, military manpower management problems are immense in the Soviet Union. Improved education, which makes it possible for present-day recruits to master military skills more quickly than their predecessors, probably prompted the Defense Ministry to cut draft age and conscript service a full year in 1967,[53] but the sacrifice in experience has been considerable. Turnover is terrific. Callups take place twice a year, spring and fall. A quarter of all draftees are discharged at the same times.[54]

Technical competence is afflicted especially by short tours. Civilians ashore, not seamen afloat, consequently serve a high percentage of Soviet shipboard equipment. Aircraft maintenance is equally aggravating. Supervisory requirements in all high skill areas are far greater than for U.S. forces.

Soviet training and operational procedures are *effective,* but commonly *inefficient.* Efforts often are excessive in terms of ends achieved, partly because uncompromising dedication sets equal priorities for all objectives. It is all very well to insist on toughness,

[51] DOSAAF stands for the All-Union Voluntary Society for Assistance to the Army, Air Force, and Navy. Its clubs and schools play an important part in pre-induction training.

[52] Telephone conversation with Frederick C. Turner, March 7, 1977.

[53] Turner, "How Tall Are the Russians?" p. 3. Callup age was cut from 19 to 18 years. Service with the Army, Air Force, naval air elements, and border/security troops was reduced from three to two years. Service on naval ships, with Coast Guard combat units, and with maritime units of border troops was cut from four years to three. For full details, see *Current Digest of the Soviet Press* 19: 45 (November 29, 1967), pp. 4-10.

[54] Goldhamer, *The Soviet Soldier,* pp. 4-5, 39.

Demographic difficulties have also developed to such an extent that Feshbach and Rapawy foresee an obligation to change the current system, perhaps by extending the length of service once again. See "Soviet Population and Manpower Trends," p. 149.

That trend may already be in motion (see Erickson, John, Soviet-Warsaw Pact Force Levels U.S.S.I. Report 76-2, Washington, U.S. Strategic Institute, 1976, pp. 17, 20). If so, and rates of induction were substantially reduced, no buildup would occur. Such action would simply confirm that shorter enlistments unduly degrade performance in several ways. However, holding three age groups under arms at current conscription rates, despite strains on the civil economy, *could* indicate plans for early aggression, according to Robert L. Goldich, a manpower analyst with the Congressional Research Service. Active duty strengths would shoot up by more than a third and stay there until reductions in force were effected. Close U.S. scrutiny thus would be commendable.

for example, but human errors increase inevitably when specialists work under abysmal conditions that could be avoided.[55]

Political indoctrination, protracted and pervasive, competes eternally with military training for time and attention, although there is no close linkage between professional competence and the Communist Party line. Conflicts of interest assume special significance in high technology units that can least afford the tradeoff.[56]

Worse yet, powerful political officers often second-guess commanders, who must keep a tight rein on subordinates to minimize "mistakes."[57] As a direct result, innovation is a rare commodity. The Soviet socialist system, which stresses collective enterprise and isolates citizens from outside contact, further inhibits initiative. Conscripts serving in East Germany cannot speak the language or even read the road signs. Movement plans and maps are matters for officers only.[58]

Discipline is stringent by U.S. standards, but when it cracks, resultant rifts are sensational. Misdoings on U.S. ships in the recent past have been minor, compared with attempted mass defections that mar the Soviet Navy's record.[59]

CONCLUDING COMMENTS

None of the shortcomings sketched above is critical. Soviet soldiers are not "10 feet tall" when compared with American counterparts, but neither are they 10 inches. Which opposing strengths outweigh which weaknesses may someday make a great difference, but judgments now are subjective. Meanwhile, man-

[55] Some Soviet radar operators reportedly work in cabin temperatures "up to 70° centigrade . . . under conditions that blunt 'all of the senses.' " Goldhamer, *The Soviet Soldier,* p. 328.

[56] *Ibid.,* pp. 255-309, 324-27.

[57] Present-day political officers lack the clout of World War II commissars. They neither countersign orders nor have an official say in operational command. All the same, their influence is still immense. DIA comments on the draft of this study, March 2, 1977.

[58] Turner, "How Tall Are the Russians?" pp. 4-5. Regimentation obviously has strong points. Communications doctrine and discipline make it possible for thirty tanks to share a single radio net in a Soviet Tank battalion. Each U.S. Army tank company with seventeen tanks operates four internal nets, plus a fifth that connects with battalion headquarters. The Soviet system not only cuts costs, but enhances communications security. There is, however, another side to the coin. It is also quite inflexible.

[59] "Russians Stung By Dissent in Armed Forces," *Baltimore News-American,* February 9, 1976, p. 4; "Soviet Mutiny Ended Swiftly," *Washington Post,* June 7, 1976, p. 18; "Mutinied Soviet Destroyer Dispatched on Long Voyage," *Christian Science Monitor,* June 29, 1976, p. 6.

power qualities on *both* sides seem in the main sufficient to support most projected courses of action. Only the test of combat could confirm the true timber of U.S. and Soviet forces.

NET ASSESSMENT APPRAISAL

The major imbalance between U.S. and Soviet military manpower which John M. Collins documents seems to be one of the most significant changes in the balance since the end of U.S. intervention in Vietnam. The importance of the trends which have led to the current strengths is not clear. Three major factors need careful consideration:

- The testimony of the Secretary of Defense and various other U.S. officials still reflect massive variations in public estimates of Soviet manpower. It seems clear from this testimony that the uncertainty in U.S. estimates of some aspects of Soviet manpower may exceed fifty percent. There is also no demonstration in recent testimony that the Intelligence Community has been able to fully coordinate its unclassified estimates of Soviet military manpower to the point where they can be meaningfully compared with U.S. manpower.
- Much Soviet military manpower is consumed in political and internal security activity which would represent an incredible waste of resources by Western standards. Similarly, much of the Soviet reserve system seems to be more a result of a need to ensure that the state can control its manpower, and of history, than of any probable contingency requirements. Certainly, Western war games and simulations rarely, if ever, include most of Soviet military manpower in the threat played for major scenarios or contingencies, even when consideration is made of the fact that the U.S.S.R. must now deploy large amounts of its manpower against China.
- Similar uncertainty exists in defining "military" and "civilian." Many Soviet military perform what are civilian functions in the West, and there are no indications that they would have a contingency value in combat in any foreseeable war. They should ideally be compared with Western civilians performing similar functions.

Much more will have to be learned about the Soviet Union before a comparison of U.S. and Soviet manpower can be more than informed guesswork.

DEFENSE TECHNOLOGY

Science and technology exert enormous influences on the U.S./Soviet military balance. Competition between the two countries is intense. Each side struggles ceaselessly to stock its arsenal with weapons and equipment that satisfy special needs under changing conditions.

Technological Warfare

Technological warfare, which connects science with strategy and tactics, is deliberately designed to outflank enemy forces by making them obsolete. Battles are won by budgeteers and men at drawing boards before any blood is shed.[60] Technological surprise thus poses special perils in critical echelons of the conflict spectrum, where sudden, one-sided supremacy in aerospace defense, antisubmarine systems, supersmart weapons, chemical warfare, lasers, and the like could create spectacular shifts.[61] The Soviet penchant for secrecy prompts "worst case" U.S. estimates in such cases.

Classic dangers develop when new systems based on new technology burst on the scene (nuclear weapons, for example), but breakthroughs that combine new systems with old technology or vice versa can also create serious shocks. Still, creativity alone confers no advantage unless tied to procedures that translate inventive ideas into tangible instruments deployed in correct combinations and sufficient strength.[62] Tactical thought outweighs scientific theory as often as not.

"Victory" is achieved when one participant unveils technological superiority so pervasive and pronounced that opponents can neither cope nor catch up. Since indicators of rival success

[60] For discussion, see Stefan T. Possony and J. E. Pournelle, *The Strategy of Technology: Winning the Decisive War* (Cambridge, Mass.: Dunellen, 1970), pp. 1-20, 55.

[61] George Heilmeier, Director of Defense Advanced Research Projects Agency (ARPA) outlines possibilities in "Guarding Against Technological Surprise," *Congressional Record,* June 22, 1976, p. S10139.

[62] Research, development, test, and evaluation (RDT&E) precedes procurement. Abstract research improves prospects for a range of practical products by broadening the base of human knowledge. Applied technology supports specific goals. Stage I in the U.S. process identifies options, sets priorities, and defines projects. Stage II establishes program characteristics, including costs and schedules. Stage III produces prototypes. Civil contractors and military project managers complete technical tests during Stage IV, which culminates in operational assessments conducted by authorities who are independent of users as well as developers.

often surface slowly, losers sometimes cherish illusions of winning until too late. Conversely, they may long be aware that they have lost, but lack any way to rally.

Unfortunately, technological forecasting is at best imprecise.[63] Surprise can be intense when enemies are fully aware of their foe's R&D schemes, if they fail to sense the significance. Serendipitous offshoots of basic or applied research often alter perceived patterns in unexpected ways. Even so, estimating future states of the enemy's art may be the *easiest* prediction. Analysts must also account for educational, economic, institutional, bureaucratic, and doctrinal constraints.[64] The will to compete can be crucial.

Since surefire predictions perhaps are impossible, given the dearth of hard data, being "ahead" is important in technological areas that really count. Substantial leads lessen chances of surprise by allowing friendly teams to explore frontiers before they are probed by enemies.[65]

Soviet Challenges

The Soviet Union, as a closed society, enjoys an R&D edge not available to the United States. Leadership, starting with Lenin, has stressed science and technology. Command emphasis and a cohesive strategy ensure an increasingly skilled cadre, sustained heavy investments in rubles and other resources, and solid continuity. Focus remains unflinchingly on military research and development, with little fear of repercussions caused by domestic demands.[66]

[63] "In 1878, Frederick Engels stated the weapons used in the Franco-Prussian war had reached such a state of perfection that further progress which would have any revolutionary influence was no longer possible. Thirty years later, the following unforeseen systems were used in World War I: Aircraft, tanks, chemical warfare, trucks, submarines, and radio communications. A 1937 study entitled 'Technological Trends and National Policy' failed to foresee the following systems, all of which were operational by 1957: Helicopters, jet engines, radar, inertial navigators, nuclear weapons, nuclear submarines, rocket powered missiles, electronic computers and cruise missiles. The 1945 Von Karmann study entitled 'New Horizons' missed ICBMs, man in space [and so on], all operational within 15 years." Heilmeier, "Guarding Against Technological Surprise," p. S10139.

[64] Klaus Knorr and Oskar Morgenstern, *Science and Defense: Some Critical Thoughts on Military Research and Development* (Princeton, N.J.: Princeton University Press, 1965), pp. 19-27.

[65] George Heilmeier sketches a seven-step process that a free society can take to forestall technological surprise. Initiative is the most important. "Guarding Against Technological Surprise," p. S10139.

[66] Background information is contained in *The Soviet Military Technological Challenge* (Washington, D.C.: The Center for Strategic Studies, Georgetown University, September 1967), pp. 13-31.

Extreme secrecy shrouds Soviet efforts, often concealing courses of action and intent until field testing starts in full view. Shortcuts are possible, because published reports of U.S. plans and progress point them out. The Kremlin consequently can concentrate on carefully chosen goals that simplify the search for superiority in selected sectors.[67]

All, however, is not advantageous. Surreptitious science carries severe penalties. It inhibits competition and free interchange of ideas (both essential stimulants) and protects poor programs from public opposition. The Socialist system excludes many incentives that generate growth. Civil and military R&D efforts are sometimes so segregated that neither sustains the other.[68] To compensate in part, the Soviets participate in exchange programs, purchase products and processes on the open market, perpetrate espionage, plagiarize ideas, and engage in technological piracy.[69]

In the past, Soviet scientists held closely to a policy of conservative incrementalism that featured slow but steady progress.[70] The R&D community designed around difficulties. Current indications, however, suggest a significant change, characterized by expansion in the scope of Soviet basic research, greater emphasis on innovation, and an increasing inclination to take technological risks on speculative projects that promise big payoffs if successful.[71]

Soviet forces already feature a smorgasbord of brand-new systems based on technology well known in the West but not fully exploited. Significant samples include intercontinental ballistic missiles (ICBMs) with "cold launch" capabilities,[72] mobile air

[67] Malcolm R. Currie, *Program of Research, Development, Test and Evaluation* (Washington, D.C.: Department of Defense, February 3, 1976), pp. I-6, II-2.

[68] Bruno Augenstein, "Military RDT&E," in *Arms, Men, and Military Budgets,* pp. 220-21; Currie, *Program of Research, Development, Test and Evaluation, FY 1977,* pp. I-6, II-2.

[69] Possony and Pournelle, *The Strategy of Technology,* p. 22. None of the techniques noted is unique to the Soviet Union, but strong emphasis is evident.

[70] Incremental improvements are important even when the main focus is on mighty leaps forward. Indeed, they serve a crucial purpose when opposing systems are essentially equal. "In an air battle conducted with air-to-air missiles . . . a two mile difference in radar ranges can result in one side being destroyed even before it detects the other. Small improvements in missile accuracy can [immensely increase] kill probabilities." Possony and Pournelle, *The Strategy of Technology,* pp. 30, 38-39.

[71] Currie, *Program of Research, Development, Test and Evaluation,* pp. I-5, II-3-4.

[72] A "pop-up" procedure that ejects ballistic missiles from silos or submarines using power plants that are separate from the missile body. Primary ignition is delayed until projectiles are safely clear of carriers/containers, preserving the launcher intact for reuse if required.

defenses, satellite intercept and surveillance craft, armored vehicles and surface ships engineered expressly to operate in chemical/biological warfare environments, rapid-fire rocket launchers, anti-ship cruise missiles, and fire-control systems unmatched either in this country or among other NATO members.[73]

From the small fraction of Soviet exploratory efforts for which U.S. intelligence analysts have sound evidence, several now stressed could bear strongly on the future balance if breakthroughs occur. Controlled thermonuclear fusion could pave the way for limitless power supplies. Wing-in-ground effect aircraft able to skim the sea's surface apparently offer great promise as part of an anti-submarine system. Techniques subjecting certain substances to pressures exceeding a million megabars could transform matter into new forms of unfathomed importance. Metallic ammonia, for example, could constitute the ideal propellant for space ships in its highly condensed stage, or furnish unstable materials for exotic munitions. High energy lasers have endless applications.[74]

U.S. Responses

The United States starts with the world's richest reservoir of scientific resources. Constant feedback between civil and military markets encourages entrepreneurism and technological chain reactions not remotely equalled by our Russian rival. As a result, options still closed to the Soviets are completely open to us.[75]

This country's predominance, however, shows signs of perishability that make many intellectuals lament our lack of momentum.[76] Causes include uncertain goals that make it troublesome to

[73] Currie, *Program of Research, Development, Test and Evaluation,* pp. I-7, 8, and II-10, 17.

[74] "Soviet R&D: No New Weapons Expected, But Surprises Still Possible," *Aerospace Daily,* August 13, 1976, P. 241. Supplemented in a telephone conversation by staff members of ARPA on December 3, 1976. See also Robert Cooke, "The Lethal Laser Is Here—Will it Be Added to US, Soviet Arsenals?" *Boston Globe,* May 23, 1976, p. 1; and "Lasers Turned to Spying on Spies," *Boston Globe,* May 24, 1976, p. 1.

[75] Augenstein, "Military RDT&E," pp. 218, 220, 221 and 244-47; Possony and Pournelle, *The Strategy of Technology,* p. 51.

[76] Senior members of the President's Committee on Science and Technology and the National Science Board, for example, recently expressed concern. Others share such sympathies. Participants at a symposium sponsored by the Massachusetts Institute of Technology (M.I.T.) agreed that serious, speedy steps are needed to prevent further erosion of the U.S. position. Jerome B. Wiesner, M.I.T.'s President, predicted that "if we don't apply our enormous, unused capacity to technology, we face a problem of whether we shall survive 30 years from now." "R&D on the Skids," *Time* (May 3, 1976), p. 56; "Loss of Innovation in Technology is Debated," *New York Times,* November 25, 1976, p. 53.

chart a sound course for defense technology. Insistence on practical products is becoming more pronounced. Fund requests for abstract research are frequently cut or cancelled. Sharp fiscal caution extends to other R&D sectors. Consequent tendencies to tolerate few failures sometimes impede rapid progress.[77]

Beyond that, "one-way street" technological transfers to Soviet competitors create sharp anxiety among some U.S. authorities. A recent RAND report, for example, cautions that uncontrolled exports "of integrated circuit manufacturing plants, machinery, and know-how cannot but improve Soviet military computers" with many potential applications.[78] Similarly, Soviet capabilities may well be enhanced by shipments of U.S. precision grinders that polish miniature bearings for missile guidance systems and MIRVs (multiple independently targetable reentry vehicles).[79] The total impact on U.S. security is difficult to assess, because no focal center studies such trends.

Nevertheless, the United States still holds unsurpassed abilities to compete technologically, and could consistently create superior products, if policies and priorities changed.

Comparative Competence

Relative standings of U.S. and Soviet defense technologies reflect a dynamic situation. Trends there, as elsewhere, are more revealing than static snapshots at any point in time. Figure 5 therefore should segregate sample comparisons into classes that show convergence, divergence, crossovers, static situations, and uncertainties,[80] but lack of information forces a simpler configuration. Confidence in categories shown is high where evidence is conclusive or Soviet leaders signal serious shortcomings by seeking U.S. assistance. It is low where clues are equivocal.

[77] Currie, *Program of Research, Development, Test and Evaluation,* p. I-6; Possony and Pournelle, *The Strategy of Technology,* pp. 23-25; "Weapons Research: Ivan's Edge Is Our Bureaucracy," *Armed Forces Journal* (July 1976), pp. 16, 18.

[78] "Soviet Computer Technology: Catching Up," *Air Force Magazine* (November 1976), p. 67. Additional details are contained in *Long-Range U.S.-U.S.S.R. Competition: National Security Implications,* proceedings of a conference conducted by the National Defense University, July 12-14, 1976, pp. 208-46; and Augenstein, in *Arms, Men, and Military Budgets,* pp. 242-44.

[79] Robert D. Heinl, Jr., "U.S. Technological Gifts to Soviets Hit," *Detroit News,* November 11, 1976, p. 7B.

[80] Augenstein, "Military RDT&E," pp. 216-17.

More importantly, some leads and lags on each side are deliberate, caused at least as much by different missions and developmental styles as by asymmetries in technological competence or failure to foresee demands. The United States could quickly close any *current* gaps if our leaders chose to do so.

Consequently, technological comparisons can be quite confusing, unless linked with strategic concepts, significant threats, and associated force requirements.

FIGURE 5 THE TECHNOLOGICAL BALANCE

General	*Specific*
UNITED STATES CLEARLY SUPERIOR	
"Black box" electronics	Aircraft
Computers	Air-to-air missiles
Integrated circuits	Artillery ammunition
Microtechnology	ECM, ECCM
Night vision	Look-down shoot-down systems
Small turbofan engines	Precision-guided munitions
Space technology	Remotely piloted vehicles
Submarine noise suppressants	Strategic cruise missiles
Target acquisition	Survivable submarines
Terrain-following radar	
SOVIETS CLOSING GAP	
Aerodynamics	MIRVs
Composite materials	Missile accuracy
Inertial instrumentation	Satellite sensors
	Tactical Nuclear Systems
SOVIET UNION CLEARLY SUPERIOR	
Cast components	Air defense missiles
Commonality of components	Anti-ship missiles
Ease of maintenance	Armored fighting vehicles
High pressure physics	Artillery/rocket launchers
Magneto-hydrodynamic power	Chemical/biological warfare
Rockets and ramjets	Cold weather equipment
Simple systems for common use	Gas turbines for ships
Titanium fabrication	ICBM payloads, yields
Welding	Mobile ballistic missiles
	Ship size vs. firepower
	Tactical bridging
STATUS UNCERTAIN	
Acoustics	Anti-ballistic missiles
Adaptive optics	Anti-submarine warfare
High explosive chemistry	High energy lasers
Inductive storage and switching systems for pulsed power control	Satellite-borne radars
Reduced drag for submarines	

NET ASSESSMENT APF

The ability to use technology to gene
a combination of factors that are difficu
U.S. or Soviet Union. These factors includ

- *The overall trend and rate of improvement in capability in each nation.* The U.S.S.R. began th with a much lower rate of capability, and some of the major bureaucratic constraints of the Stalin era. It is unclear at what rate it is now improving relative to the United States.
- *The interaction between tactics, strategy, technology, production efficiency, the quality of planning, and the ability to pursue consistent and well chosen force improvement strategies over time.* The Soviet Union seems to have a major advantage over the U.S. in the quality and consistency of its force improvement strategies. Regardless of the comparative quality of its technology or tactics, the U.S.S.R. does not seem to go through the endless political revisions of its force plans which destabilize much of the U.S. force planning.
- *The level of resources each nation expends over time.* Regardless of the theoretical merits of Western society in permitting more freedom for research, the ultimate quality of U.S. and Soviet defense technology depends on what resources each side devotes to defense. While DOD estimates of these trends sometimes seem more advocacy than analysis, they still merit close attention. One estimate of these trends is shown in Chart Four. Although uncertain, it illustrates the best data the U.S. now has available, as well as the importance of such comparisons.
- *The weight of effort given to things which do not contribute to military effectiveness.* A great deal more U.S. technology reflects a concern with human comfort and survival than does Soviet technology. This channels resources into U.S. R&D and production which are not always important to combat effectiveness.
- *The quality of production technology.* Soviet military production capabilities are now much larger than those of the United States, as shown in Chart Five. The efficiency of Soviet facilities is unclear, but they should permit major economies of scale. Further, a great deal of Soviet equipment exploited after the October War revealed that it was much better designed for rapid low cost production than its U.S. counterpart. Israeli experts were particularly struck by the ease with which even a T-62 tank could be produced relative to the U.S. M-60.

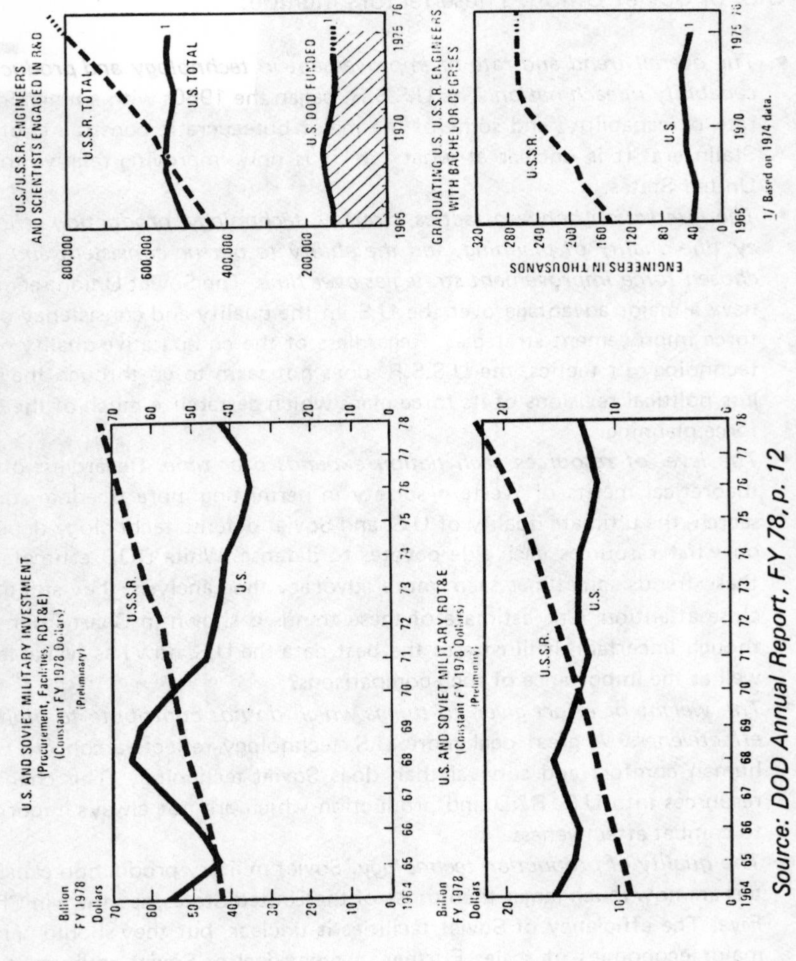

Source: DOD Annual Report, FY 78, p. 12

CHART FIVE

A. ESTIMATED US/USSR RELATIVE PRODUCTION RATES (1972–1976)

	USSR 1972-76 AVG	U.S. 1972-76 AVG	USSR/U.S. RATIO 1972-76
	2,770	469	5.9:1
	4,990	1,556	3.2:1
	1,310	162	8:1
	1,090	573	1.9:1
	666	733	0.8:1
1/	27,000	27,351	1:1

1/ Ground launched antitank missiles

B. TRENDS IN US/USSR PRODUCTION OF GROUND FORCE EQUIPMENT (1966–1976)

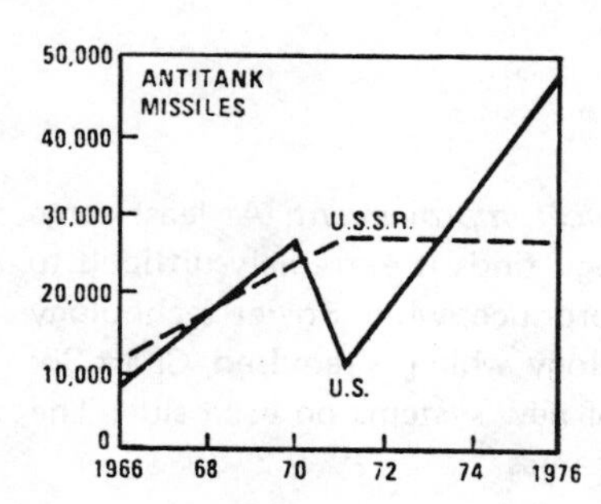

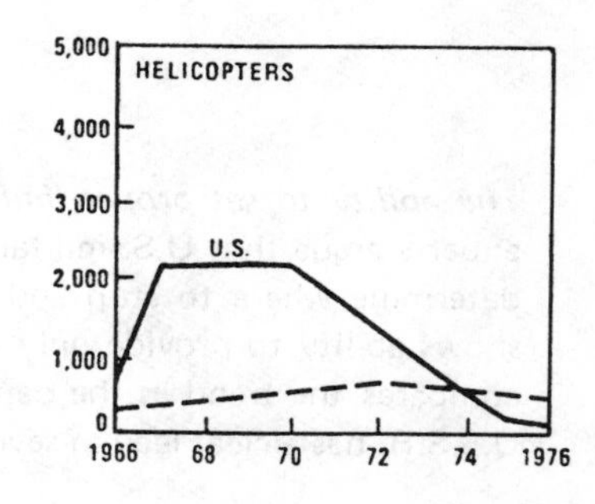

CHART FIVE (Continued)

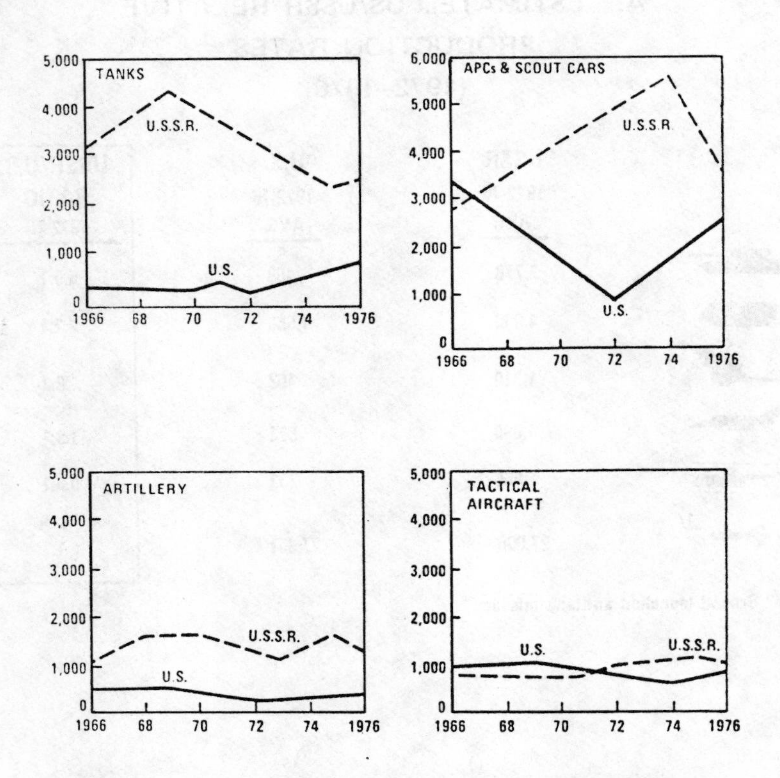

Source: DOD Annual Report, FY 78, p. 27

- *The ability to set proper limits to technical improvement.* At least some experts argue that U.S. military technology finds it extremely difficult to determine where to stop and when to produce, while Soviet technology shows ability to provide only the technology which is required. Chart Six compares the trend in the deployment of new systems on each side. The U.S.S.R. has a clear lead in several critical areas.

CHART SIX

Comparative U.S. and Soviet Defense Procurement Effort

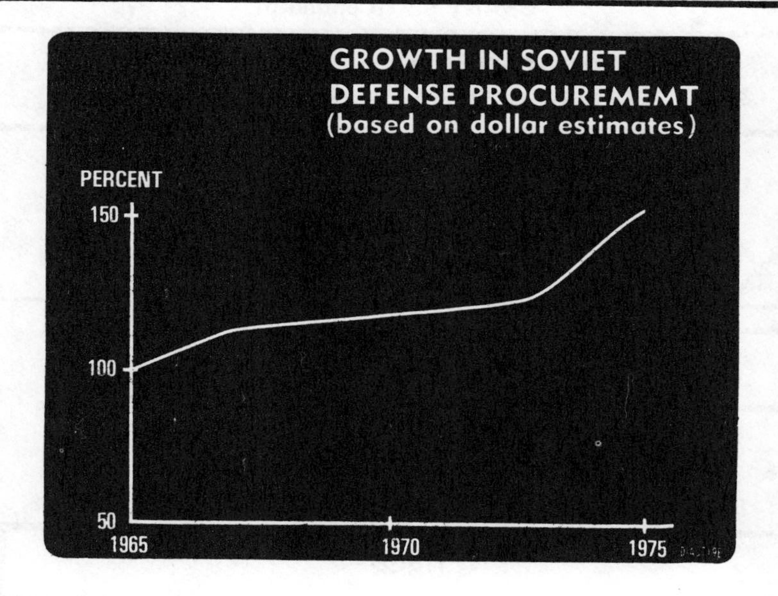

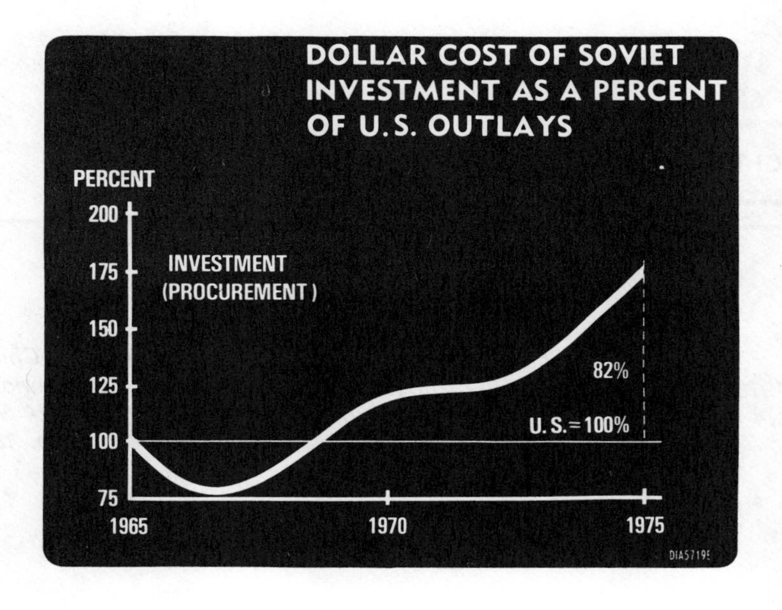

CHART SIX (Continued)

Comparative U.S. and Soviet Defense Procurement Effort

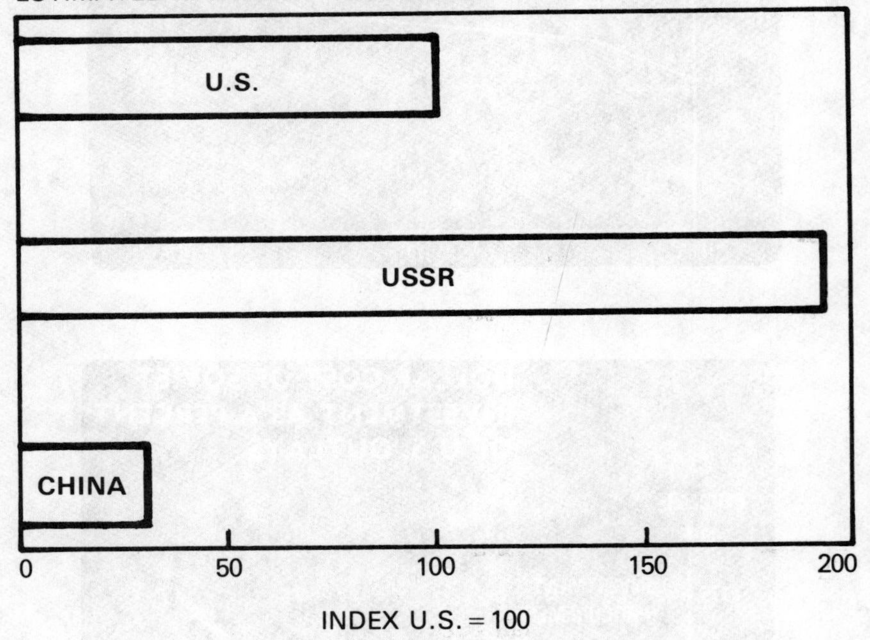

Source: "Allocation of Resources in the Soviet Union and China, 1976", Hearings Before the Subcommittee on Priorities and Economy in Government, Joint Economic Committee, U.S. Congress, 24 May and 15 June, 1976

CHART SEVEN

COMPARISON OF NUMBERS OF NEW SYSTEMS DEVELOPED DURING 1965-1976 BY U.S. AND U.S.S.R.

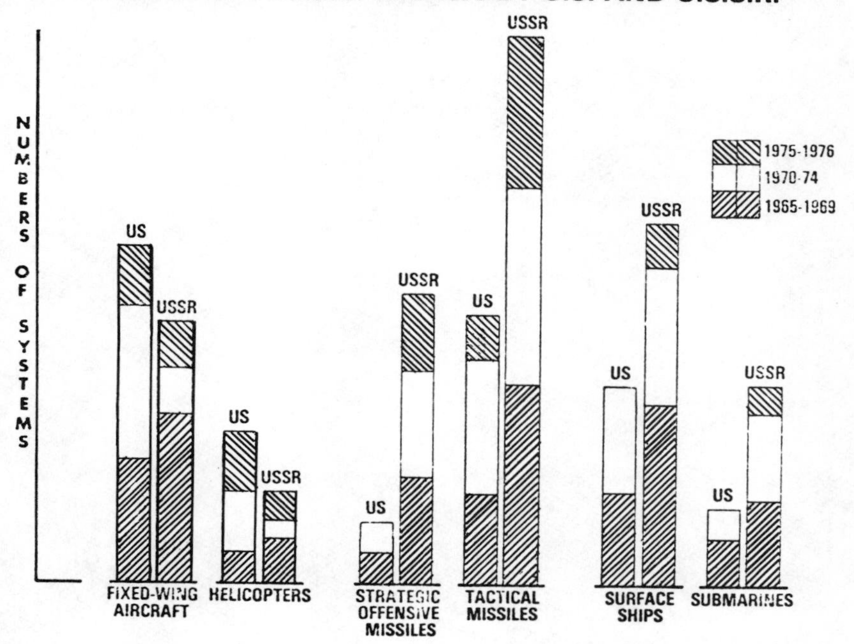

*Sharply understates rate of Soviet deployment of new types because does not count modifications of aircraft like the Mig-21. This favors the U.S. which tends to deploy new types rather than make major changes in existing types.

Source: DOD Annual Report, FY 78, p. 113

[Chapter 3]

Strategic Nuclear Forces

STRATEGIC OFFENSIVE COMPARISONS

United States nuclear deterrent strategy presently depends almost entirely on a retaliatory triad of manned bombers, ICBMs, and submarine-launched ballistic missiles (SLBMs) coupled with a sophisticated command/control system.[1] Those components, in combination, are designed to survive a savage first strike by any enemy and still retain infallible Assured Destruction abilities. According to current concepts, the mix also must afford a range of flexible responses within the framework of our second-strike strategy, minimize probabilities that all U.S. systems could be compromised concurrently by technological surprise, and contribute to stability. This section relates U.S. and Soviet strategic offensive forces to those requirements.[2] (See Graph 3).

[1] A sizable number of Air Force and Navy tactical aircraft, routinely earmarked for strategic nuclear missions, reinforce the U.S. triad. Four new families of nuclear weapons are now under development: air- and sea-launched cruise missiles (ALCMs, SLCMs), air-launched ballistic missiles (ALBMs), and land-mobile ICBMs.

This study uses the terms strategic offensive and strategic retaliatory interchangeably, although U.S. forces are for retaliatory purposes only, according to pronounced policy.

[2] For basic characteristics and implications of delivery systems by functional class, including those now in R&D stages (such as mobile ICBMs), see John M. Collins, *Strategic Nuclear Delivery Systems: How Many? What Combinations?* (Washington, D.C.:

Intercontinental Ballistic Missiles

All U.S. and Soviet ICBMs currently are cased in subterranean concrete silos, although both sides are experimenting with land-mobile models, and U.S. Minutemen have been test launched from aircraft. American efforts, however, are still in early stages, whereas Soviet SS-16s reputedly are ready to join using units. (See Figure 6.)

U.S. TRENDS

The total number of U.S. ICBMs and the "heavy-light" mix of 54 Titan IIs and 1,000 Minutemen has stayed static since 1967. Converting half the force to MIRVed Minuteman IIIs[3] drastically reduced weapons in the megaton range (only 504 remain), but concurrently doubled our warheads from 1,054 to 2,154.

Titans have ample explosive power to crack hard structures, such as Soviet silos, but lack the accuracy to "kill" point targets consistently. Minuteman missiles, which mount smaller warheads, still lack sufficient yield, although that condition is being corrected.[4] Consequently, U.S. ICBMs at this stage are essentially designed to fulfill Assured Destruction functions against Soviet population/production centers and other soft targets.[5]

Congressional Research Service, October 7, 1974), pp. 1-84. Updated and supplemented by *Strategic Force Options Related to SALT II,* Issue Brief 77046 (Washington, D.C.: Congressional Research Service, March 1, 1977), 25 pp. Future trends are addressed in A. A. Tinajero, *Projected Strategic Offensive Weapons Inventories of the U.S. and U.S.S.R.: An Unclassified Estimate* (Washington, D.C.: Congressional Research Service, March 24, 1977), 180 pp.

[3] Multiple reentry vehicles (MRVs) on any ballistic missile are similar to the pellets in a shotgun shell. They saturate a single target. Multiple independently targetable reentry vehicles (MIRVs) also are carried by a single missile, but engage several separate targets.

[4] Yields expressed in kilotons or megatons and accuracies expressed in circular errors probable (CEPs) are subject to considerable speculation. Unclassified documents universally agree on estimates for some U.S. and Soviet systems, but disagree on others. Thomas A. Brown cites three sources (Robert A. Leggett, Kosta Tsipis, and Edward Luttwak) in "Missile Accuracy and Strategic Lethality," *Survival* (March/April 1976), pp. 52-59. See also Thomas J. Downey, "How to Avoid Monad—and Disaster," *Foreign Policy* (Fall 1976), pp. 172-201.

[5] *Ibid.,* pp. 54-55. Warhead lethality, derived from accuracy and yield, is expressed as K. More than 30 K reputedly is required to destroy a silo hardened to resist overpressures of 330 pounds per square inch (psi). About 50 K could crack one hardened to 500 psi. A 1,000-psi silo could survive a shock of almost 80 K. Titan II's largest warhead is rated at less than 20 K. Each MIRV on Minuteman III exerts about 8 K. Several such weapons would be needed to neutralize the weakest Soviet silo.

Graph 3

STRATEGIC OFFENSIVE FORCES
Statistical Summary

(Note Different Scales)

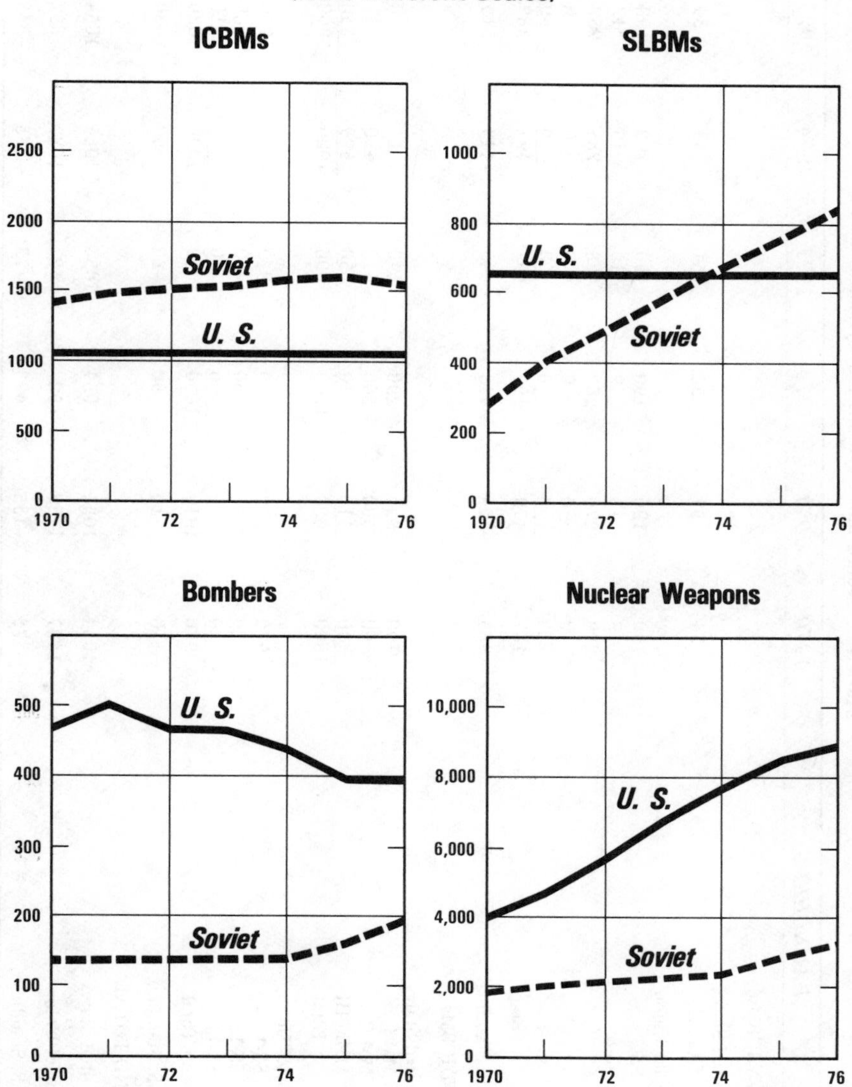

FIGURE 6 INTERCONTINENTAL BALLISTIC MISSILES:
Statistical Trends and System Characteristics

LAUNCHERS	*1970*	*1971*	*1972*	*1973*	*1974*	*1975*	[illegible]
HEAVY ICBMs							
United States							
Titan II	54	54	54	54	54	54	[illegible]
Soviet Union							
SS-7	190	190	190	190	190	190	140
SS-8	19	19	19	19	19	19	19
SS-9	228	270	288	288	288	288	264
SS-18[1]	0	0	0	0	0	10	36
SS-19	0	0	0	0	0	60	100
Total	437	479	497	497	497	567	559
U.S. Standing	-383	-425	-443	-443	-443	-513	-505
LIGHT ICBMs							
United States							
MM I	490	390	290	140	21	0	0
MM II	500	500	500	510	450	450	450
MMM III	10	110	210	350	529	550	550
Total	1000	1000	1000	1000	1000	1000	1000
Soviet Union							
SS-11	970	970	970	970	1030	960	910
SS-13	20	40	60	60	60	60	60
SS-17	0	0	0	0	0	10	20
Total	990	1010	1030	1030	1090	1030	990
U.S. Standing	+10	-10	-30	-30	-90	-30	+10
GRAND TOTAL							
United States	1054	1054	1054	1054	1054	1054	1054
Soviet Union	1427	1489	1527	1527	1587	1597	1549
U.S. Standing	-373	-435	-473	-473	-533	-543	-495

[1] Soviet SS-18 Mod 2 ICBMs have been flight tested with 8-10 MIRVs; a few reportedly have been deployed, but the count is classified.

Figure 6 (continued)

WARHEADS	*1970*	*1971*	*1972*	*1973*	*1974*	*1975*	*1976*
United States							
MIRV	30	330	630	1050	1587	1650	1650
Other	1044	944	844	704	525	504	504
Total	1074	1274	1474	1754	2112	2154	2154
Soviet Union							
MIRV	0	0	0	0	0	400	680
Other	1427	1489	1527	1527	1587	1528	1429
Total	1427	1489	1527	1527	1587	1928	2109
U.S. Standing	-353	-215	-53	+227	+525	+226	+45

SYSTEM CHARACTERISTICS

	First Deployed	*Number RVs*			*Warhead Yield*	*CEP (Nautical Miles)*	*Range (Miles)*	*Cold Launch*
		Single	*MRV*	*MIRV*				
United States								
Titan II	1963	1			10 MT	0.8	7250	No
MM I	1962	1			1 MT	0.5	7500	No
MM II	1965	1			1 MT	0.3	8000	No
MM III	1970			3	170 KT ea	0.2	8000	No
Soviet Union								
SS-7	1962	1			5 MT	1.5	6900	No
SS-8	1963	1			5 MT	1.5	6900	No
SS-9	1967	1			18-25 MT	0.7	7500	No
Mod 4	1971		3		4-5 MT	0.5	7500	No
SS-11	1966	1			1-2 MT	0.7	6500	No
Mod 3	1973		3		500 KT ea	0.5	6500	No
SS-13	1969	1			1 MT	0.7	5000	No
SS-17	1975			4	200 KT ea	0.3	6500	Yes
SS-18	1974	1			18-25 MT	0.3	7500	Yes
Mod 2	N/A			8-10	2 MT ea	0.25	7500	Yes
SS-19	1974			6	200 KT ea	0.25	6500	No

NOTE: The three MRVs on each SS-9 Mod 4 and SS-11 Mod 3 count as single warheads in this summary.

)VIET TRENDS

Moscow has stressed land-based ballistic missiles since the early 1960s. Deployments soared during that decade, then slowed when the tally outstripped our own by almost a third. U.S. intelligence confirms just 122 new Soviet silos since 1970, but all are in the "heavy" category, which includes five systems (SS-7, SS-8, SS-9, SS-18, and SS-19).[6]

Twenty percent of all Soviet warheads are in the 5 to 25 megaton range, versus fewer than five percent for our ICBM force (469 to 54). Except for "small" MIRVs on SS-17s and SS-19s, the least lethal tips have three times the yield of those on Minuteman III. That disparity derived from a conscious U.S. decision to emphasize precision when the Soviets stressed raw power. The Kremlin now strives for both. Before long, its land-based missiles therefore could boast the best accuracy/yield combinations.[7]

Systems with MIRV capabilities are now supplanting those with single weapons. The Soviets already have twice as many ICBM *warheads* as we have *missile silos.* Our count still outnumbers theirs slightly (2,154 to 2,109), but larger Soviet missiles with greater MIRV capacities put Moscow in position to pass us quickly, and Soviet craftsmen have the competence to achieve impressive accuracies if ruling councils so choose.

CONSEQUENCES

The U.S. second-strike strategy, professed by every President since Truman, imposes constraints on this country that the Soviets do not share.

Bigger ICBMs and warheads, combined with better accuracy, would enhance U.S. abilities to smash rival *silos,* but not before the *missiles* therein took flight. At best, such improvements could

[6] A unilateral U.S. statement associated with the 1972 SALT I interim agreement on the limitation of selected strategic offensive systems identified "heavy" ICBMs as those having "a volume significantly greater than that of the largest light ICBM," which then was the SS-11. Since no different definition has been formally adopted, this study considers SS-19s to be "heavies." They exceed SS-11s by about 60 percent in volume and 350 percent in payload capacity, which enables them to handle half a dozen MIRVs rated at roughly 200 KT each.

[7] Lethality (K) is directly proportional to yield$^{2/3}$ and inversely proportional to CEP^2. Thus, increasing any weapon yield by a factor of eight produces just four times more lethal power. Reducing the same weapon's CEP by a factor of eight multiplies K 64 times. Improving accuracy *and* yield by factors of eight increases K 256 times.

check the release of some Soviet reserves, including rapid refirings from cold launch facilities (poor comfort in a cataclysmic war and needless in lesser conflicts).

Far more importantly, present trends bode badly for the prelaunch survival of undefended U.S. ICBMs, whose power to perform Assured Destruction tasks is by no means permanent. Former Defense Secretary Schlesinger, for example, once predicted that Soviet SS-18s with MIRV warheads "could pose a serious threat to our ICBMs."[8] The Federation of American Scientists concurred, with the comment that "our fixed land-based missiles in silos will look more and more vulnerable to attack if MIRV and increases in accuracy cannot be prevented."[9]

Corrective courses of action are subject to question. Installing more U.S. ICBMs would prove impractical, because the Soviets could add hard target warheads much faster than we could build silos and fill them with missiles, at a fraction of the cost.

Prospects for reinforcing U.S. silos face finite limitations (the ultimate compressive strength of cement approximates 3,000 psi, to cite just one criterion). After maximum practical hardness has been achieved, an advanced generation of Soviet warheads with appropriate yields and pinpoint precision could strike each silo in a closely spaced ICBM field with far less fear of "fratricide" than saturation attacks would currently cause.[10]

Launch-on-warning policies would help preserve U.S. ICBMs against surprise assaults only if adequate information were immediately accessible to our National Command Authorities (NCA), who reserve exclusive rights to sanction retaliatory strikes.[11] As a minimum, rational response would depend on sufficient data to determine the magnitude and impact areas (cities or silos) of a Soviet missile attack. Should the U.S. alert apparatus fail to func-

[8] James R. Schlesinger, Report on the FY 1975 Defense Budget, p. 46.

[9] *Solution to Counterforce: Land-Based Missile Disarmament,* Public Interest Report, Federation of American Scientists (February 1974), p. 1.

[10] The simultaneous explosion of two or more nuclear weapons over any target is almost impossible to plan. "Fratricide" occurs when blast, heat, or radiation from the first detonation destroys or deflects other warheads in the salvo. If successive shots delay until adverse conditions dissipate, slightly damaged silos can launch missiles through the resultant "window." Joseph J. McGinchley and Jakob W. Seelig, "Why ICBMs Can Survive a Nuclear Attack," *Air Force Magazine* (September 1974), pp. 82-85.

[11] U.S. National Command Authorities are limited to the President, the Secretary of Defense, and their duly deputized alternates or successors.

tion for any reason, including early enemy action, chances are slim that *any* decision to launch could be made, much less implemented, in the few minutes available.

Undefended U.S. ICBMs in concrete silos may serve deterrent purposes for an extended period of time, despite such disadvantages.[12] In the final analysis, however, the question is not *whether* the Soviets could eventually crush them with a first strike, only *when* they could achieve that capability. The consequent uncertainty makes that leg of our triad an increasing source of instability.

Ballistic Missile Submarine Systems

All U.S. and Soviet sea-launched ballistic missiles (SLBMs) are carried by submarines.[13] They are "sitting ducks" in port, but on station at sea are the most survivable of all strategic nuclear systems. "Hair-trigger" action would not be needed to avoid heavy losses from surprise attacks. Decisions to fire against Assured Destruction type targets could be delayed indefinitely without degrading deterrence or countervalue capabilities. (See Figure 7.)

U.S. TRENDS

The total number of U.S. nuclear-powered ballistic missile submarines has stayed constant at 41 since 1967. Each, regardless of class, still carries 16 SLBMs. The missile count remains at 656, but the mix is decidedly different than it was early in this decade.

All single-warhead Polaris A-2s have been phased out. MRVs on the remaining A-3s have larger yields than most U.S. ICBMs,

[12] General David C. Jones, current Air Force Chief of Staff, contends that "it will be a long time before [the Soviets] could disarm the Minuteman force with any great assurance." His Director of Plans amplified that statement. "Any reasonable cautious Soviet planner or policy-maker contemplating a nuclear strike on the Minuteman force (which—if it failed—could result in the destruction of the power base of the Communist Party of the Soviet Union) would entertain the following sorts of doubts . . . . Would the Soviet missiles work with the reliability estimated from limited peacetime tests? Would there be previously undetected bias errors which degrade accuracy? Would the U.S. silos be 'harder' than anticipated? Would the surviving U.S. ICBMs in silos still possess sufficient destructive power to pose a significant threat to the Soviets because of their multiple warheads?" Testimony before Senate Budget Committee, Task Force on Defense, November 13, 1975.

[13] Soviet ships, not counting coasters, carry more than 350 SS-N-3/SS-N-12 cruise missiles (SLCMs) that can be armed with nuclear warheads. Surface ships carry 56, submarines the remainder. Their effective range of 150 to 250 nautical miles makes them a potential threat to U.S. targets close to ocean shores, but their main mission seems to be anti-shipping.

FIGURE 7 BALLISTIC MISSILE SUBMARINE SYSTEMS
Statistical Trends and System Characteristics

SUBMARINES	*1970*	*1971*	*1972*	*1973*	*1974*	*1975*	*1976*
NUCLEAR POWER							
United States							
Poseidon	1	7	12	22	24	28	28
Polaris	40	34	29	19	17	13	13
Total	41	41	41	41	41	41	41
Soviet Union							
D-II	0	0	0	0	0	0	4
D-I	0	0	0	1	5	11	13
Y	13	20	26	31	33	34	34
H	7	7	7	7	7	7	7
Total	20	27	33	39	45	52	58
U.S. Standing	+21	+14	+8	+2	-4	-11	-17
DIESEL POWER							
United States	0	0	0	0	0	0	0
Soviet Union							
G-II	11	11	11	11	11	11	11
G-I	9	9	9	9	9	9	8
Total	20	20	20	20	20	20	19
U.S. Standing	-20	-20	-20	-20	-20	-20	-19
TOTAL							
United States	41	41	41	41	41	41	41
Soviet Union	40	47	53	59	65	72	77
U.S. Standing	+1	-6	-12	-18	-24	-31	-36

NOTE: U.S. Franklin, Madison, and Lafayette Class submarines are armed with Poseidon SLBMs. George Washington and Ethan Allen class carry Polaris missiles. Three Polaris boats presently are being converted to Poseidon. When that process is complete, the count will be 31 Poseidons and 10 Polaris submarines, total 41.

Soviet D-II submarines are armed with 16 SS-N-8 SLBMs. D-Is have 12. Y-Class submarines have 16 SS-N-6 missiles. H and G-II models carry 3 SS-N-5 missiles. G-I submarines have 3 SS-N-4s.

FIGURE 7 (continued)

SLBMs	*1970*	*1971*	*1972*	*1973*	*1974*	*1975*	*1976*
ON NUCLEAR SUBS							
United States							
Polaris A-2	128	128	128	128	64	32	0
Polaris A-3	512	416	336	176	208	176	208
Poseidon	16	112	192	352	384	448	448
Total	656	656	656	656	656	656	656
Soviet Union							
SS-N-5	21	21	21	21	21	21	21
SS-N-6	208	320	416	496	528	544	544
SS-N-8	0	0	0	12	60	132	220
Total	229	341	437	529	609	697	785
U.S. Standing	+427	+315	+219	+127	+47	–41	–129
ON DIESEL SUBS							
United States	0	0	0	0	0	0	0
Soviet Union							
SS-N-4	27	27	27	21	21	21	18
SS-N-5	33	33	33	39	39	39	39
Total	60	60	60	60	60	60	57
U.S. Standing			–60	–60	–60	–60	–57
GRAND TOTAL							
United States	656	656	656	656	656	656	656
Soviet Union	289	401	497	589	669	757	842
U.S. Standing	+367	+255	+159	+67	–13	–101	–186

FIGURE 7 (continued)

WARHEADS	*1970*	*1971*	*1972*	*1973*	*1974*	*1975*	*1976*
United States							
MIRV	160	1120	1920	3520	3840	4480	4480
Other	640	544	464	304	272	208	208
Total	700	1664	2384	3824	4112	4688	4688
Soviet Union							
MIRV	0	0	0	0	0	0	0
Other	289	401	497	589	669	757	842
Total	289	401	497	589	669	757	842
U.S. Standing	+411	+1263	+1887	+3235	+3443	+3931	+3846

	MISSILE CHARACTERISTICS						
		Number RVs					
	First Deployed	*Single*	*MRV*	*MIRV*	*Warhead Yield*	*CEP (Nautical Miles)*	*Range (Miles)*
United States							
Polaris A-2	1962	1			800 KT	0.5	1750
Polaris A-3	1964		3		200 KT ea	0.5	2880
Poseidon	1971			10	40 KT ea	0.3	2880
Soviet Union							
SS-N-4	1961	1			? MT	2.0	350
SS-N-5	1963	1			? MT	2.0	750
SS-N-6	1968	1			1 MT	1.5	1750
Mod 3	1974		3		? KT ea	1.0	2000
SS-N-8	1973	1			1 MT	0.8	4800

NOTE: The three MRVs on each Polaris A-3 and SS-N-6 Mod 3 count as single warheads in this summary.

but poorer CEPs.[14] The preponderance of power now lies with MIRVed Poseidon missiles, which may carry as few as 6 or as many as 14 warheads. Most are tipped with 10, for an average of 160 per boat. Accuracy, however, at a third of a mile, is still less than the best ICBMs, and 40 KT yields are only one-third as much as Minuteman's MIRVs. Their lethality against hard targets is commensurately less.

U.S. submarine missile systems therefore afford an effective Assured Destruction force for use against most surface structures in Soviet cities, but several warheads would be needed to cripple a single silo. Flight times to bomber bases deep in Soviet territory allow enemy aircraft on alert ample opportunity to scramble.

Switching to many small MIRVs had other side effects. The current Poseidon complement concentrates 4,480 weapons on 28 boats, of which about a third are tempting targets immobilized in port for maintenance at any given time.[15] Each submarine at sea, however, poses deterrent threats of immensely greater magnitude than its 1970 predecessors.

SOVIET TRENDS

Antiquated submarines armed with three short-range SLBMs each accounted for two-thirds of Moscow's sea-launched missile strength as late as 1970. Half were diesel powered. Since than, 38 new nuclear subs have hoisted the total to almost twice our own. The number of tubes has nearly tripled (from 289 to 842). Only 78 short-range missiles remain. Other types could obliterate U.S. bomber bases (but not alert aircraft) from firing positions on our continental shelf and in the Caribbean, using yields measured in megatons. Short flight times would allow fewer than ten minutes warning. SS-N-8s, recently introduced, have a 4,800-mile range

[14] The last three Polaris A-3 submarines being converted to Poseidon are still in shipyards. Those boats, with a total of 48 missile tubes aboard, count as Polaris A-3 in this study, no matter what their state of completion on December 31, 1976.

[15] In the past, U.S. ballistic missile submarines averaged 60 days on patrol and 30 days in port. Transit times to launch stations and return still take 2 to 10 days, depending on base locations. Improved boats, better maintenance procedures, and longer range missiles will enable each Polaris/Poseidon boat to stay on station for longer periods, but a sizable percentage must always be in port for repair and crew rest. The same will be true for Trident.

that this country's missiles will not match for several more years.[16] Immense launch areas, which complicate U.S. search procedures, enhance survivability.

Even so, Soviet submarine systems are inferior to U.S. counterparts in most crucial respects. Diesel-powered types still comprise a third of the count. Late-model nuclear boats, being noisier than Polaris and Poseidon, are more susceptible to compromise. On-station time is substantially less. Total target coverage is only a fifth that afforded by our force (842 to 4,688), since none of their missiles are MIRVed.[17] SS-N-8s must sacrifice security for counterforce capabilities. Should they rely on remote firing points, flight times to U.S. bomber bases would triple, and threats would be much reduced.

CONSEQUENCES

The expanding strength of Soviet submarine-launched missile systems in no way endangers U.S. counterparts, except for those in port, nor does it degrade their deterrent capacities. Matching Moscow's buildup with more submarines, SLBMs, and MIRVs would therefore serve little purpose in terms of essential missions.

Reinforcing SLBMs with bigger warheads and better accuracy would improve U.S. hard target kill capabilities from firing positions close to Soviet shores, since flight times would be shorter than those for ICBMs. Each submarine, however, would run serious risks in enemy coastal waters, and chances of catching Soviet missiles in silos would be slight, given our second-strike strategy.[18]

Qualitative changes centered on continued pre-launch survivability therefore seem to proffer the best prospects for preserving the deterrent powers of American SLBMs, despite Soviet ASW (anti-submarine warfare) efforts.

[16] George S. Brown, *United States Military Posture for FY 1976* (Washington, D.C.: The Joint Chiefs of Staff, 1976), pp. 23-24.

[17] Soviet SS-N-X18s are being tested in a MIRV mode with about three warheads. Their estimated range is 4,600 miles. The SS-N-X17, with a range of roughly 2,000 miles, has not yet been tested with MIRVs, but may have such a capability. Donald H. Rumsfeld, news conference at the Pentagon, September 27, 1976, p. 4. See also "Russians Test New Submarine Missile," *Chicago Tribune* (November 24, 1976), p. 2.

[18] Related problems were reviewed earlier in the subsection on ICBMs, including Notes 5 and 7.

Strategic Bombers

The United States and Soviet Union both based strategic nuclear strength on manned bombers until ballistic missiles were deployed en masse, beginning early in the 1960s. Thereafter, the U.S. accent on air power stayed comparatively strong, while Soviet stress has been slight (see Figure 8).[19]

Assured Destruction is the principal capability in each case, because U.S. and Soviet alert forces (aircraft and ICBMs) have adequate time to launch before bombers could arrive. That characteristic, which makes aircraft a poor first-strike system, enhances strategic stability.

U.S. TRENDS

B-52s assigned to Strategic Air Command (SAC) as unit equipment have decreased 30 percent since the start of this decade, from 465 to 330. The oldest airframes have been flying for 20 full years, the newest for 15,[20] but they still pack a powerful wallop.

Multimegaton gravity bombs come in various sizes, one of which offers assorted yields. B-52s carry up to four of either as their basic load.[21] They could also be fitted for up to 20 nuclear-tipped Short-Range Attack Missiles (SRAMs), although only 1,500 were produced, an average of four per plane. Those weapons are designed to assist aircraft penetration by suppressing enemy defenses and/or engaging main targets.[22]

Sixty-six FB-111s in the "medium" category supplement U.S. "heavy" bombers. Payloads are roughly half that of B-52s, and ranges are less than one-third, but each can carry six slim bombs, six SRAMs, or some combination (an average of two SRAMs per aircraft is in stock).[23]

SAC tanker squadrons furnish aerial refueling support that

[19] Statistics in this section refer only to unit equipment (UE) aircraft. Those in storage, being cannibalized, or used for training purposes do not count. Significant numbers of U.S. land- and carrier-based tactical aircraft that act as strategic nuclear auxiliaries are also excluded. Soviet forces have no counterparts.

[20] About 75 B-52Ds, delivered to SAC in 1957, still were assigned in December, 1976, according to Air Force staff officers. The last B-52H models entered service in 1962.

[21] Craig Covault, "B-52 Training Stresses Timing, Realism," *Aviation Week and Space Technology* (May 10, 1976), p. 127.

[22] *Annual Air Force Almanac Issue, Air Force Magazine* (May 1976), p. 123.

[23] *Ibid.*, p. 112.

FIGURE 8 STRATETIC NUCLEAR BOMBERS
Statistical Trends and System Characteristics

AIRCRAFT	*1970*	*1971*	*1972*	*1973*	*1974*	*1975*	*1976*
HEAVY BOMBERS							
United States							
B-52	465	435	397	397	372	330	330
Soviet Union							
Bear	100	100	100	100	100	100	100
Bison	40	40	40	40	40	35	35
Total	140	140	140	140	140	135	135
U.S. Standing	+325	+295	+257	+257	+232	+195	+195
MODERN MEDIUM BOMBERS							
United States							
FB-111	4	66	66	66	66	66	66
Soviet Union							
Backfire	0	0	0	0	0	25	60
U.S. Standing	+4	+66	+66	+66	+66	+41	+6
GRAND TOTAL							
United States	469	501	463	463	438	396	396
Soviet Union	140	140	140	140	140	160	195
U.S. Standing	+329	+361	+323	+323	+298	+236	+201
NUCLEAR WEAPONS							
United States	2226	1762	1842	1206	1426	1658	
Soviet Union	140	140	140	140	140	185	
U.S. Standing	+2086	+1622	+1702	+1066	+1286	+1473	

Note: Bomber force loads vary according to assigned missions. U.S. weapons figures above were derived by subtracting ICB heads from total loads published in DOD posture statements for FY 1971-1977. Soviet figures reflect one large bomb or ASM p and two ASMs per Backfire. No reserve weapons are counted on either side.

FIGURE 8 (continued)

	BOMBER SYSTEMS						
	First Deployed	*Unrefueled Combat Radius (Miles)*	*Bomb Load (Lbs)*	*ASM*	*Max Speed (Mach)*	*Engines*	
						Nr	*Type*
AIRCRAFT							
United States							
B-52 G	1959	3,385	60-70,000	SRAM	0.95	8	Jet
FB-111	1969	1,550	13,500	SRAM	2.5	2	Jet
Soviet Union							
Bear	1956	3,900	40,000	AS-3, AS-4	0.78	4	Piston
Bison	1956	3,250	20,000		0.87	4	Jet
Backfire	1974	2,500	20,000	AS-4	2.5	2	Jet

	First Deployed	*Warhead Yield*	*Range (Miles)*
SUPERSONIC AIR-TO-SURFACE MISSILES			
United States			
SRAM (W-69)	1972	200 KT	100
Soviet Union			
AS-3[1]	1961	1 MT	400
AS-4[2]	1962	? KT	450
AS-6	1975	200 KT	155 lo, 500 hi

[1] AS-3 is commonly called Kangaroo.

[2] AS-4 is commonly called Kitchen.

[3] B-52 G combat radius reflects maximum high altitude mission armed with average load of bombs and SRAMs. B-52H radius under those conditions is 4060 nautical miles. FB-111 radius is with SRAMs only. The 1,500-mile combat radius would be reduced if the load were 13,500 lbs of nuclear bombs.

assures intercontinental range for B-52s and FB-111s under combat conditions.

Flexible force loadings for U.S. bombers are adjusted to suit changing missions and target assignments, as Figure 8 shows. As currently armed, SAC's air wings account for almost a fourth of all allocated U.S. weapons and more than half of all megatonnage.[24] Fewer aircraft lift larger loads than in the recent past.

SOVIET TRENDS

Soviet heavy bomber strength peaked at about 210 turboprop Bears and jet-powered Bisons in 1966, then steadily dropped to 135, the current tally. Something like 80 tankers serve those antiques, whose penetration prospects would be poor against any determined defense. Both types probably average just one large gravity bomb as the basic load, although B-Model Bears may carry a single AS-3 Kangaroo missile, which could be released 400 miles from its target.[25]

Supersonic Backfire, the first new Soviet bomber deployed in the past fifteen years, is smaller but much more sophisticated. That modern aircraft probably threatens sea lanes and NATO more than North America, but could strike some U.S. cities without resorting to tanker support, then recover in Cuba or another "neutral" country. In-flight refueling is technically possible, because all Backfires are fitted with receptacles. Standoff missiles extend their range by as much as 500 miles.[26] (Medium-range Badgers, which could attack undefended targets in the United States on one-way missions, are covered in sections concerning theater nuclear and land-based naval aircraft.)

CONSEQUENCES

Backfire bombers may have some bearing on U.S. needs for

[24] U.S. Congress, House, *Hearings by the Research and Development Subcommittee of the Armed Services Committee on FY 1977 Authorization*, Part 5 (Washington, D.C.: U.S. Government Printing Office, 1976), p. 239.

[25] "Second Annual Soviet Aerospace Almanac," *Air Force Magazine* (March 1976), pp. 94-95, 105.

[26] *Ibid.*, pp. 95-96, 105. See also A. A. Tinajero, *The Soviet Backfire Bomber* (Washington, D.C.: Congressional Research Service, November 24), 1975, 3 pp.

Backfire's capabilities and limitations still cause controversies that will not likely be soon resolved. (See Chapter I for one example.)

improved air defenses, but they are completely unrelated to offensive force requirements. Backfire squadrons, which currently contribute less to Soviet strategic nuclear capabilities than forward-based fighters add to ours, could double in number or disappear without diluting any advantages that accrue from SAC's aircraft. Manned bombers may indeed be a legitimate leg of the U.S. triad, but maintaining superiority, essential equivalence, or any other military balance with Backfire would serve some cosmetic or symbolic purpose, nothing more.

Triads Assessed in Tandem

Separate assessments of triad components supply incisive insights, but only a survey of interactions on competing sides can measure complete implications.

COMPARATIVE PATTERNS

The basic composition of both triads has stayed constant during this decade. Only sizes and shapes have changed. (See Graph 4 and Figure 9.)

America maintains a balanced structure of ballistic missiles and bombers. The Soviet Union does not. Most of its might will remain with large, land-based missiles, even if Moscow MIRVs SLBMs, because bigger, more numerous ICBMs can carry many more warheads.[27] The importance of manned bombers is minuscule in comparison.

Soviet delivery systems are somewhat more numerous than in 1970, while our count is slightly smaller, but MIRV programs caused U.S. weapon holdings to rise at a rapid rate until they reached a plateau not yet approached by our rival. America's transcendence, however, may be transitory. The Soviets have established a solid expansion base, and growth could be just beginning.

COGENT IMPLICATIONS

U.S. bombers and ICBMs are more vulnerable than ever before. That condition causes instability, even though no current combi-

[27] ICBMs presently carry 70 percent of all Soviet warheads. That share could increase to 90 percent or more if SS-19 deployments are large and SS-18s mount maximum MIRVs. By way of contrast, U.S. ICBMs carry fewer than a fourth of our strategic nuclear weapons.

Graph 4

TRIADS COMPARED STATISTICALLY

(Note Different Scales)

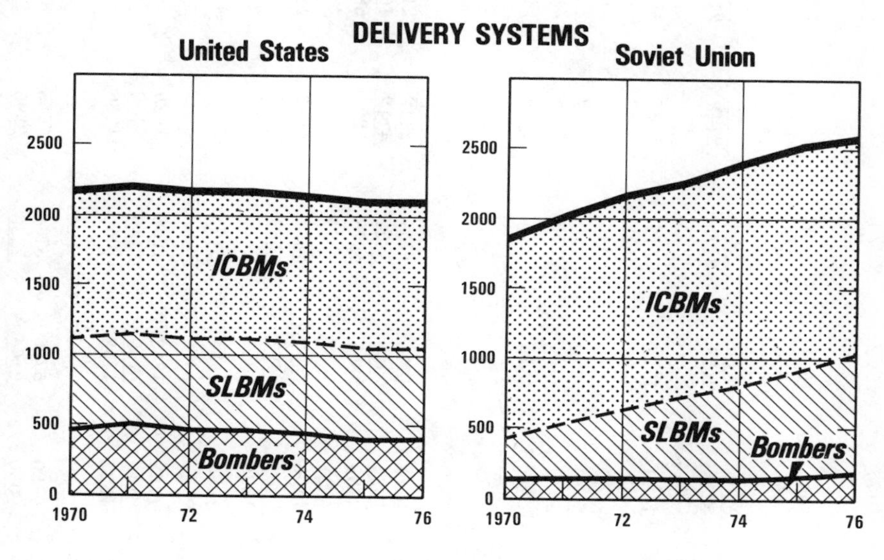

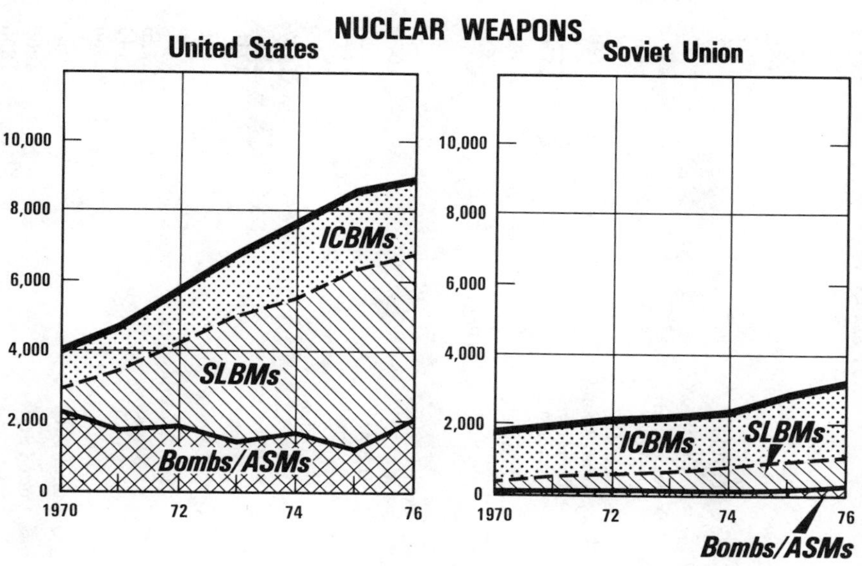

FIGURE 9 STRATEGIC OFFENSIVE FORCES
Statistical Recapitulation

SYSTEMS	*1970*	*1971*	*1972*	*1973*	*1974*	*1975*	*1976*
United States							
ICBM	1054	1054	1054	1054	1054	1054	1054
SLBM	656	656	656	656	656	656	656
B-52	465	435	397	397	372	330	330
FB-111	4	66	66	66	66	66	66
Total	2179	2211	2173	2173	2148	2106	2106
Soviet Union							
ICBM	1427	1489	1527	1527	1587	1597	1549
SLBM	289	401	497	589	669	757	842
Bear/Bison	140	140	140	140	140	135	135
Backfire	0	0	0	0	0	25	60
Total	1856	2030	2164	2256	2396	2514	2586
U.S. Standing	+323	+181	+9	-83	-248	-408	-480
WEAPONS							
United States							
ICBM	1074	1274	1474	1754	2112	2154	2154
SLBM	700	1664	2384	3824	4112	4688	4688
Total	1774	2938	3858	5578	6224	6842	6842
Soviet Union							
ICBM	1427	1489	1527	1527	1587	1928	2109
SLBM	289	401	497	589	669	757	842
Total	1716	1890	2024	2116	2256	2685	2951
U.S. Standing	+58	+1036	+1834	+3462	+3968	+4157	+3891
Bombs/ASMs							
U.S.	2226	1762	1842	1206	1426	1658	2058
Soviet	140	140	140	140	140	185	255
U.S. Lead	+2086	+1622	+1702	+1066	+1286	+1473	+1803
Grand Total							
United States	4000	4700	5700	6784	7650	8500	8900
Soviet Union	1856	2030	2164	2256	2396	2870	3206
U.S. Standing	+2144	+2670	+3536	+4528	+5254	+5630	+5694

nation of assaults could smother the two systems simultaneously.[28] Soviet counterparts are comparatively secure, because of our second-strike strategy. SLBMs are still safe at sea.

Both sides consequently display awesome Assured Destruction abilities, but neither could neutralize the other by adding offensive power under current conditions. Defensive measures may endanger U.S. survival to a great degree, as discussed in the following section.

[28] Soviet SLBMs, with flight times of 6 to 10 minutes if fired close to U.S. coasts, might catch some of SAC's aircraft on strip alert, but still lack sufficient accuracy/yield combinations to crush concrete silos. Soviet ICBMs, which take about 30 minutes to reach targets, pose greater threats to American land-based missiles, but would allow authorities ample time to launch bombers. Scheduling problems of that sort are close to insoluble.

NET ASSESSMENT APPRAISAL

In many ways, there are three strategic balances. An "invisible" balance which is only seen by those U.S. and Soviet strategic planners who have unique access to the classified information available to each side, and who model and plan strategic nuclear forces. A "visible" balance reflected in public discussion and unclassified studies. And, a "phantom" balance that will only become real if SALT II is negotiated.

THE "INVISIBLE" BALANCE: THE IMPORTANCE OF UNCERTAINTY AND DYNAMIC ANALYSIS

It may initially seem difficult to believe that little is publicly known about actual U.S. and Soviet capability to fight strategic wars. The decades-long debate over the size of U.S. strategic forces has led to an incredible amount of formerly classified data reaching the public. This data has been supplemented by countless estimates of the balance by various experts. The resulting flood of figures and studies sometimes gives the impression that "all is known," that an analyst outside the highly classified staffs who do real strategic planning can validly measure the strategic balance.

THE IMPACT OF UNCERTAINTY

There are two major reasons why this is not true. The first reason is uncertainty. Some of the major areas of controversy and uncertainty in strategic force analysis are shown below:

- Number, purpose, and individual system capabilities of U.S. and Soviet MIRVing programs.
- Range, reliability, re-targeting capability, total system accuracy, yield, burst attitude, simultaneous or staggered launch capability, and time on target capability, of U.S. and Soviet ICBMs, SLBMs, and air to surface missiles.
- U.S. and Soviet silo hardness, and total system survivability including missile to silo coupling effects in given types of attacks.
- Bomber penetration capabilities. A few of the uncertainties include actual effectiveness of Soviet SAMS and fighters, the effectiveness of active penetration aids, and effectiveness of electronic warfare systems.
- Alert rates under attack conditions, and response or reaction capabilities. For example, how many bombers could take off on warning or how many SSNBs will be operational.
- Pre-strike targeting, ability to achieve warning that a strike is occurring and analyze its nature, post-strike damage assessment capability, post-strike target acquisition and retargeting capability.
- Operational capability to minimize radar cross section, minimize submarine noise, rely on terrain following radar, achieve MIRV release, and countless other factors.

It should be clear from this list that no single unclassified estimate of the balance can be fully objective or accurate. It should be equally clear that no expert outside government could analyze all the credible possibilities, and that even an analyst inside government cannot plan without uncertainty.

THE IMPORTANCE OF DYNAMIC ANALYSIS

The U.S. and Soviet military planning staffs who try to consider the impact of all these uncertainties can only do so by using large scale computer models, and by gaming or simulating the interactions between as many major contributors to the outcome of a strategic exchange as they can possibly consider. They must also model uncertainty and consider "high," "low," and "best" estimates of virtually every number used in their model. Accordingly, as complex as "static" counts of strategic warheads or delivery systems may seem, they are only a small part of the analysis that goes into strategic planning. As one USAF expert notes:

> "Hundreds of other force parameters and capabilities exist which should be compared when examining the Strategic Balance. Most of these describe complex technical or force interactions which are not the domain of the layman.
>
> ". . . Dynamic battle management includes those capabilities important for flexible response options and limited nuclear wars, such as: pre-impact missile retargeting; launch-under-attack for ICBMs; empty-silo information on enemy missiles to avoid wasting weapons; bomb damage assessment for one's own weapons to know if second attack waves are required; attack damage assessment on one's own forces to know what retaliatory capability still exists. Invulnerable or hardened command, control and communication assets such as airborne or underground command posts, satellite relays, high data rate communication and computer systems could easily spell the difference in a nuclear engagement.
>
> ". . . Most of these factors are best treated in a dynamic balance context where interactions such as bombers vs air defense can be quantified depending on the scenario being considered."[1]

Strategic war planning is concerned with "outcomes" of a war and not with force "inputs." The number of total forces available to each side is only important in terms of its ultimate effect, along with all other factors, in killing opposing strategic systems, other military targets, and civilian populations or economic capabilities. Even this output in total casualties is only important relative to the number that the planner is seeking to kill to achieve a given outcome; overkill is useless.

MEASURING THE "VISIBLE" BALANCE: PUBLIC PERCEPTION OF STRATEGIC FORCES

It is not surprising, given these uncertainties and complexities, that experienced analysts can publicly defend such radically different views of the "true" balance. The differences in viewpoint are even greater when an analyst adds his own ideological, political, or moral judgments to his analysis. For example, some analysts find

[1]*Source:* Lt. Col. Gerald T. Rudolph, "Assessing the Strategic Balance," DOD Report, June 1976.

it impossible to accept anything less than clear U.S. superiority in all measures of force strength, in all contingencies, and prolonged indefinitely into the future. Other analysts find it equally impossible to justify more U.S. strategic forces than a token capability to destroy Soviet civilian populations.

MANIPULATING THE "VISIBLE" BALANCE

Collins has, quite correctly, chosen to challenge some comparisons which have broad official sanction. He has avoided the more controversial or partisan ways of portraying statistics. From a net assessment perspective, however, it is the range of public perceptions that shape the "visible" balance, and not the best or most accurate set of figures.

Two themes emerge from Collins's study: One concerns the inflexibility of U.S. forces, some of which are poorly structured to accomplish their assigned missions, and are chronically spread too thin. Numbers *do* count, when the imbalance favors one side too heavily. The second concern is the inability of bigger defense budgets to cure connected ills, unless concurrent steps are taken to correct shortsighted strategies.

A closer look, then, at the recent trends which John Collins's study highlights, offers new insight into the thinking and analysis which shape that "visible" balance he details so clearly.

COMPARING U.S. AND SOVIET DELIVERY SYSTEM STRENGTH

Delivery vehicle strengths tell three specific things about the U.S. and Soviet balance. First, they provide an approximate index of the number of counter-force targets; second, they summarize the differences in the structure of the triad on each side; and third, they provide a gross measure of the level of effort going into each side's strategic forces. They do not tell much about kill capability because the number of warheads, yield, and accuracy is more critical in determining relative ability to destroy given types of targets.

Graph number 4 in Collins's force comparison has shown the broad trend in delivery system numbers during 1970-1976, but there are other ways such numbers can be portrayed. Chart One, for example, shows perhaps the best available unclassified estimate of how delivery system numbers are projected to change between 1970 and 1980. It shows that the Soviets will probably double their delivery vehicles during the decade, and overcome a major

CHART ONE

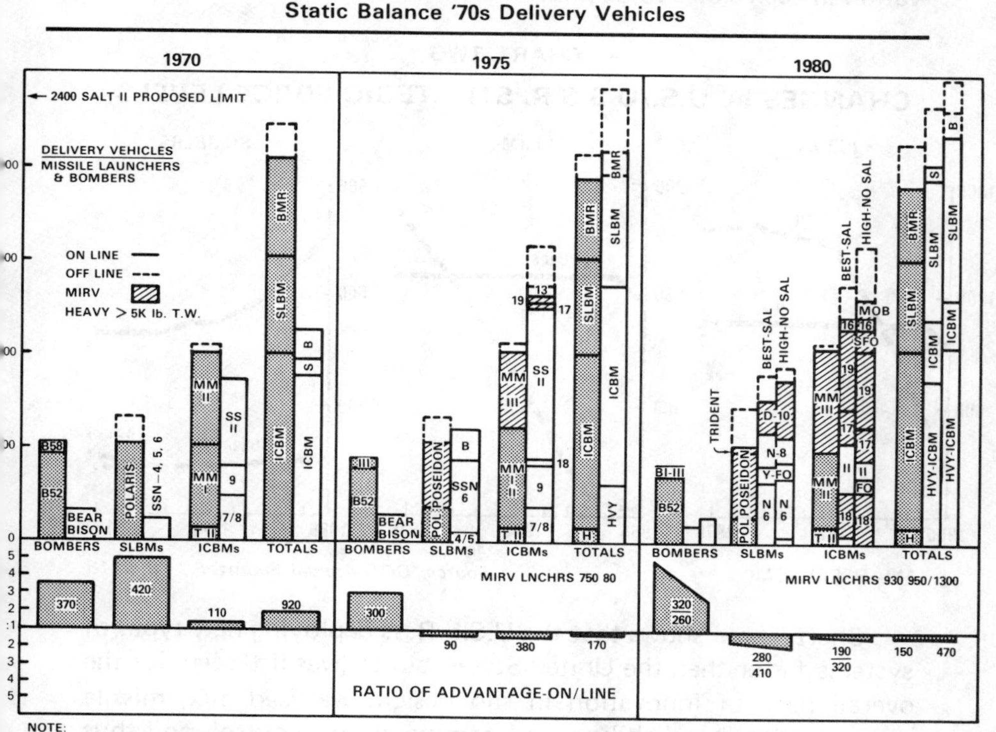

NOTE:

1. Ratio of advantage is indicated by the height of the graph; absolute difference (rounded off) is shown by the number in the ratio graph.
2. Symbols used are:

HVY – Heavy ICBMs	MOB – Mobile missile	C3, C4 – Trident SLBMs	17, 18, etc. SS17, 18...ICBM
SFO – Small follow-on	H.D. – Hound Dog ASM	Y, D – Yankee, Delta SSBNs	MM – Minuteman
LFO – Large follow-on	T II – Titan II ICBM	N – SSN 4, 5, 6, 7, 8 SLBM	(hatched) – MIRVED Systems

3. A launcher is an SLBM tube, an ICBM silo, or equivalent soft missile launcher. This would not reflect available missiles if launchers are capable of being reloaded after firing.

Source: Lt. Col. Gerald T. Rudolph, "Assessing the Strategic Balance," DOD Report on Measuring the Strategic Balance, 24 June, 1976

initial inferiority unless the SALT I ground rules change. The U.S. will reduce its force numbers slightly during the same period. The U.S.S.R., however, will rely on ICBMs and SLBMs which can be upgraded to counter-force capability. The U.S. will rely on a balanced triad with a large bomber force. Neither side will have striking advantage in numbers.

Chart Two provides a variation of this analysis which highlights the shift in the balance in ICBMs and SLBMs in favor of the U.S.S.R., and illustrates the dependence the U.S. continues to place upon manned bombers. It also shows the rate of change and innovation in each side's force mix.

CHART TWO

CHANGES IN U.S./U.S.S.R. STRATEGIC FORCE LEVELS

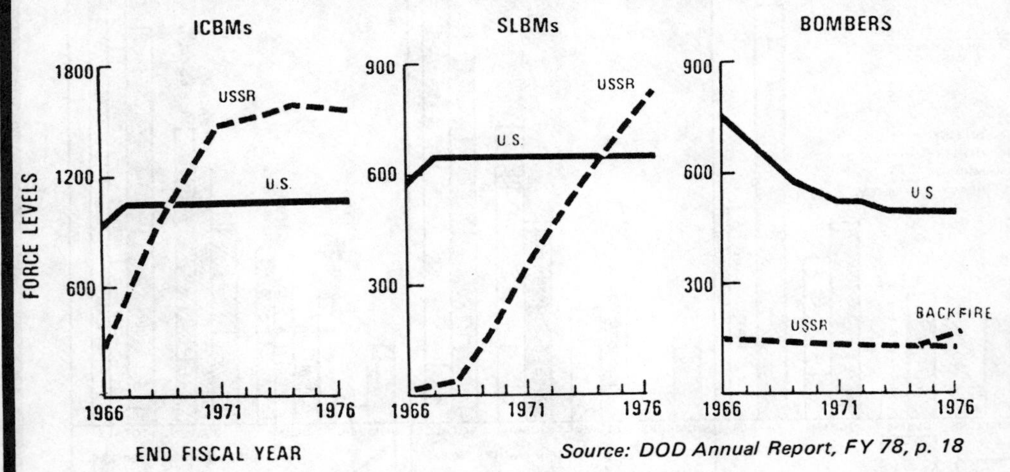

Source: DOD Annual Report, FY 78, p. 18

Chart Three shows that the U.S.S.R. is deploying new types of systems faster than the United States, but it gives little feel for the overall pace of innovation in silo design, warhead mix, missile range, guidance, reliability, and command and control, and thus tends to sharply understate the rate of overall change in both U.S. and Soviet forces. For example, the U.S. has deployed 550 new ICBMs since 1970 versus 330 for the U.S.S.R. The U.S. ICBM force carries roughly 2,100 warheads versus 1,050 in 1967. U.S. silo hardness has been increased significantly, and U.S. missile accuracy now approaches about 300 yards or 0.15 miles. U.S. missiles now have 100-200 targeting plans in their computer memories, and have nearly 100% alert rate. Soviet systems have also improved in accuracy, but most experts feel they do not yet have the accuracy to strike at U.S. missiles with high lethality. They also are largely liquid fueled, and their alert rate and reliability may be lower than that of U.S. systems.

However, four new types of Soviet ICBMs not shown on Chart Three are known to be under development, and the Soviets may now have the capability to deploy the solid fueled SS-16 in both silo and mobile form. A mobile SS-16 would be difficult or impossible to distinguish from an SS-20. Similar trends exist in the modernization of SLBMs, and recent press reports indicate that Soviet SLBMs may be able to strike at U.S. targets from Soviet ports, and that the low Soviet SSBN deployment rate may no longer be a critical measure of Soviet capability, assuming a Soviet first strike.

Chart Four provides a relatively speculative estimate of projected changes in U.S. and Soviet strategic forces. It gives the

CHART THREE

US AND SOVIET ICBM DEVELOPMENTS [1]

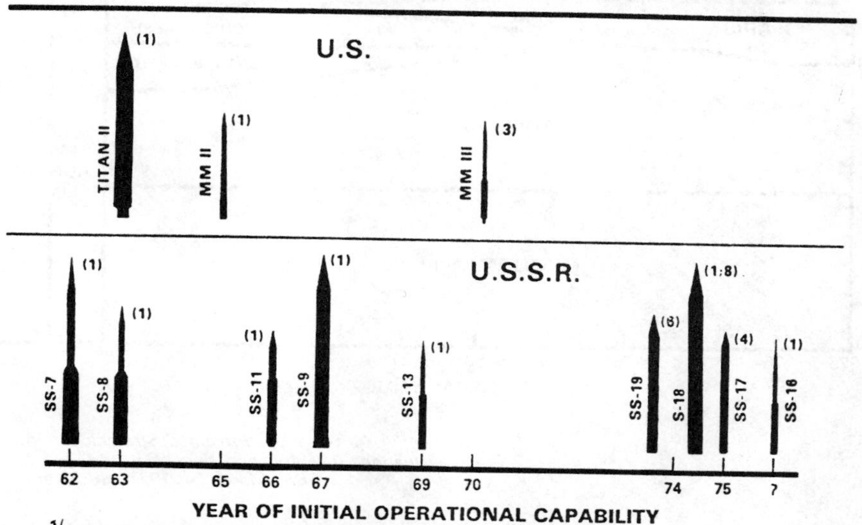

[1] The numbers in parentheses represent the number of independently targetable re-entry vehicles associated with each missile

Source: DOD Annual Report, *FY 78, p. 12.*

CHART FOUR

Changes in U.S. and Soviet Offensive Forces, by Strength and Weapon: 1975-1985

U.S. STRATEGIC OFFENSIVE FORCES
(ESTIMATED LEVELS)

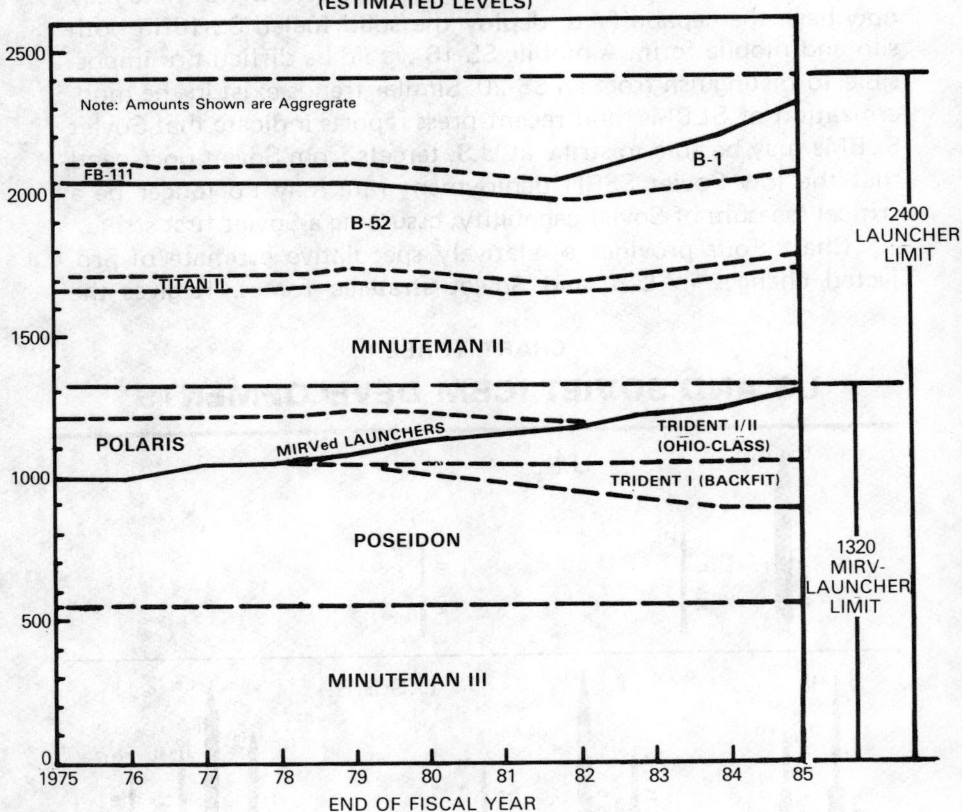

Source: A.A. Tinajero, "Projected Strategic Offensive Weapons Inventories of the U.S. and U.S.S.R.," CRS, March 24, 1977, pp. 28 and 77

CHART FOUR (Continued)

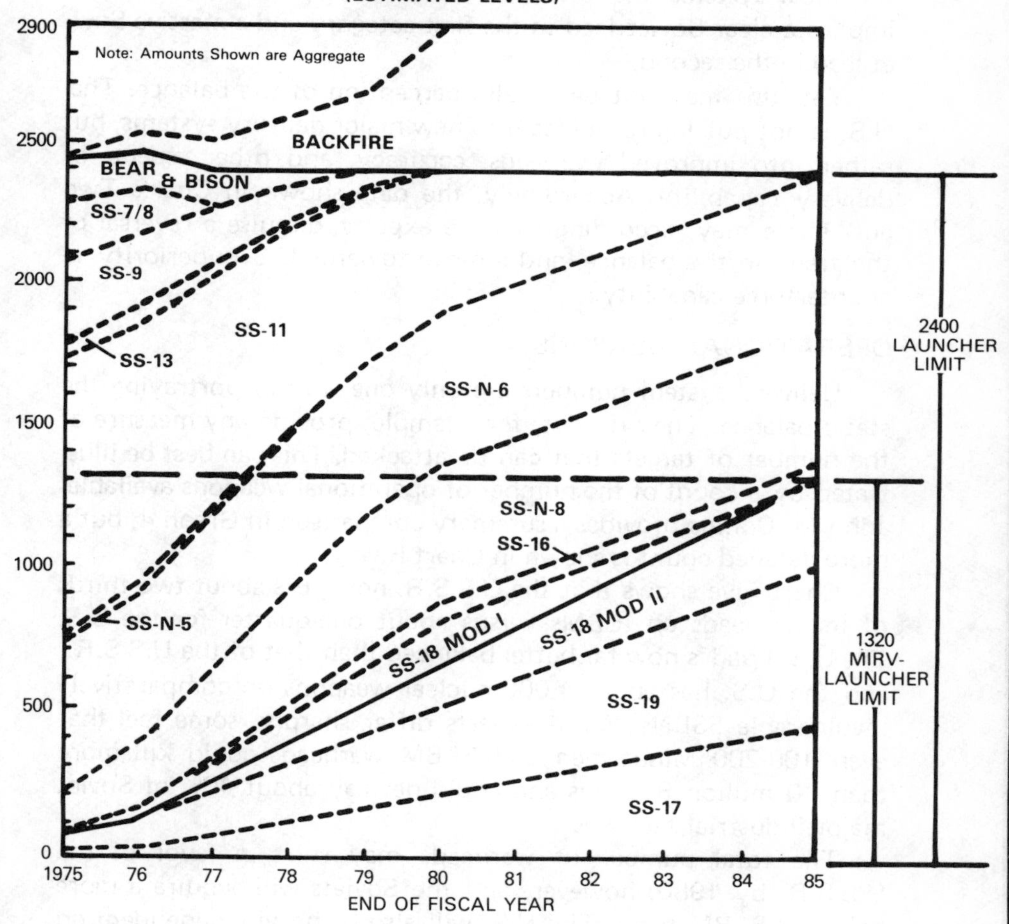
USSR STRATEGIC OFFENSIVE FORCES
(ESTIMATED LEVELS)
Note: Amounts Shown are Aggregate
2900
2500
2000
1500
1000
500
0
BACKFIRE
BEAR
& BISON
SS-7/8
SS-9
SS-11
SS-13
SS-N-6
SS-N-8
SS-16
SS-N-5
SS-18 MOD I
SS-18 MOD II
SS-19
SS-17
2400
LAUNCHER
LIMIT
1320
MIRV-
LAUNCHER
LIMIT
1975
76
77
78
79
80
81
82
83
84
85
END OF FISCAL YEAR

impression of extremely rapid improvement in Soviet forces while only marginal change is taking place in U.S. forces.

Much of the public perception of the balance in western nations is shaped by two basic aspects of strategic forces: who has the most systems and who has the newest systems. Chart Four implies a clear Soviet lead in the first category and a massive Soviet lead in the second.

Yet, this may not be a valid perception of the balance. The U.S. is not putting resources into new major delivery systems, but rather into improved warheads, accuracy, and other aspects of delivery capability. Accordingly, the data shown in Charts Two and Three may, according to some experts, disguise a reversal of the trend in the balance, and a move towards U.S. superiority in counterforce capability.

OPERATIONAL WEAPONS

Delivery system numbers are only one way of portraying the static balance. They do not, for example, provide any measure of the number of targets that can be attacked. This can best be illustrated by a count of the number of operational weapons available. John M. Collins provides a summary comparison in Graph 4, but a more detailed count is shown in Chart Five.

Chart Five shows that the U.S.S.R. now puts about two-thirds of its warheads on ICBMs versus about one-quarter for the U.S. The U.S. Triad is now far better balanced than that of the U.S.S.R., and the U.S. has about 4,000 nuclear weapons on comparatively invulnerable SSBNs. While experts differ sharply, some feel that even 100-200 Minutemen and SLBM warheads could kill more than 40 million Russians and could destroy about 40% of Soviet major industrial facilities.

The total number of warheads may shift in favor of the U.S.S.R. by 1980, however, and the Soviets will acquire a more balanced SLBM force. The U.S. will also be heavily dependent on the B-52, and will not have the ALCM until the mid or late 1980s.

COMPARATIVE "THROW WEIGHT"

Missile throw weight is important because it reveals the potential capability each side has to increase the yield of its weapons, to increase MIRVing or MARVing, to add penetration aids, or to add

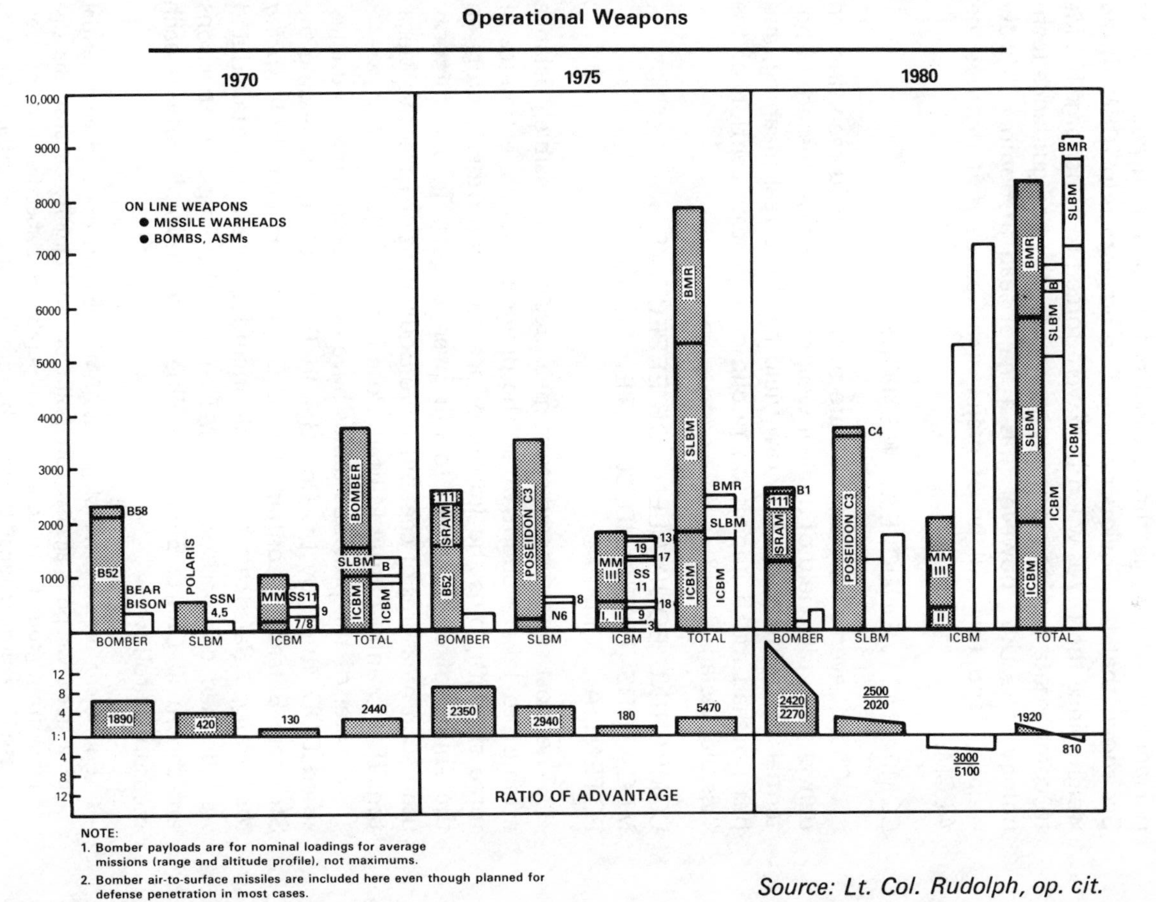

NOTE:
1. Bomber payloads are for nominal loadings for average missions (range and altitude profile), not maximums.
2. Bomber air-to-surface missiles are included here even though planned for defense penetration in most cases.

Source: Lt. Col. Rudolph, op. cit.

counterforce or surgical targeting guidance and navigation equipment. Advances in technology may, however, be reducing the importance of very large throw weights for certain purposes. Chart Six shows that the Soviets have had an advantage in ICBM throw weight since the 1970s when they were forced to build large ICBMs to compensate for inferiority in warhead, RV, and guidance technology. The U.S., however, has a major lead in bomber "throw weight," and this will be of major significance if it deploys the ALCM.

COMPARATIVE NUCLEAR WEAPONS YIELD

Chart Seven shows an estimate of the total yield each side can deliver. The raw yield of nuclear weapons is a gross measure of potential destructiveness. But raw yield has little real meaning, and has not had much impact on public or expert perception of the "visible" balance.

COMPARING EQUIVALENT OR EFFECTIVE MEGATONS (EMT) AND COUNTER MILITARY POTENTIAL (CMP)

The most sophisticated static measures of the "visible" balance attempt to take account of all the previous factors, estimates of force strength, plus the impact of accuracy and other factors on the ability of each system to kill given targets. Two commonly used such measures of capability are Equivalent or Effective Megatons (EMT) and Counter Military Potential (CMP).

Several of the previous tables have been taken from an unclassified DOD study by Lt. Col. Gerald T. Rudolph, "Assessing the Strategic Balance," contained in a DOD report called *Measuring the Strategic Balance.*[1] His article includes unusually detailed and well defined comparisons of EMT and CMP. These comparisons are shown in Charts Eight and Nine, and Rudolph defines each measure as follows:

> "Equivalent or Effective Megatons (EMT) reflects the damage potential of a side's weapons against soft point or area targets. Because the distance from ground-zero to a point vulnerable to a specified level of overpressure is proportional to the cube-root of yield, and because the area

[1] Anthony H. Cordesman, editor, DOD, 1976. Available through the Defense Documentation Center.

CHART SIX

U.S. and Soviet Throw-Weight (mlbs): 1970 to 1980

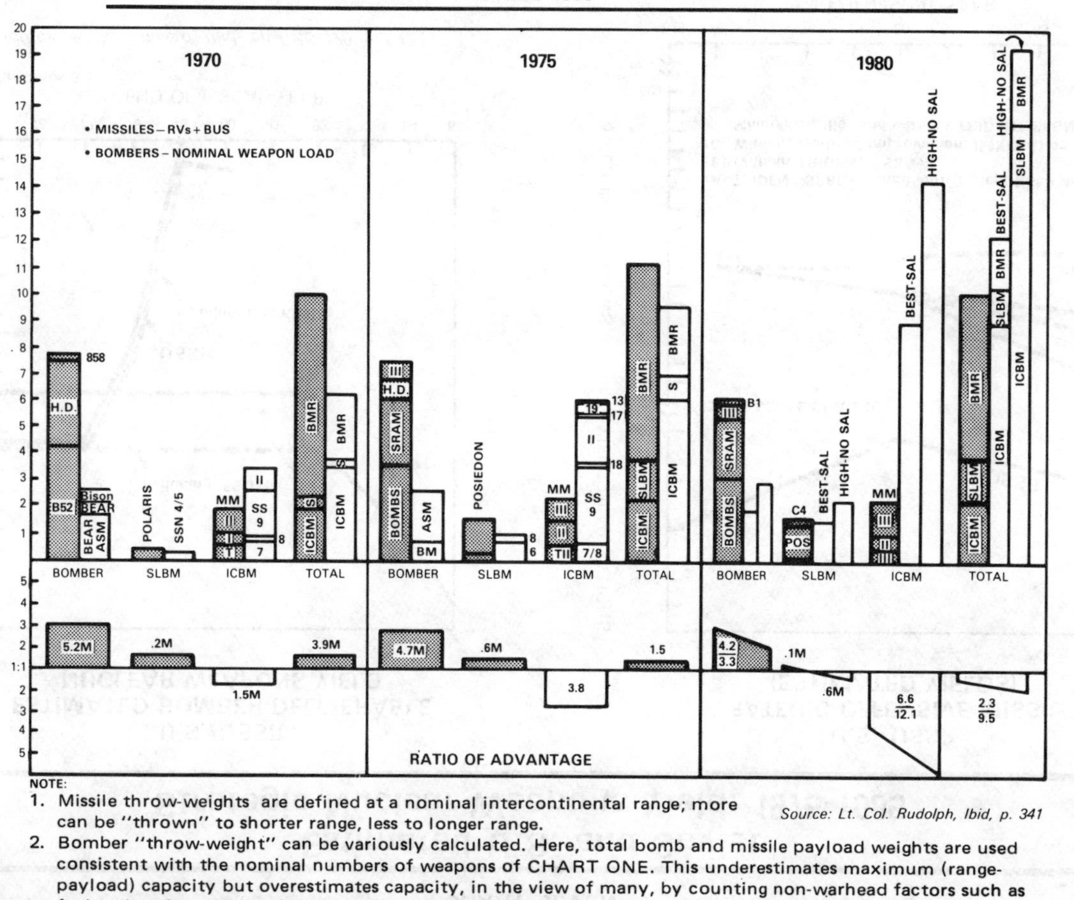

NOTE:

1. Missile throw-weights are defined at a nominal intercontinental range; more can be "thrown" to shorter range, less to longer range.
2. Bomber "throw-weight" can be variously calculated. Here, total bomb and missile payload weights are used consistent with the nominal numbers of weapons of CHART ONE. This underestimates maximum (range-payload) capacity but overestimates capacity, in the view of many, by counting non-warhead factors such as fuel and engine weight of air-to-surface missiles (ASMs).

Source: Lt. Col. Rudolph, Ibid, p. 341

CHART SEVEN

Estimated U.S. and Soviet Strategic Nuclear Weapons Yield: 1975-1985

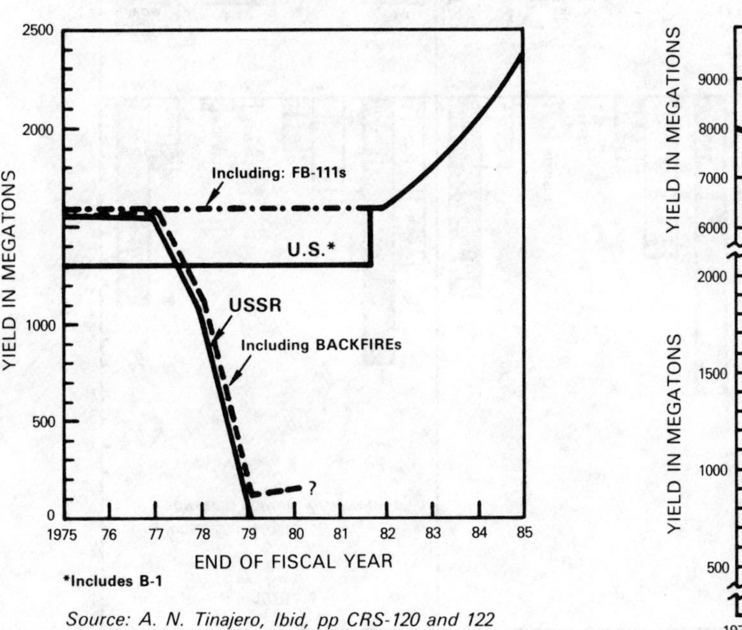

*Includes B-1

Source: A. N. Tinajero, Ibid, pp CRS-120 and 122

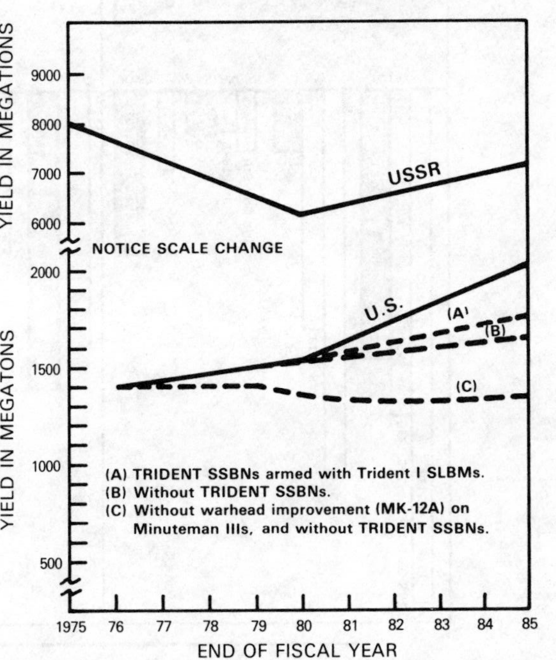

affected is proportional to distance squared, a two-thirds power of raw yield is used to better describe the effect of yield on the ground. Stated another way, equivalent megaton units can be converted to units of damageable area by multiplying EMT by a constant for each overpressure value of interest.

"A more complex static measure combines weapon delivery accuracy and yield into a single measure of hard target damage potential. Because delivery accuracy is stated in probabilistic terms, usually Circular Error Probable (CEP), a weapon's ability to destroy a hard target is mathematically formulated as a Probability of Kill (Pk). If the weapon dependent terms in this Pk equation are grouped together, a new parameter is formed, herein called Counter Military Potential (CMP). These weapon dependent terms are: the number of weapons per target, the yield per weapon, and the CMP per weapon. CMP can be thought of as the probability of applying a given over-pressure to a target. The other term of the Pk equation, target hardness, represents the resistance to all nuclear weapon effects but is usually discussed in units of overpressure. The CMP of a total force is simply the sum of individual weapon CMPs, just as throw weight or yield are added.

". . . values of CMP are extremely sensitive to accuracy assessments which, themselves, are fairly uncertain. Some published articles, based on a similar analysis conclude the opposite: U.S. superiority. This conclusion can be reached by taking extremes of the CEP uncertainty range; that is, upper bounds for U.S. accuracy (best CEP) and lower bounds for Soviet accuracy (worst CEP). The displayed CMP values are based on nominal or central value CEP and yield estimates. The U.S. must necessarily keep its true understanding of hard target capability classified to protect both its own accuracy capability and its estimates of Soviet accuracy."

Charts Eight and Nine represent the practical current limit of what can be explicitly compared without making use of classified information. They also represent the point at which the uncertainties in the data available begin to multiply to the point where analysts begin to draw radically different public conclusions about the balance. It is interesting to note, however, how important U.S. bombers are to the balance in both charts. Any chart which did not include U.S. bombers would greatly reduce the estimate of U.S. counterforce capabilities. At the same time, however, a chart that includes them disguises the fact that bombers make poor counterforce weapons. They provide ample warning to the U.S.S.R., and most Soviet strategic systems would probably launch before they were attacked by them.

A different view of the balance using a measure similar to CMP,

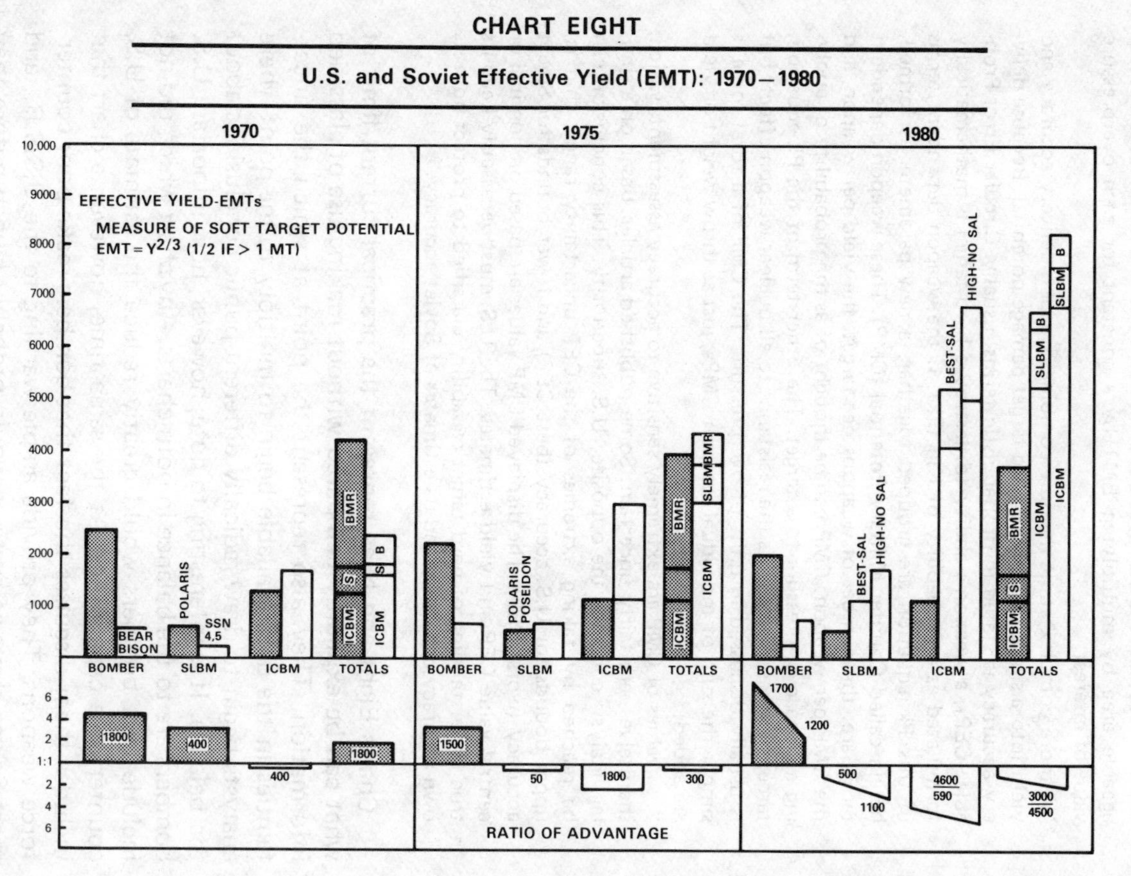

Source: Lt. Col. Gerald T. Rudolph, Ibid, p. 343

CHART NINE

U.S. and Soviet Hard Target Potential (CMP): 1970-1980

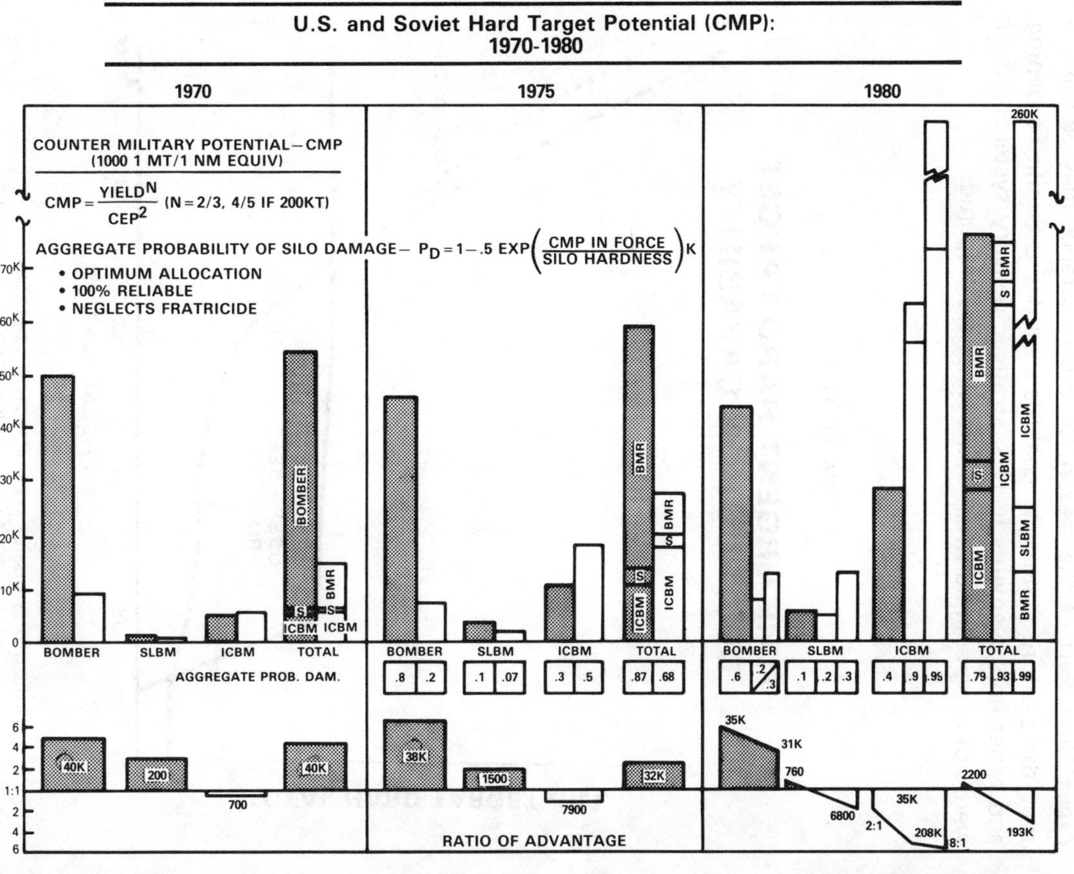

Source: Lt. Col. Gerald T. Rudolph, Ibid, P. 344

but which is never explicitly defined, is contained in the Secretary of Defense's posture statement for FY 1978. The comparison is shown in Chart Ten. It draws radically different conclusions from Chart Nine. Part of the reason for this difference may be the fact that the posture statement analysis includes U.S. silo hardening, and other improvements in U.S. strategic delivery systems, which were not programmed when Chart Nine was prepared.

CHART TEN

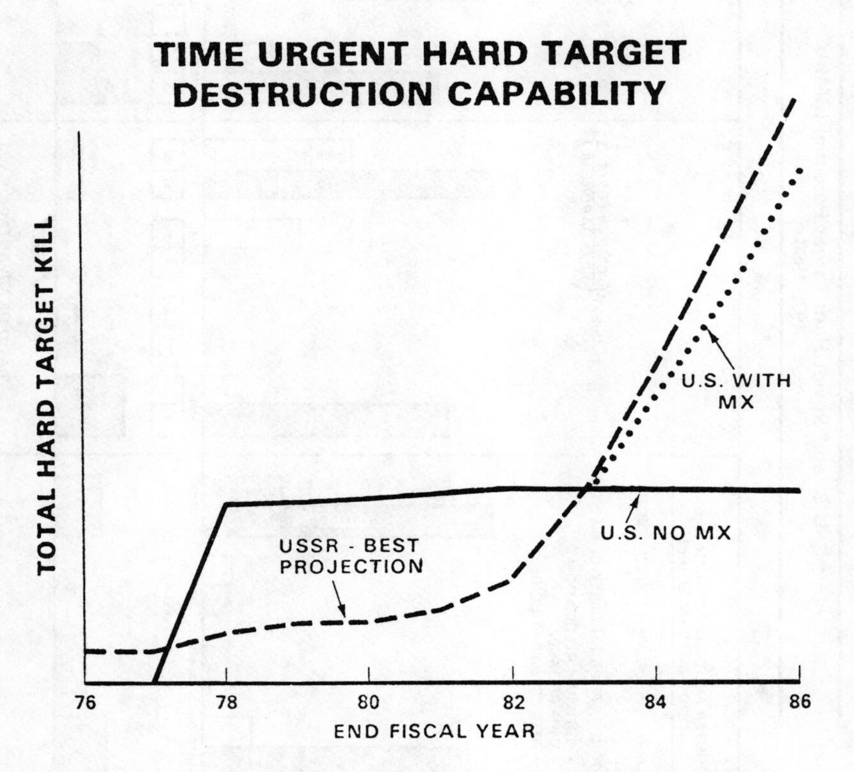

NOTE: ONLY MISSILE SYSTEMS WITH A HIGH PROBABILITY OF HARD TARGET KILL (0.7 OR BETTER) ARE INCLUDED.

Source: DOD Annual Report, *FY 78, p. 125*

COMPARING RESIDUAL STRIKE CAPABILITY AND CASUALTIES

A few analysts have attempted to go further, and to actually game a strategic war using unclassified data. The simplest of these comparisons to portray this was presented in Paul H. Nitze's article, "Deterring Our Deterrent" in *Foreign Policy.* The overall results of his analysis are shown in Chart Eleven, and indicate that a Soviet first strike might be effective enough in the mid-1980s to deprive the United States of a meaningful capability to retaliate. However, Mr. Nitze's analysis is definitely not the official view of the Department of Defense. Further, most of the other analysts who have attempted such analysis have drawn still other conclusions.

CHART ELEVEN

The Outcome of a Soviet First Strike on U.S. Strategic Forces: 1962-1992

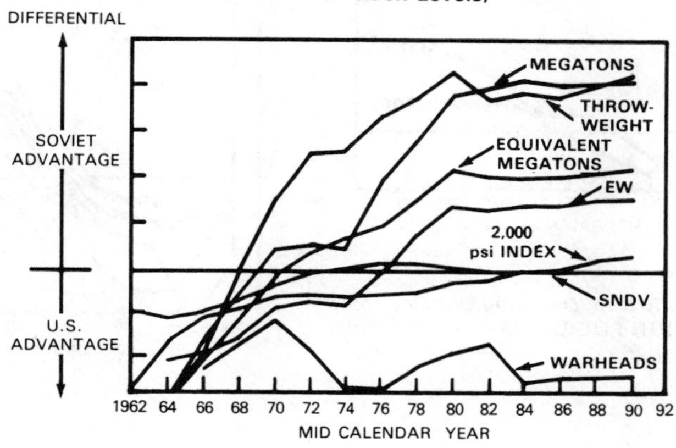

CHART ELEVEN (Continued)

CAPABILITIES AFTER SOVIET FIRST STRIKE

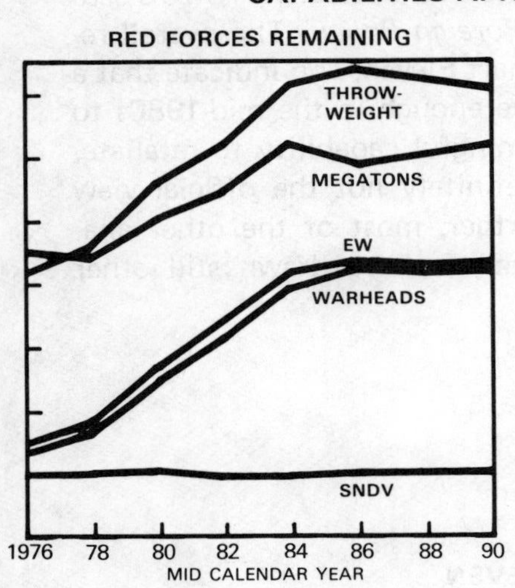

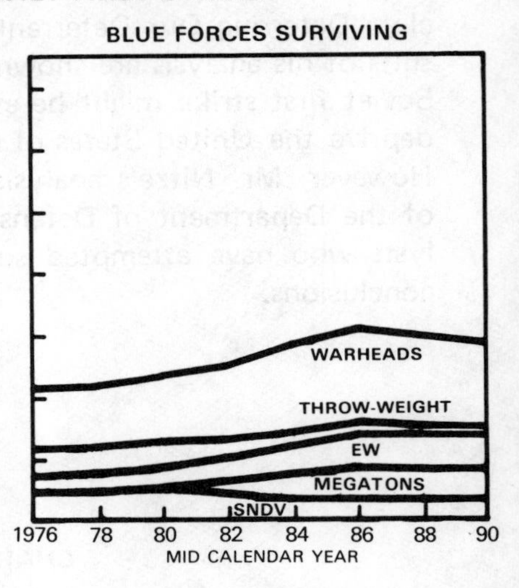

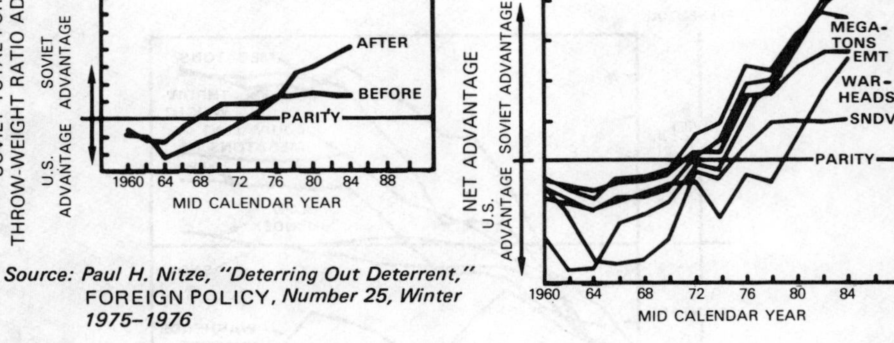

Source: Paul H. Nitze, "Deterring Out Deterrent," FOREIGN POLICY, *Number 25, Winter 1975–1976*

CHART TWELVE

Soviet—U.S. Throw-Weight Differentials

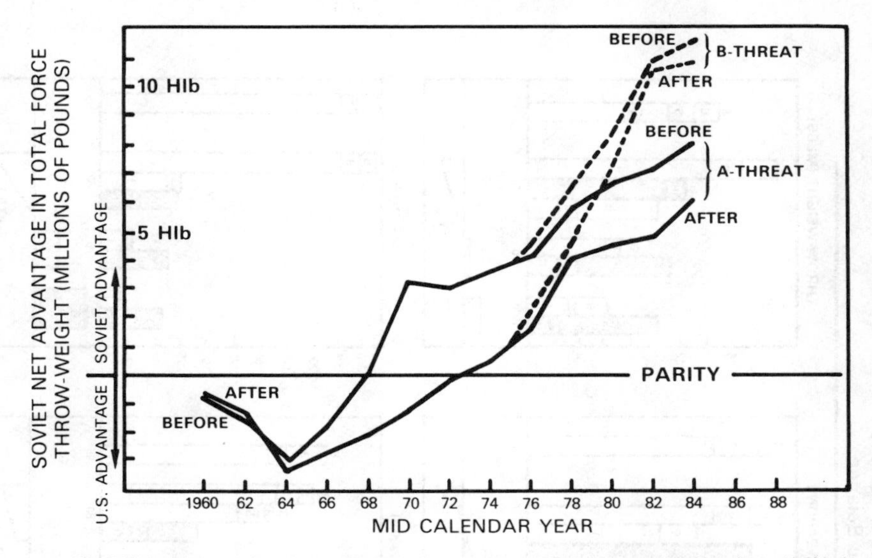

[16]A B-52 has been assigned an equivalent throw-weight of 10,000 lbs. and a B-1 about 19,000 lbs. The SRAM air-to-surface missile has a yield about equal to that of a Minuteman III warhead; hence, for every three SRAMs carried by a bomber, that bomber is given a throw-weight equivalent equal to the throw-weight of one Minuteman III. Laydown bombs are assumed to have roughly the yield of Minuteman II; hence, for each laydown bomb carried by a bomber it is given a throw-weight equivalent equal to the throw-weight of a Minuteman II. The alert bomber force is assumed to be 40 percent of the B-52 inventory and 60 percent of the B-1 inventory, degraded to incorporate penetration factors.

Source: "Assuring Strategic Stability in an Era of Detente," Paul H. Nitze, FOREIGN AFFAIRS, *Vol. 54, No. 2, January 1976*

THE "VISIBLE BALANCE": DRAWING NO CONCLUSIONS

Virtually all analysts would agree that the Soviet Union moved from a position of quantitative inferiority in 1970, to a position of parity by 1975-78. Many would agree that the U.S. is now inferior in some aspects of the "visible" balance, and will grow more inferior during the early 1980s. At this point, however, consensus stops.

This is illustrated in the next three charts. Chart Thirteen integrates all of the comparisons taken from the study of Lt. Col. Rudolph. Chart Fourteen integrates all of the major measures of the strategic balance in Secretary of Defense Donald H. Rumsfeld's FY 1978 Posture Statement. Chart Fifteen provides the data

CHART THIRTEEN

Summary of Static Measures of the Strategic Balance

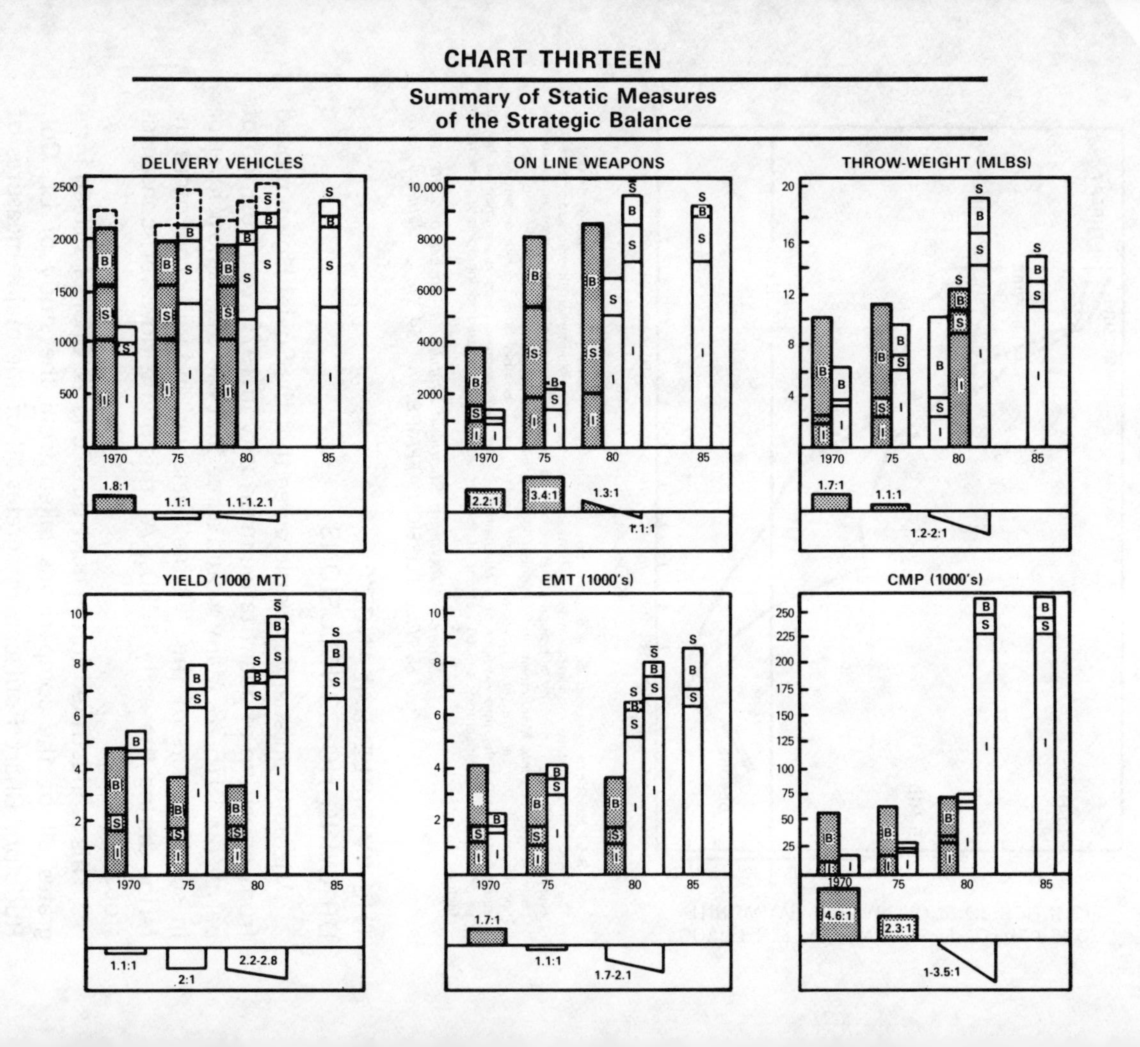

CHART FOURTEEN
THE SECRETARY OF DEFENSE'S PORTRAYAL OF THE BALANCE

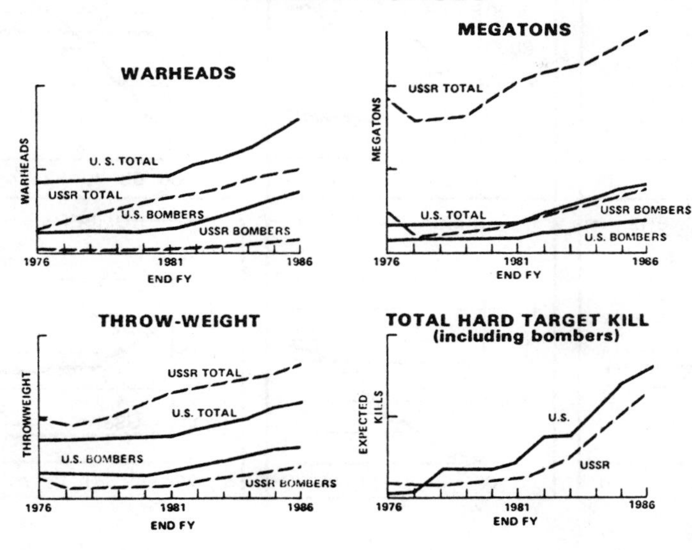

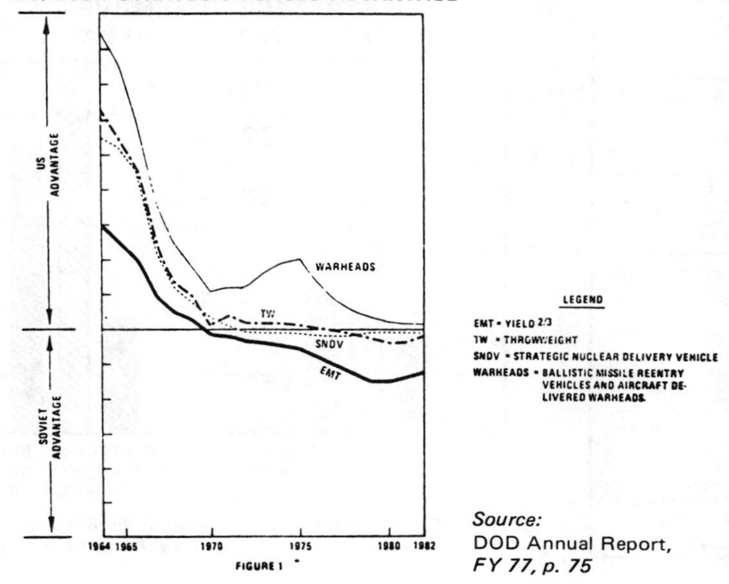

Source:
DOD Annual Report,
FY 77, p. 75

CHART FIFTEEN

THE PORTRAYAL OF THE BALANCE

A. HISTORICAL FACTORS (1966–1976, End of Fiscal Year)

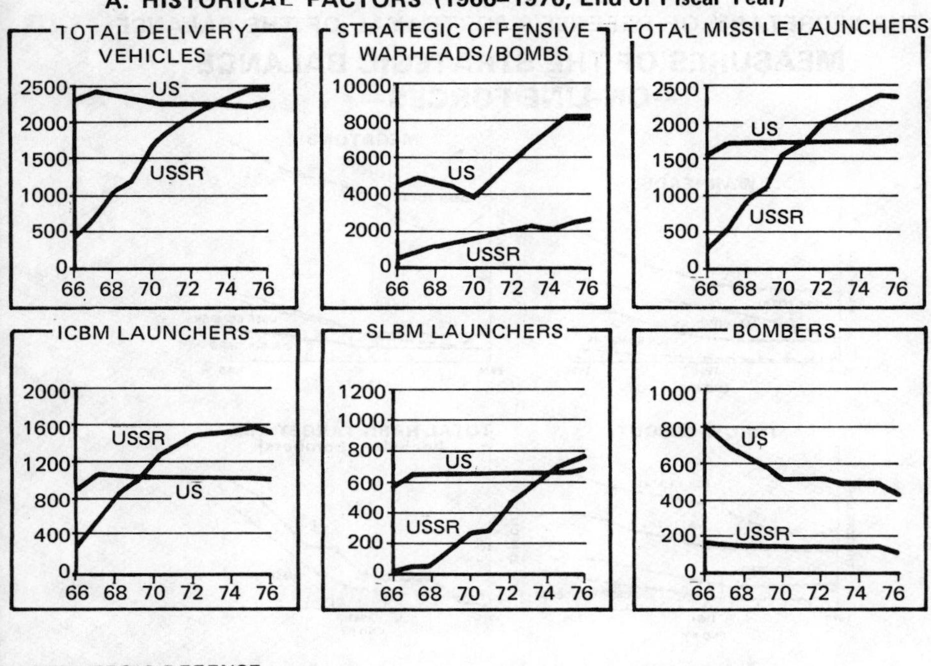

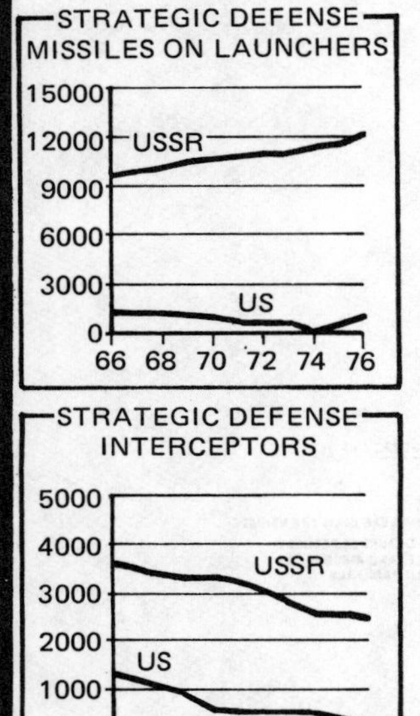

B. US AND USSR STRATEGIC OFFENSIVE THROW-WEIGHT/PAYLOAD

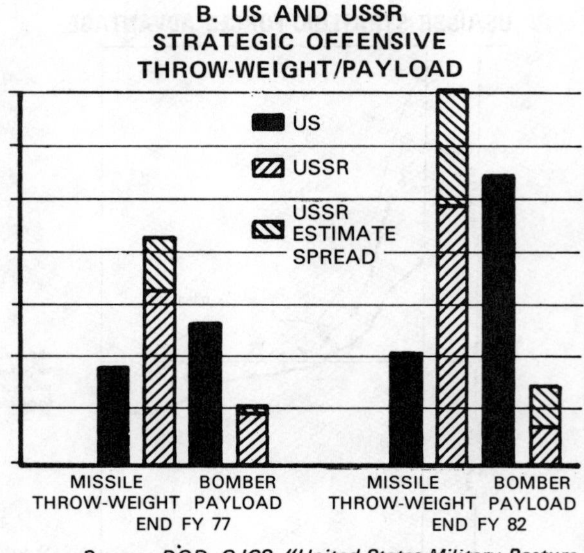

Source: DOD, OJCS, "United States Military Posture for FY 78," Charts 10 and 14

presented by the Joint Chiefs of Staff in their FY 1978 Military Posture Statement.

The resulting range of views and perceptions—all of which are current—should give pause to any one who is seeking to select a single view of the balance or a set of clear conclusions. Even the Secretary of Defense and Joint Chiefs appear to be describing roughly similar forces located on different planets.

GUESSING AT THE IMPACT OF SALT ON THE BALANCE

In addition to previous complexities, there is another balance which must ultimately be considered in addition to the "visible" and "invisible" balances. SALT presents the constant possibility that the U.S. and U.S.S.R. might suddenly agree on arms control measures which could radically change the balance, and even the importance of strategic forces as a measure of U.S. and Soviet great power competition.

THE ABM TREATY AND SALT I

The ABM treaty and interim agreement on strategic arms (SALT I) were signed in 1972. The ABM treaty limited deployed ABM strength on both sides to token levels. The interim SALT agreement set numerical ceilings on Soviet and U.S. forces at 2,400 bombers and missiles on each side, of which 1,320 can be MIRV'd. No precise definition exists of what medium bombers, cruisers, or IRBM missiles should be included in this ceiling. The agreement also allows both sides almost endless options for qualitatively improving their forces. For example:

- No limitation exists on the number of warheads in a MIRV'd missile.
- No limitation exists on the number of weapons a bomber can carry.
- No limitation exists on improvements in passive strategic defenses, in SAM or fighter defenses, or on high energy ABM devices. ABM missiles and sites are limited, but it is questionable whether the limit on ABM radars is meaningful or enforceable with current technology.
- Changes in the yields and accuracy of ICBMs and SLBMs are not affected. For example, all ICBMs and SLBMs could be given counterforce kill capabilities in the 1980s.
- No limitation exists on improvements in command and control, intelligence, or conflict management capabilities.

SALT I thus gave both the U.S. and U.S.S.R. almost unrestricted ability to continue the strategic arms race indefinitely into the future. It did not limit strategic arms, but meant that neither side would have to worry about ABM deployments, and that competition had to be shifted in areas other than the number of restricted delivery systems.

SALT II: THE "PHANTOM" BALANCE

Since 1972, the strategic arms limitation talks have been in their second phase or SALT II. Ideally, this should have culminated in a permanent treaty by October, 1977. At this point, however, no one can predict when, if ever, such an agreement will be reached. As a result, there is now a "phantom" balance of strategic arms: the balance that will exist if (a) a new SALT agreement is reached, or (b) if either side chooses to radically react if an agreement is not reached.

The effects of the U.S. proposal of Spring 1977, are summarized in Chart Sixteen. It is not surprising, given their overall effect, that the U.S.S.R. did not rush to accept this U.S. proposal.

THE NEW SALT II PROPOSAL

During the last six months, however, the U.S. and U.S.S.R. have reportedly evolved a very different proposal, and seem to be on the edge of developing a new SALT II treaty. The exact nature of this proposal has not been made public, but though experts differ, their various reports indicate the following likely features:

Possible Details of the SALT II Treaty
(now being negotiated by the U.S. and U.S.S.R. as derived from a working paper by Paul H. Nitze, 1 November 1977)

- *Strategic Nuclear Launch Vehicles (SNLVs) limited to 2,400.* Would limit number of ICBM and SLBM launchers and bombers.
- *Further reduction to 2,160 SNLVs by October 3, 1980 (U.S.) or 2,250 by 1982 (U.S.S.R.).*
- *Sublimit on MIRV'd ICBMs (5,500 km. + range) and heavy bombers carrying cruise missiles to 1,320 launchers/aircraft.*
- *No Mobile ICBMs.* The U.S.S.R. cannot deploy the SS-16, and the U.S. could not deploy a mobile M-X, for the three year life of the treaty.
- *Sublimit on total number of MIRV'd ICBM and SLBM launchers. U.S. proposes 1,200; U.S.S.R. proposes 1,250.*

CHART SIXTEEN

The "Phantom Balance": The Potential Effect of the U.S. SALT II Proposa on U.S. and Soviet Strategic Forces

A. NATURE OF THE U.S. COMPREHENSIVE PROPOSAL

Restrictions	Impact on: U.S.	U.S.S.R.
1. Limit Strategic Launchers to 1800-2000	Cut 62-262 launchers.[1]	Cut 540-740 launchers.[1]
2. Limit MIRV Launchers to 1100-1200.	Could add 54-154 MIRV launchers.[2]	Could add 900-100 MIRV launchers.[2]
3. Limit MIRV ICBMs to 550.	No change	Could add 350 MIRV ICBMs.[2] Force Soviets to push MIRV SLBM program.
4. Limit Modern Large Ballistic Missile Launchers to 150, (such as Soviet SS-9/18)	No effect on U.S.	Would not deploy about 150 heavy ICBMs.[3]
5. Ban Development and Deployment of new ICBMs.	Would not deploy MX ICBM.	Stop deployment of new ICBMs.
6. Ban Mobile ICBMs.	Would not deploy mobile ICBMs.	Prevent deployment of SS-16.
7. Limit ICBM and SLBM Flight Tests to six per year of each.	Slow Trident I missile program, eliminate Trident II Program. Reduce confidence in ability to launch successful 1st strike.	Slow MIRV SLBM programs, slow operational silo test firings, slow MIRV development. Reduce confidence in ability to launch successful 1st strike.
8. Ban Modifications on existing ICBMs.	Stop construction of MK 12A warheads, and other ICBM modifications.	Stop changes in existing missiles.
9. Limit Deployment of Backfire bomber.	No effect.	Prevent deployment of Backfire in certain parts of U.S.S.R.
10. Ban Cruise Missiles with ranges of more than 2500 km. Only long-range bombers could carry cruise missiles with ranges of 600 to 2500 km.	Cruise missiles up to 2500 km. range could be deployed in unlimited numbers.	No immediate effect as Soviets do not have long-range cruise missiles.

NOTES:
[1]Assumes 2062 U.S. and 2540 Soviet launchers.
[2]Assumes Soviets now have approximately 200 MIRV ICBMs. U.S. has 1046 MIRV missiles.
[3]Assumes approximately 50 SS-18 currently in place.

B. IMPACT ON THE BALANCE

	1985 WITHOUT U.S. SALT II			1985 WITH U.S. SALT II Limits of 2200 Delivery Vehicles, 1200 MIRV		
U.S. FORCES				**U.S. FORCES**		
Delivery System	**number**	**MIRV**	**number of weapons**	**number**	**MIRV**	**number of weapons**
Titan II	54		54	54		54
Minuteman III	650	650	1950	550	550	1650
Minuteman II	350		350	450		450
MX	100	100	1000	0		
Poseidon (31 submarines)	496	496	4960	400	400	4000
Trident (10 submarines)	240	240	1920	240	240	1920
B-52 (with ALCM and SRAM)	265		3180	265		3180
Air Launched Cruise Missile Carriers	100		1200	0		
TOTALS	2205	1436	14614	1965	1190	10154
U.S.S.R. FORCES				**U.S.S.R. FORCES**		
SS-18 ICBM	400	400	3200	308	308	2464
SS-17 ICBM	600	600	2400	340	340	1360
SS-19 ICBM	600	600	3600	550	550	3300
SS-16 ICBM	100		100	60		60
SS-N-6 SLBM	544		544	464		464
SS-N-8 SLBM	412		412	476		476
SS-N-18	64	64	192	0		
TOTALS	2720	1664	10448	2198	1198	8124

Source: Adapted from THE DEFENSE MONITOR, *"SALT: A Race Against the Arms Race," Vol. VI, No. 5, July, 1977, Center for Defense Information, Washington, D.C.*

- *Sublimit of 820 on number of MIRV'd ICBM launchers.*
- *U.S.S.R. can keep 308–326 modern large ICBMs (MLBMs) as part of 820 sublimit.*
- *ICBMs with payloads greater than SS-19 are MLBMs; no missiles permitted with payloads greater than SS-18.*
- *Backfire bomber not counted as strategic bomber; U.S.S.R. will not employ Backfire in role threatening to U.S.*
- *90–100 Soviet heavy bomber variants used primarily as recce and ASW aircraft not counted as SNLVs; status of U.S. B-52s in storage uncertain.*
- *Definition of MIRV'd missile undecided; affects number of SS-18s limited by treaty.*
- *Testing and deployment of air launched cruise missiles being negotiated.* The Soviets have evidently backed off a prior position that ALCMs with ranges above 600 km. could only be tested on heavy bombers and that all ALCMs could not exceed 2,500 km. in range and have agreed to ranges of up to 3,250 km. The U.S. is now seeking to expand the range to 3,500 km. No agreement has been reached on the definition of "range."
- *Definition of heavy bomber undecided.* It is unclear what future bombers are heavy, and whether FB-111 H and wide-body jets will be able to carry ALCMs.
- *Prohibition of reload ICBMs and testing of rapid silo reload systems.*
- *Prohibition of transfer of strategic equipment and technology to third countries.*
- *Reciprocal disclosure of key data on developments in strategic forces.* The U.S.S.R. does not favor this position.
- *Obligation not to interfere with opposing sides' unilateral verification techniques.* The details are being negotiated. Would prohibit measures to prevent or mislead satellite and signals intelligence.
- *Would be in effect for three years.* U.S. and U.S.S.R. negotiating over starting date.
- *Ban on testing of all new SLBMs except the Trident I and its new Soviet equivalent, the SSNX-18.* Would kill the Trident II program.
- *Issue statement in principle for SALT III which would:*
  - Reduce SNDVs to 1,800–2,000 (U.S. position).
  - Reduce MIRV'd launchers to 1,000–1,100 and set lower MIRV'd ICBM sublimit (U.S. position).
  - Further restrict ICBM and SLBM R & D and flight testing.
  - Further restrict ABM development, and introduce restrictions on air defenses and civil defense.
  - Strengthen verification.
  - Limit all NATO TNW systems capable of reaching the U.S.S.R. (Soviet position).

The Risks Inherent in the New Proposal

Such a SALT II treaty would be less favorable to the U.S. than the proposal in Chart Sixteen and some experts feel that the Soviet Union could use the treaty to achieve a significant strategic superiority over the United States.

Paul H. Nitze and the Committee on the Present Danger have made the following criticisms of the SALT II proposals:

THE MIRV'D ICBM BALANCE

- Could leave the U.S. limited to Minuteman III missiles, with 2,200 pound payloads until at least 1986. Might delay or prevent deployment of the more accurate and higher yield MK-12A warhead. The U.S. would thus be limited to 550 MIRV'd ICBMs, and 1,650 relatively low yield RVs.
- The U.S.S.R. might be able to deploy 200 SS-18s with 16,000 pound payloads, and 220 SS-17s and 400 SS-19s with 7–8,000 pound payloads. The SS-18 can hold 8–10 RVs, the SS-17 4 RVs, and the SS-19 6 RVs.
- This would give the U.S.S.R. 8 million pounds of throw weight by 1985 versus 1¼ million for the U.S., and up to 5,000 RVs versus 1,650 for the U.S. Each Soviet RV would have much higher yield.
- By 1985, ongoing improvements in Soviet ICBM accuracy to 0.15 NM, would allow the Soviets to use this force to kill 90 percent of U.S. ICBMs with an attack using two Soviet RVs per U.S. silo and only half the total number of Soviet RVs. The U.S. could only destroy less than 60 percent of the Soviet ICBM silos with the entire U.S. ICBM force even if all were upgraded to MK-12A warheads.

THE SINGLE WARHEAD ICBM BALANCE

- By 1985, the U.S. will still have 450 Minuteman II missiles with a single RV of less than 2,000 pounds. This will give the U.S. a million pounds of throw weight and 450 MT of yield. The U.S. might also maintain 54 Titans with 500,000 pounds of payload and 400 MT.
- The U.S.S.R. may keep 100 of the SS-18s with single warheads. This would provide 1,600,000 pounds of payload and 2,500 MT. They will also retain 300–400 SS-11 missiles or some successor with 1,000,000 pounds of payload and 1,000 MT.
- The total U.S. force of un-MIRV'd missiles could at most provide 1,500,000 pounds of throw weight and 850 megatons of yield. The Soviet force could provide 2,600,000 pounds and 3,500 MT. The Soviets would have more than four times the total U.S. yield for use against population and civilian targets.

CHART SEVENTEEN

The Impact of the Current SALT II Proposal on the Strategic Balance

A. Time Urgent Counter-Military and Soft Target Destruction Potential

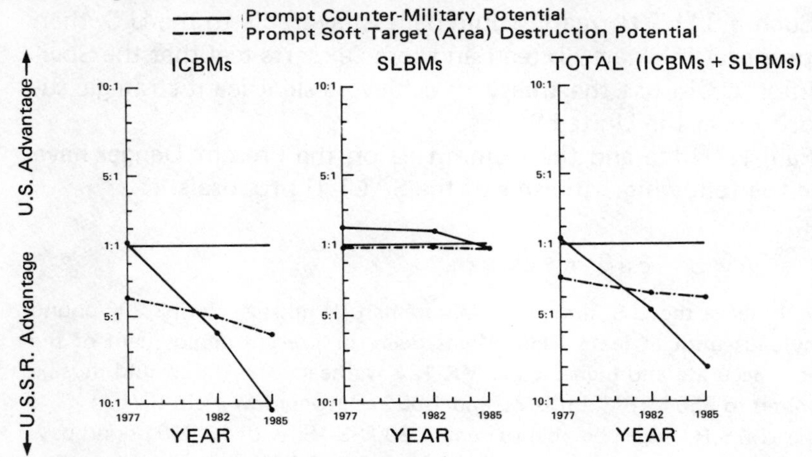

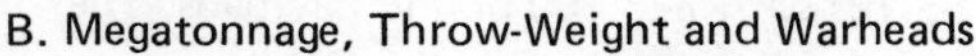

B. Megatonnage, Throw-Weight and Warheads

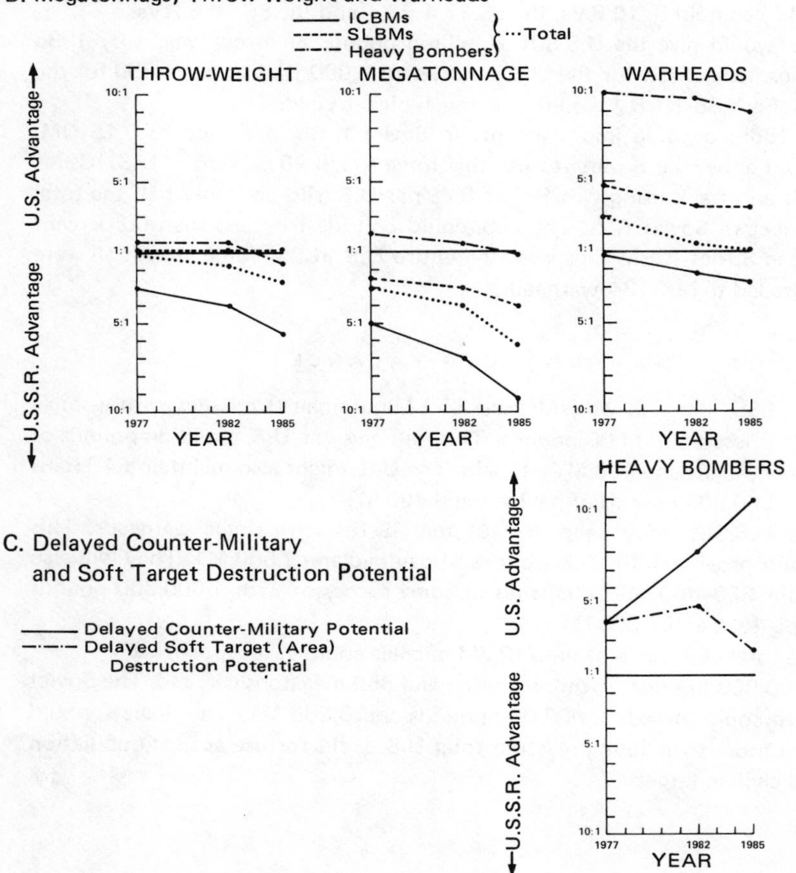

THE SLBM BALANCE

- By 1985, the U.S. can have 31 Poseidons with 16 missiles and 8–10 RVs of 40 KT per missile, plus 10 Tridents with 24 missiles and 8 RVs of 80–120 KT per missile. The Trident would have a 4,000 NM range. This would be a force of 740 launchers and 6,000 RVs of which 60 percent would normally be operational, and of which 80 percent of the missiles would be reliable. This is an effective force of 2,900 RVs or 300 MT, although it might be peaked to 3, 625 RVs at 375 MT.
- The Soviets would probably have only 400 MIRV'd SLBMs, although the new Soviet Typhoon SSBN would have the new SSNX-18 missile and 20–24 launchers.
- All U.S. single RV Polaris & SLBMs would be dropped from U.S. forces. The Soviets would have up to 500 un-MIRV'd SLBM tubes.
- Soviet SLBM forces have low operational rates but Soviet forces might be peaked before an attack.
- The U.S. will thus retain the lead in SLBM forces, but the Trident cannot be used against Soviet ICBMs. Accordingly, the U.S. SLBM lead has no impact on the Soviet lead in ICBM counterforce capability.

THE BOMBER/CRUISE MISSILE BALANCE

- The cancellation of the B-1 means that U.S. bombers can only achieve a high probability of penetrating Soviet air defenses in the mid-1980s if ALCMs have ranges greater than 2,500 km. Further, the new agreement only permits 120 B-52s equipped with ALCMs, of which only 60 could be kept on continuous alert and 84 could be brought to readiness with fully-generated warning. The FB-111 H lacks secure penetration capability.

  In contrast, the U.S.S.R. can peak the Backfire to full readiness before it launches an attack and faces essentially no threat from NORAD.

The Other Major Problems in SALT II

- The definitions in the treaty permit a steady unverifiable upgrading of all elements of Soviet forces while the treaty will probably halt such U.S. initiatives.
- Soviet ALCM, GLCM and SCLM capability and deployment is unverifiable.
- MIRVing of un-MIRV'd Soviet ICBMs is unverifiable.
- No limits are placed on improvements in Soviet civil or passive defense.
- Improving Soviet SAMs and warning radars allow rapid upgrading to an ABM defense capability.
- The three year ban on mobile ICBMs will probably kill U.S. development of the M-X missile. The Soviet SS-20 is permitted as a "theater weapon," although it can be unverifiably modified to have ICBM range. The Soviets will also have an SS-16 mobile ICBM ready for production and fully tested.

Such experts argue that the overall effect of the proposed SALT II treaty is a massive shift in the balance in favor of the U.S.S.R. This is shown in Chart Seventeen. Regardless of current Soviet intentions, this shift is so great that it greatly weakens U.S. political influence, and would drastically weaken U.S. deterrent and war fighting capabilities in a crisis.

Other Views of the New SALT II Proposal

Though experts differ, the following points have been made:

- The "leaks" of the treaty are incomplete and pre-judge the outcome of key negotiating points. Many may be more favorable to the U.S. than the leaks indicate.
- The Nitze calculations are a "worst case" view.
- U.S. ICBMs would be equally vulnerable to a Soviet counterforce strike without SALT II.
- The U.S. bomber and SSBN force will remain survivable.
- U.S. bombers will retain high penetration capability with 2,500 km. ALCM.
- Trident I missiles can be upgraded to counterforce capability.
- The U.S. could go to a launch on warning posture and negate Soviet first strike capabilities.
- The only war fighting advantage the Soviets would have would be in an ICBM counterforce exchange so politically incredible as to be unimportant for planning.
- SALT III negotiations will either prevent the balance from eroding in favor of the U.S.S.R. as Nitze predicts, or the failure of SALT III will give the U.S. time to build up its bomber force with ALCMs, and the U.S. can deploy the new M-X ICBM.
- The "true" figures on the post SALT II balance cannot be released without destroying the negotiations or greatly weakening the U.S. position in seeking further concessions.

Nevertheless, Mr. Nitze's analysis obviously merits serious attention. It is also clear that the "phantom balance" of SALT II may be more important in future international politics than the "visible" and "invisible" balances.

STRATEGIC DEFENSIVE COMPARISONS

American forces, people, and production base are naked to nuclear attack. A "vulnerability gap" of disputed proportions

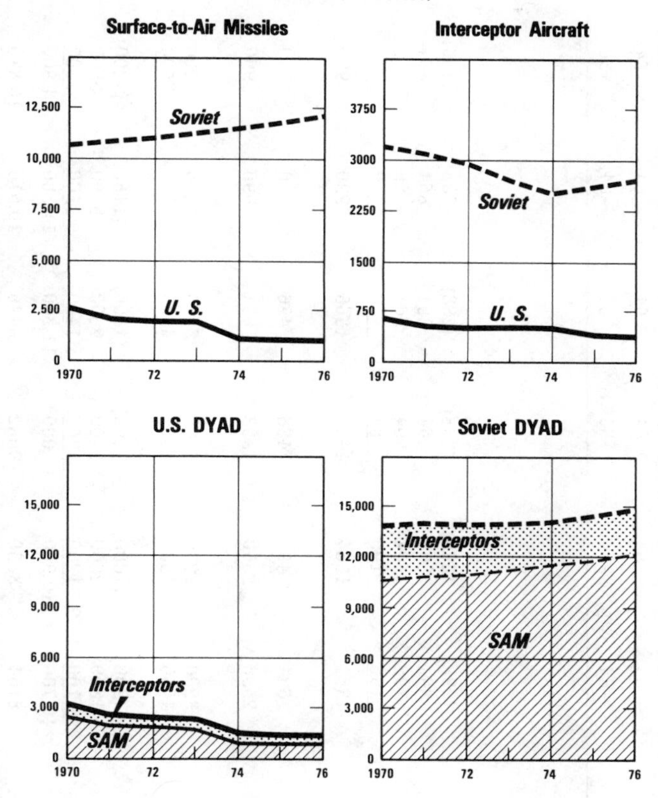

grows, because Soviet leaders stress defense, while U.S. leaders do not. End results eventually could erode our deterrent.[29]

Active Defense

Statistical summaries of U.S. and Soviet active defenses are so nearly self-explanatory that a few words suffice. (See Graph 5 and Figures 10 and 11.)

[29] Relationships between deterrence and defense, in principle and practice, are described in U.S. Congress, Senate Document 94-268, *United States and Soviet City Defense: Considerations for Congress,* a study prepared by the Congressional Research Service, Washington, D.C.: U.S. Government Printing Office, 1976), pp. 3-26.

FIGURE 10 STRATEGIC DEFENSIVE MISSILE SYSTEMS
Statistical Trends and System Characteristics

ABM LAUNCHERS	*1970*	*1971*	*1972*	*1973*	*1974*	*1975*	*1976*
United States	0	0	0	0	0	100	0
Soviet Union	64	64	64	64	64	64	64
U.S. Standing	-64	-64	-64	-64	-64	+36	-64
SAM LAUNCHERS							
United States							
Active							
Hawk							
Launchers	(288)	(288)	(288)	(288)	(288)	(288)	(288)
3 Arms Each	864	864	864	864	864	864	864
Nike Herc	792	504	504	504	126	126	126
Bomarc	196	196	84	0	0	0	0
Total	1852	1564	1452	1368	990	990	990
National Guard							
Nike Herc	684	486	486	486	0	0	0
Grand Total	2536	2050	1938	1854	990	990	990
Soviet Union							
SA-1	3200	3200	3200	3200	3200	3200	3200
SA-2	4600	4500	4300	4100	3700	3500	3400
SA-3							
Launchers	(900)	(1000)	(1100)	(1100)	(1150)	(1200)	(1300)
Rails	1800	2000	2200	2600	3100	3500	3700
SA-5	1100	1200	1300	1400	1500	1600	1800
Total	10,700	10,900	11,000	11,300	11,500	11,800	12,100
U.S. Standing	-8164	-8850	-9062	-9446	-10,510	-10,810	-11,110

NOTES: All Hawk launchers have three arms. SA-3s originally had 2 rails, but some were deployed with 4 rails, beginning in 1973. Each arm/rail holds one missile.

Fifty-four Nike-Hercules launchers now are in Alaska. All other U.S. SAMs are in Florida, on call, but out of position to deal with surprise attacks.

FIGURE 10 (continued)

	SYSTEM CHARACTERISTICS					
	First Deployed	*Nr Rails, Arms*	*Type Warhead*	*Slant Range (miles)*	*Combat Ceiling (Feet)*	*Launch Site*
United States						
Hawk	1960	3	HE	25	Lo-Med	Mobile
Nike Herc	1958	1	HE, Nuke	100	100,000	Fixed
Bomarc	1958	1	HE, Nuke	200-400	70,000	Fixed
Soviet Union						
ABM						
Galosh	1964	1	Nuke	200		Fixed
SAM						
SA-1	1956	1	HE			Fixed
SA-2	1958	1	HE	25	Med-80,000	Fixed
SA-3		2,4	HE	18	Lo-40,000	Mobile
SA-5	1963	1	HE	50-150	95,000	Fixed

NOTES: An improved version of SA-3, first displayed in 1967, may have a nuclear warhead.

A 4-rail version of SA-3 began replacing the standard 2-rail system beginning in 1973.

SA-4 and SA-6 through SA-9 are all tactical missiles associated with battlefield air defense. So are many Hawk and Nike Hercules. Those weapons are excluded from this summary, although some Soviet launchers could contribute to strategic defense if properly positioned at appropriate times. So could air defense artillery.

FIGURE 11 STRATEGIC DEFENSIVE INTERCEPTORS
Statistical Trends and System Characteristics

	1970	*1971*	*1972*	*1973*	*1974*	*1975*	*1976*
United States							
Active							
F-106	207	199	162	126	120	115	114
F-102	58	14	16	5	0	0	0
F-101	56	1	4	3	0	0	0
Total	321	214	182	134	120	115	114
National Guard							
F-106	0	0	33	67	68	90	90
F-102	255	192	157	172	167	44	19
F-101	45	110	102	107	117	122	134
Total	300	302	292	346	352	256	243
Grand Total	621	516	474	480	472	371	357
Soviet Union							
MIG-17	1000	800	650	400	200	150	100
MIG-19	350	350	350	300	200	200	150
MIG-23	0	0	0	0	0	0	100
MIG-25	0	50	100	150	200	200	300
SU-9	750	750	750	750	750	700	650
SU-15	400	550	550	600	650	850	850
TU-28	150	150	150	150	150	150	150
YAK-25	200	100	50	0	0	0	0
YAK-28	350	350	350	350	350	350	300
Total	3200	3100	2950	2700	2500	2600	2700
U.S. Standing	-2579	-2584	-2476	-2220	-2028	-2229	-2343

NOTE: MIG-23 and MIG-25 are the only interceptor aircraft presently being produced for Soviet air defense squadrons.

Three squadrons of Canadian F-101s, which total about 44 aircraft, are assigned to North American Air Defense Command (NORAD). They complement U.S. capabilities by covering northern approaches to the United States, which strengthens defense-in-depth.

Thirty-six U.S. F-4s are dedicated to strategic air defense, but none are in CONUS; 24 are in Alaska, 12 in Iceland.

Figures above reflect Unit Equipment (UE) aircraft only.

FIGURE 11 (continued)

	AIRCRAFT CHARACTERISTICS					
	First Deployed	*Nr Jet Engines*	*Combat Radius (Miles)*	*Max Speed (Mach)*	*Typical Armament*	
					Guns	*Missiles*
United States						
F-106	1959	1	600	2.0	0	4 AIM-4F/G, 1 AIR-2
F-102	1959	1	450	1.5	0	3 AIM-4C/D, 1 AIM-26
F-101	1958	2	400	1.8	0	2 AIM-4D, 2 AIR-2
Soviet Union						
MIG-17	1953	1	360	Subsonic	0	4 Alkali
MIG-19	1955	2	425	1.0	3X30mm	4 Alkali
MIG-23	1972	1	550	2.5	Gatling	or 4 Missiles
MIG-25	1965	2	700	3.2	0	4 Acrid
SU-9	1959	1	685	2.2	0	4 Alkali
SU-15	1967	2	450	2.5	0	2 Anab
TU-28	1966	2		1.7	0	4 Ash
YAK-25	1953	1		Subsonic	2X37mm	2 Missiles
YAK-28	1961	2	575	1.1	0	2 Anab

FIGURE 11 (continued)

AIR-TO-AIR MISSILE/ROCKET CHARACTERISTICS

	First Deployed	*Guidance*	*Range (Miles)*	*Speed (Mach)*	*Warhead*	
					Type	*Yield*
United States						
Rockets						
AIR-2	1957	None	6	3.0	Nuclear	1.5 KT
Missiles						
AIM-4C,D	1956	Infrared	6	2.0	HE	
AIM-4F,G	1960	Radar	7	2.5	HE	40 lbs
AIM-26A	1960	Radar	5	2.0	Nuclear	
AIM-26B	1963	Radar	5	2.0	HE	
Soviet Union						
Acrid	1975	Radar, IR	23	2.2	HE	
Alkali		Radar	3.7-5	1-2	HE	
Anab	1961	Radar, IR	5-6.2		HE	
Ash		Radar, IR	18		HE	

NOTE: Combat radius is with external fuel tanks.

ANTI-BALLISTIC MISSILE DEFENSE

Ballistic missiles on both sides are essentially unopposed, because neither country has ever erected an extensive ABM shield. The Soviet Galosh system, which comprises 64 launchers in four sites around Moscow, could be easily saturated. A single U.S. Safeguard installation opened operations with 100 missiles in an ICBM field near Grand Forks, North Dakota in October 1975, but shut down one month later.[30]

Scientists on both sides still pursue credible ABM capabilities within confines imposed by the SALT I Treaty and subsequent Protocol. Those documents restrict development and deployments, but not basic research.[31] The stakes are high, because a breakthrough by either side could suddenly shift the strategic balance. There is no consensus in the U.S. intelligence community concerning Soviet *progress* in that regard, but their *purpose* appears unswerving.[32] Certainly, there is no conviction comparable to the belief in some influential U.S. circles that defense degrades deterrence by making nuclear conflict seem a sensible choice, and true "victory" seem attainable.[33]

AIR DEFENSE

Cutbacks in U.S. interceptor aircraft, begun in the 1960s, accelerated sharply after ABM was excised, on the supposition that "a major anti-bomber defense of CONUS [the Continental United States] without a comparable anti-missile defense . . . would not be a sound use of resources."[34] Surface-to-air missile (SAM) batteries, which once defended U.S. cities, have all but disappeared.[35] Point defenses are not presently possible.

[30] Rumsfeld, *Annual Defense Department Report for FY 1977,* pp. 70, 91.

[31] In accord with the 1972 SALT I ABM Treaty, the United States and Soviet Union renounced rights to erect area defenses of respective homelands, and agreed that point defenses should comprise no more than two complexes of 100 ABM launchers and missiles each, sited to cover an ICBM field and the capital city. A Protocol signed in 1974 reduced that authorization to one complex in each country. For verbatim texts, see Senate Document 94-268, *United States and Soviet City Defense,* pp. 66-71.

[32] Personal conversation between the author and DIA officials on April 22, 1976.

[33] John Erickson, *Soviet Military Power,* Report No. 73-1 (Washington, D.C.: United States Strategic Institute, 1973), pp. 42, 45, 47, 49; Thomas W. Wolfe, *Soviet Power and Europe, 1945-1970* (Baltimore: The Johns Hopkins Press, 1970), pp. 186, 439-40.

[34] Rumsfeld, *Annual Defense Department Report for FY 1977,* p. 88.

[35] Four Nike-Hercules and eight Hawk batteries in Florida currently are under operational command of ADCOM, but are available for overseas deployment. Three additional Nike-Hercules batteries are positioned in Alaska.

To compensate, our Air Defense Command (ADCOM) now is compelled to supplement dedicated interceptors with F-4 fighters from the general purpose pool. Even so, the attrition of aging F-106s, still our first line of defense, will make it impossible to maintain even the present minimum number of alert sites in the late 1970s, unless F-4s join the Air National Guard as planned.[36]

Consequently, U.S. air defenders find it difficult to meet requirements of a watered-down mission, which merely demands capabilities sufficient for "limited day-to-day control of U.S. airspace in peacetime," warning of possible bomber attacks, and enough surge strength to "deny any intruder a free ride."[37]

The Soviet Union, in stark contrast, has amassed the world's most impressive array of air defenses, which currently includes 2,700 interceptor aircraft and 12,000 SAMs. Numerical strengths are slightly smaller than they were in 1970, but sheer mass serves a useful purpose, even though half the inventory consists of items outmoded according to U.S. standards. U.S. bombers fighting their way to targets therefore would face serious competition that cuts penetration prospects considerably.

[36] Rumsfeld, *Annual Defense Department Report for FY 1977,* p. 70; J-5 comments on the draft of this study, March 4, 1977.
[37] *Ibid.*

NET ASSESSMENT APPRAISAL:

The overall interaction between the trends in U.S. and Soviet bomber forces and strategic air defenses are shown in Chart Eighteen. These trends show a fundamental difference in Soviet and U.S. force structures and capabilities. Yet, the immense Soviet superiority in strategic air defenses may have little real meaning.

SOVIET STRATEGIC AIR DEFENSES

While Soviet air defenses are steadily improving, they now have only limited effectiveness against low flying U.S. bombers with advanced penetration aids and air-to-surface missiles. The current Soviet strategic air defense system has the following major defects:

CHART EIGHTEEN

THE INTERACTION BETWEEN TRENDS IN U.S. AND SOVIET BOMBER FORCES AND AIR DEFENSES

A. OFFENSIVE FORCES

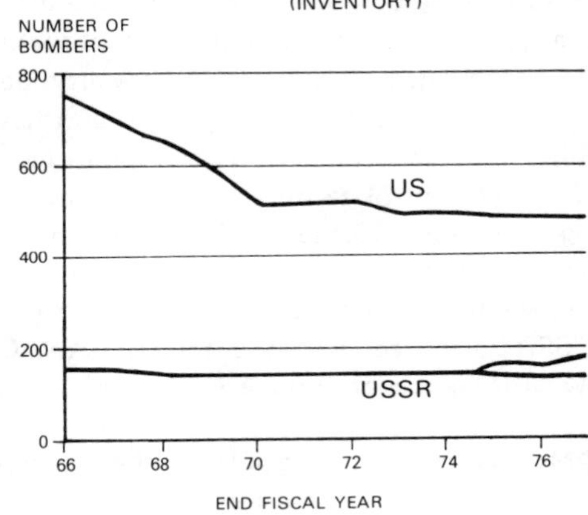

B. DEFENSIVE FORCES

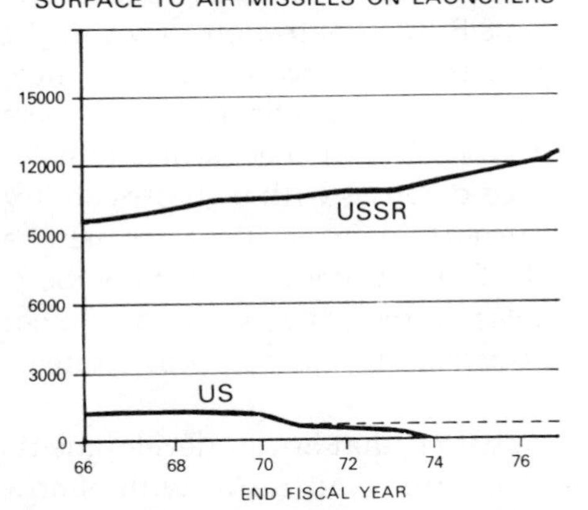

*DOTTED LINE INDICATES GENERAL PURPOSE FORCE MISSILES ASSIGNED TO CINC NORAD FOR USE IN STRATEGIC DEFENSE ORDER OF BATTLE

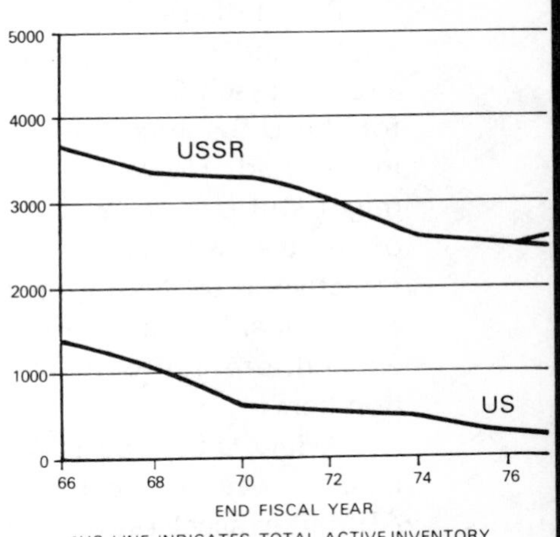

*US LINE INDICATES TOTAL ACTIVE INVENTORY

Source: OJCS, "Military Posture, FY 78," Charts 7, 11, 12.

- They are highly vulnerable to electronic warfare.
- Fighter forces still lack an effective overall combination of radar and missiles into a system which is effective for long range medium to high altitude intercepts. The Soviets have no equivalent to the F-15, or F-14 with the Phoenix missile, although the Foxbat has the potential to be upgraded to such capability.
- The Soviets do not yet have a real "look down/shoot down" fighter system to find and kill low flying bombers, and it will take the USSR some years to deploy such a system after it is tested and put into production.
- Current Soviet SAMs are highly vulnerable to low altitude penetration, still leave important gaps in low altitude coverage, and have no meaningful capability to kill bomber launched air-to-surface missiles.
- Soviet command and control technology is still limited by Soviet computer technology, and this places severe limitations on certain key capabilities. This however, may be rapidly changing.
- Although Soviet warning and air defense sensor systems are steadily improving—the USSR has no equivalent to the U.S. AWACs, and still is vulnerable to low altitude penetration even by aircraft without advanced ECM.
- The Soviet SAMs used in defense against bombers cannot be rapidly relocated, and must actively emit or transmit their location to fire. This makes them vulnerable to suppression, and it is unclear that the tactical option of pulsing units in different locations would be technically feasible.

All of these defects could be corrected by some time in the 1980s, and some experts feel that B-52 penetration losses might already reach 50%. Such a shift in effectiveness would be critical for the U.S., since the U.S. places such heavy reliance on bombers in its triad. However, advanced U.S. air launched cruise missiles might still penetrate such improved defenses with high probability of survival. Accordingly, while current Soviet air defenses may be effective in protecting against NATO and Chinese fighter-bombers, and probably against French nuclear forces, they are a vast Soviet expenditure for only token kill capability against U.S. strategic bombers.

A few U.S. experts do feel, however, that Soviet developments in laser weapons, command and control, and SAMs with almost ABM-like acceleration and very advanced sensors, may give the USSR radically improved future defenses even against bombers with advanced long range cruise missiles. Such Soviet improvements would be extremely costly, and probably require far greater investment in new Soviet defenses than the U.S. has or will have made in its bombers during 1950 to 1985, but they may be technically possible. It is premature, however, to draw any conclusions as to whether the Soviets will really try to develop such capability.

U.S. STRATEGIC AIR DEFENSES

Different problems have shaped the sizing of U.S. continental air defenses. It would require a massive U.S. investment to develop a modern NORAD to stop a comparatively small number of Soviet bombers. The U.S. might, for example, have to fund six to eight U.S. fighters per Soviet bomber killed. Yet, the overall effect of such a defense might be zero in reducing damage to the U.S. population, other military targets, and U.S. strategic forces.

Soviet ICBMs and SLBMs can almost certainly now kill all critical non-time sensitive U.S. targets. As Soviet missile forces improve, Soviet bombers, even under pessimistic U.S. assumptions about Backfire, will have a negligible incremental counter-force capability over Soviet ICBMs. The growth in Soviet warhead strength will also soon allow the USSR to target all major U.S. civilian and non-strategic military targets with its ICBM and SLBM forces even in a second strike.

Given the force ratios involved, a strong U.S. strategic air defense would have no real contingency value, and would be a major waste of resources, unless the U.S. first developed and deployed extremely capable ABM defenses against Soviet ICBMs and SLBMs. This is forbidden by the 1972 SALT I ABM treaty, which sets rigid limits on ABM numbers. It also would require successful U.S. development of ABMs to kill maneuvering re-entry vehicles or MaRVs. There are insufficient unclassified data available to estimate whether this is possible by the mid-1980s, or how the relative advances in ABM and warhead technology will affect the cost-feasibility of an ABM defense over the next ten years.

Passive Defense

DEFENSE FOR DELIVERY SYSTEMS

Pre-launch survival of U.S. nuclear delivery systems is predicated completely on passive defense. Fixed-site ICBMs, required to ride out any enemy first strike before retaliating, rely on hard silos to reduce initial attrition. Our bombers and SLBMs depend on mobility and dispersion.

Soviet passive protective measures are comparable, but security for land-based components in that country is enhanced considerably, because active defenses assist and U.S. second-strike concepts create a small threat.

CIVIL DEFENSE

Civil defense (CD) has received scant attention in the United States since the Cuban missile crisis.[38] Soviet stress on city evacuation and shelter programs reputedly is quite strong. Comprehensive programs provide incomplete but significant protection for the production base, as well as selected members of the population.[39]

Consequent asymmetries in U.S. and Soviet susceptibilities to nuclear attack cause increasing controversy. Cassandras at one end of the spectrum contend that emerging Soviet abilities, abetted by detailed plans, psychological conditioning, and physical preparations, already degrade U.S. deterrence and place this country in peril.[40]

Princeton's Nobel Prize-winning nuclear physicist, Dr. Eugene P. Wigner, and Joanne S. Gailar, a Soviet civil defense specialist

[38] Senate Document 94-268, *United States and Soviet City Defense,* pp. 88-91 contain a concise summary.

[39] *Ibid.,* pp. 15-17.

Representative Soviet writings on the subject include P. T. Yegorov, I. A. Shlyakhov, and N. I. Albin, *Civil Defense: A Soviet View,* translated and edited by Oak Ridge National Laboratory, published under auspices of the U.S. Air Force (Washington, D.C.: U.S. Government Printing Office, no date), 374 pp.; and N. M. Titov, P. T. Yegorov, B. A. Gayko, and others, *Civil Defense,* translated and edited by G. A. Cristy, Oak Ridge National Laboratory, Document ORNL-TR-2845, (July 1975), 118 pp.

Unclassified U.S. studies include Leon Gouré, *War Survival in Soviet Strategy* (Center for Advanced International Studies, University of Miami, 1976), 218 pp. Boeing Aerospace Company, *Industrial Survival and Recovery After Nuclear Attack: A Report to the Joint Committee on Defense Production, U.S. Congress* (Seattle: November 18, 1976), 81 pp.; and U.S. Congress, House, *Civil Defense Review, Hearings by the Civil Defense Panel of the Subcommittee on Investigations of the Committee on Armed Services,* 94th Congress, 2d Session (Washington, D.C.: U.S. Government Printing Office, 1976), 428 pp.

[40] Conrad V. Chester, Leon Gouré, T. K. Jones, Paul H. Nitze, and Harriet Fast Scott are among authoritative students of Soviet civil defense. All generally concur (as Jones put it before the Joint Committee on Defense Production in November, 1976) that "Soviet preparations substantially undermine the concept of deterrence that forms the cornerstone of U.S. security."

with Oak Ridge National Laboratory, speculate that crisis relocation procedures would limit Soviet fatalities to four or five percent during a general war, under worst-case conditions.[41] Official estimates indicate that almost half the American people would die under similar circumstances. Another 35 million would demand medical attention.[42] If those casualty ratios are reasonably correct, Wigner would be right in claiming, "Assured Destruction has become a myth."[43]

Skeptics, whose ranks reportedly include the present Secretary of Defense,[44] draw less drastic conclusions. Most concede that the Kremlin stresses city defense, but doubt that U.S. deterrence is in danger. Followers of one faction, for example, see the so-called civil defense gap as a spurious issue, because they believe that nuclear blasts can break through the best protection.[45] Others, whose opinions are widely shared, suspect that Soviet CD capabilities, while significant, are overstated.[46] U.S. overreaction, they contend, could be just as ruinous as complacency.

Which claims are correct is still not clear. The U.S. intelligence community accorded such a low priority to Soviet civil defense for so many years that crash efforts to estimate current effectiveness are inconclusive.[47] Classified studies as well as open assessments thus lack sufficient hard data and depth to support solid conclusions concerning Soviet strengths and weaknesses.

All the same, Cassandras and skeptics seem to agree that Soviet

[41] Joanne S. Gailar and Eugene P. Wigner, "Civil Defense in the Soviet Union," *Foresight* (May-June 1974), p. 10; and "Will Soviet Civil Defense Undermine SALT?" Human Events, July 8, 1972, pp. 497-98. See also Wigner's original estimate in "The Myth of Assured Destruction," *Survive: An American Journal of Civil Defense* (July-August 1970), pp. 2-4.

[42] U.S. Congress, Senate, *Analyses of Effects of Limited Nuclear Warfare,* Committee Print, prepared for the Subcommittee on Arms Control of the Committee on Foreign Relations (Washington, D.C.: U.S. Government Printing Office, 1975), pp. 112, 119; *Post Nuclear Attack Study (PONAST) II,* briefing prepared by Studies Analysis and Gaming Agency, Joint Chiefs of Staff, May 23, 1973.

[43] Wigner, "The Myth of Assured Destruction," p. 4.

[44] Robert Gillette, "Incoming Defense Chief Skeptical of Soviet Civil Shelter Reports," *Los Angeles Times,* December 27, 1976, p. 15.

[45] Gene R. La Rocque, "Danse Macabre in a Divided Ballroom," *New York Times,* October 14, 1976, p. 37; see also "The New Nuclear Strategy: Battle of the Dead?" *Defense Monitor* (Washington, D.C.: Center for Defense Information, July 1976), 8 pp.

[46] Les Aspin, "Soviet Civil Defense: Myth and Reality," *Arms Control Today* (September 1976), pp. 1-4.

[47] Henry S. Bradsher, "Civil Defense Plans Compared," *Washington Star,* November 9, 1976, p. 2.

CD capabilities would considerably exceed our own even if the Wigner-Gailar calculations were overstated by several hundred percent and U.S. casualty statistics were off by, say, half.

The Upshot

Active and passive defenses in combination are beginning to create a survivability imbalance that favors the Soviet Union. Assertations that the Soviets soon could survive a general war appear premature, but long-term consequences could be severe if the trend proceeds too far, particularly if accompanied by surprised Soviet breakthroughs in ASW or ABM (which presently seem possible, but not soon probable). *Any* amalgam that allowed the Soviets to evade Assured Destruction while America still could not would discredit this country's deterrent strategy based on *mutual* dangers.[48] Strategic defense thus seems to merit close and continuous attention by individuals and agencies responsible for U.S. national security.

[48] A case could be built for more and bigger U.S. ballistic missiles, more MIRVs, and MaRV if the Soviets deployed a credible ABM system and/or secured major elements of the population and production base in hard shelters. Deterrence would be well served, because U.S. weapons that survived a first strike would be numerous enough and possess sufficient lethal power to saturate defenses and ensure Assured Destruction.

NET ASSESSMENT APPRAISAL:

Experts now have very different views of Soviet civil defense capabilities. They disagree sharply on whether the Soviet civil defense system can now, or will become able, to accomplish a combination of three major objectives in the event of strategic nuclear war:

- Protect the Soviet government and military, and retain the general purpose forces necessary to achieve strategic objectives such as the seizure of Western Europe.
- Protect the Soviet population from nuclear attack.
- Protect Soviet economic and technological facilities and resources.

There is no doubt that the Soviets can do much to achieve all three objectives if they wish to. Analysts like Herman Kahn, Glenn

Kent, and others pointed out in the late 1950s that a U.S. shelter program could reduce U.S. casualties in a nuclear war to a percentage of the total population lower than the percentage losses the Soviet Union suffered in World War II. The U.S. eventually rejected this civil defense option, but it is clear that *if* the Soviets built suitable hard shelters, *if* they evacuated and dispersed other parts of their population from major target areas before launching a first strike, and *if* they then put that population into fall-out shelters, Soviet prompt casualties might be reduced to as little as five to fifteen million lives. Soviet death rates from long-term radiation effects would be uncertain, but might be an acceptable risk.

The debate over Soviet civil defense is essentially over whether the U.S.S.R. is doing this, and not over whether it is theoretically possible.

SOVIET SUPERIORITY IN THE PROTECTION OF SENIOR OFFICIALS AND OTHER MILITARY TARGETS

There is no debate over whether the Soviet Union has already largely achieved the first objective. It has provided massive, secure protection and shelter facilities for its civilian and military leaders, and for its command and control structure, which it may be impossible for present U.S. ICBMs and SLBMs to destroy.

Unlike the United States—which has only a few airborne facilities, a sheltered SAC headquarters, and a few additional obsolete and vulnerable hardened command centers—the Soviets have protected a large cadre of their senior military and civilians. The United States considered, but rejected, this option in the late 1950s and early 1960s; it limited protection of its National Command Authority to avoid presenting the U.S.S.R. with the threat of a U.S. first strike capability.

A few experts also feel that the Soviets could now evacuate or disperse many of their other military forces before a first strike, and that NATO and the United States then could not target enough of these forces to destroy their effectiveness. This is much more questionable, but it is technically possible, and such large scale evacuation might be accomplished with no, or only ambiguous, warning.

POPULATION CASUALTIES AND CIVIL DEFENSE IN COUNTERFORCE WARS

Soviet ability to achieve the second and third objectives is far less clear and depends upon the type of war to be fought. Requirements for counterforce wars are different from those for general war.

Although estimates of the long-term effects of thermonuclear weapons are controversial, unclassified DOD studies indicate that a Soviet attack on the 200 ICBM silos at Malmstrom Air Force Base (roughly 20% of U.S. silos) would produce 120,000 - 310,000 casualties. An attack on all U.S. silos with one megaton warheads would produce about 800,000 to 1,000,000 casualties. A comprehensive Soviet attack on all SAC ICBMs, SAC bombers, and U.S. Navy SSBN bases, would produce 3.2 to 16.3 million casualties, with 6 - 10 million as the most probable range.

Similar U.S. attacks on the U.S.S.R. would produce fewer Soviet casualties even without Soviet civil defense, because of differences in population density, U.S. yields, and fallout patterns. If the Soviets conducted covert or limited evacuation and shelter preparation before a U.S. retaliatory counterforce strike, they could probably reduce population losses to a level which would only be a fraction of that suffered by the U.S.

Both sides, therefore, have the capability to conduct limited counterforce strikes that would produce so few enemy casualties, even without civil defense, that it would not be profitable for the other side to escalate to countervalue (population and economic recovery) strikes.

Some experts argue that the Soviets would have a significant advantage in such exchanges since they would launch the first strike and could take suitable civil defense measures. They might then be able to kill large numbers of U.S. strategic systems without risking major U.S. civilian casualties. The U.S. could not strike back efficiently at dispersed Soviet bombers, hit deployed Soviet submarines, or know how to target Soviet ICBM silos which were not empty holes. The U.S. might then have no clear way of responding with a counterforce second strike without expending more U.S. systems than it would kill Soviet systems, and could not escalate to a selective strike on countervalue targets without risking general war and far more U.S. lives than Soviet lives.

However, there are several reasons why other experts question whether the Soviets would ever launch such counterforce attacks:

- the great risks and uncertainties inherent in such attacks.
- the uncertain accuracy, reliability, and lethality of Soviet ICBMs against U.S. ICBM silos.
- the problems of raising normally low Soviet alert rates, and deploying Soviet submarines and bombers, without raising U.S. alert rates and reaction times to the point where the U.S. might launch a warning attack against still targetable Soviet bombers, submarines, and IRBM/MRBMs.
- the fact that U.S. might seek second strike equivalence in Soviet other military targets and not more escalatory countervalue targets.
- the possibility the U.S. will acquire the capability to determine which Soviet ICBM silos are empty after a Soviet strike, and be able to efficiently kill remaining Soviet ICBMs, before the U.S.S.R. can improve the capability of its ICBMs to kill U.S. ICBMs with enough probability to make counterforce attacks credible options.

Although the data are lacking to make any clear judgment about which group of experts is right, it does not seem likely that Soviet civil defense would have much impact on Soviet assessment of the risk of launching such attacks, or U.S. retaliatory capabilities, although it might cut Soviet population losses significantly. The military risks of such attacks to the Soviets also seem likely to remain so great at least into the late 1980s, that such wars do not seem a major threat.

CASUALTIES AND CIVIL DEFENSE IN GENERAL WAR

A major Soviet advantage in civil defense might be more significant if the Soviets should try to win a general strategic nuclear war against the United States. Presumably, the Soviets would be willing to accept large scale casualties, and losses of military forces, and economic recovery capability if they felt they could achieve a decisive advantage over the U.S. in destroying its strategic forces, other military targets, and countervalue targets.

Various experts argue the Soviets would have significant advantages in population kill capability in such wars. They note that the Soviets have much larger yields available on their ICBMs and could blanket the U.S. with fallout. This would be the critical factor in "kill" capability since both sides would retain enough weapons in a second strike to make individual yield ultimately more significant than warhead numbers.

Depending upon the estimate, recent studies indicate the Soviets might be confident of killing 60 to 140 million Americans, with 80 to 120 million being the more probable range. In contrast, with no Soviet civil defense, the unclassified data in DOD PONAST II briefings indicate the U.S. could kill 50-100 million Soviets with no civil defense, and perhaps only 30-80 million with minimal civil defense. The smaller U.S. yields limit U.S. counter population kill capabilities, and the Soviet population is more dispersed. Accordingly, the Soviets might suffer only half as many casualties as the U.S.

Most experts would agree, however, that this would scarcely inspire the U.S.S.R. to attack since it would still lose millions of lives and the U.S. would have large numbers of warheads available even in a second strike. Given superior U.S. accuracy, the U.S. might well be able to strike a larger percentage of Soviet economic recovery targets with more accuracy and lethality in its second strike, which could be largely limited to countervalue targets, than the U.S.S.R. could hit U.S. targets in its combined counterforce and countervalue first strike. Such an outcome is uncertain, and economic recovery capability is notoriously hard to define, but the yield effects important in giving the Soviets superior population kill capability are probably less important in killing key plant and economic facilities than accuracy and warhead numbers, and nominal civil defense would scarcely make such attacks attractive to the Soviets. Accordingly, the real issue becomes one of whether Soviet civil defense might reach a far higher level of effectiveness than the minimal level discussed earlier.

The uncertainties which these experts are debating are to what extent Soviet civil defense can protect far more of the Soviet civilian population, and economic recovery capability. Some experts feel that the Soviets are developing comprehensive civil defense which could reduce Soviet casualties to as low as 5 million lives. This would mean that the U.S. could lose 12 to 24 times as many lives in a strategic exchange as the U.S.S.R. Some experts feel, therefore, that there is a potential "genocide gap" in the force postures of both sides, and that the growing protection of the Soviet population could make the gap hard to close. A few experts also argue that the U.S.S.R. does not merely lead the U.S. in civil defense, but already can cut its population losses to a strategically acceptable level. They see such a "genocide gap" as a current reality.

It is questionable, however, whether there really is such a gap, and whether it would give the Soviets a useful lever against the United States in any practical contingency without guaranteed survival of Soviet economic recovery capability. Most experts also feel that there is as yet no firm evidence that the Soviet Union has any intention of developing such a costly passive defense program.

SOVIET ABILITY TO PROTECT ECONOMIC RECOVERY CAPABILITY

The same uncertainty exists over current and future Soviet capability to protect the Soviet economy. Secretary McNamara stated in 1968 that 440 Minuteman III ICBMs (1,320 warheads) or 340 Poseidon SLBM (3,400) warheads could destroy 30% of the Soviet population, and 75% of Soviet industry. This is a comparatively limited proportion of the total U.S. strategic warhead force.

However, analysts like T. K. Jones of the Boeing Corporation have recently argued that the relatively low cost forms of protection set forth in Soviet civil defense manuals, coupled with dispersal, would allow the U.S.S.R. to protect enough of its industry and technology capacity so that the Soviet economy could rapidly recover from a U.S. attack. They argue that the massive short term losses in plant and facilities would then be acceptable to the Soviet leaders as the price of ultimate strategic world dominance.

This is a still very controversial conclusion, but Herman Kahn's writing in *On Thermonuclear War,* and in *Thinking About the Unthinkable,* suggests that such recovery may theoretically be feasible. It cannot be stressed too firmly however, that these possibilities have been raised by only a few experts, and are far from being verified in terms of either their ultimate effectiveness or in terms of evidence that the Soviets are developing such capabilities or have such goals.

There are also several basic problems the U.S.S.R. faces in achieving such capabilities. Under optimal circumstances, the U.S.S.R. would still have to accept five to fifteen million prompt deaths, an uncertain amount of long term or genetic deaths, and trillions of rubles worth of losses in plant and facilities. They would risk early detection of any evacuation measures by U.S. intelligence, and this might theoretically allow the U.S. to selectively re-target warheads against dispersal areas. The U.S.S.R. would have to make massive expenditures on civil defenses which have not yet been detected. And, the U.S.S.R. would have to believe that such a "war winning" strategy would really buy ultimate security far into the future.

U.S. OPTIONS FOR RESPONSE

Further, many experts feel the U.S. should be able to confirm the existence of such Soviet effort long before it occurs and take appropriate countermeasures. While the U.S. does not have an active civil defense program in the sense of building shelters, it does maintain an extensive planning capability through the Federal Preparedness Agency, and can rapidly improve its own passive defense. Further, the U.S. can both increase its yields over a period of time, and may ultimately improve guidance and targeting to counter Soviet actions. Experts have also discussed "super" strategic weapons which might raise Soviet casualty levels back to at least the level that would result from no Soviet civil defense level.

In summary, with the exception of a few analysts, most experts feel that a critical shift in Soviet civil defense capabilities could not occur before the mid-1980s, and that the Soviets would still not have a reasonable probability of "winning" such a war. They also agree there is no firm evidence that the Soviets have any intention of pursuing such goals, and that Soviet success even after a massive effort would be problematic and incredibly costly.

Most also feel the U.S. will have ample warning of such a combination of developments, and that the cost of the improvement in U.S. strategic offensive forces necessary to counter Soviet improvements in strategic defenses would be substantially lower than the costs to the Soviets of the improved defenses.

There is no way at this point in time to reach any conclusion about these arguments. There has been much more vehemence from the experts who take any position on these issues than either analysis or evidence.

[Chapter 4]

General Purpose Forces

IMPORTANT U.S. MISSIONS

The most important single mission of U.S. general purpose forces is to help NATO allies deter, and if need be defeat, Soviet armed aggression in Europe. Our Army and tactical air power are tailored primarily to meet that threat. Marine and Navy needs are more global in nature, but preserving open sea lanes for NATO takes a high priority.

Simultaneously, without undercutting deterrent capabilities in Europe, America's armed forces should be sufficient to discourage Soviet aggression elsewhere, if it endangers U.S. security; deal with such ventures if they do develop; and cope with selected contingencies caused by other countries, when U.S. decision-makers deem armed force advisable. Robbing Peter to pay Paul, as we did during the Vietnam and Yom Kippur Wars, intensifies risks in Western Europe, where we can ill afford it.

Coverage herein consequently assesses total U.S. and Soviet general purpose forces, service by service, with the focus on flexibility. Regional interactions are reviewed in Chapter 6, which includes NATO and Warsaw Pact partners.

Graph 6

SELECTED GROUND FORCE STRENGTHS COMPARED
Statistical Summary
(Note Different Scales)

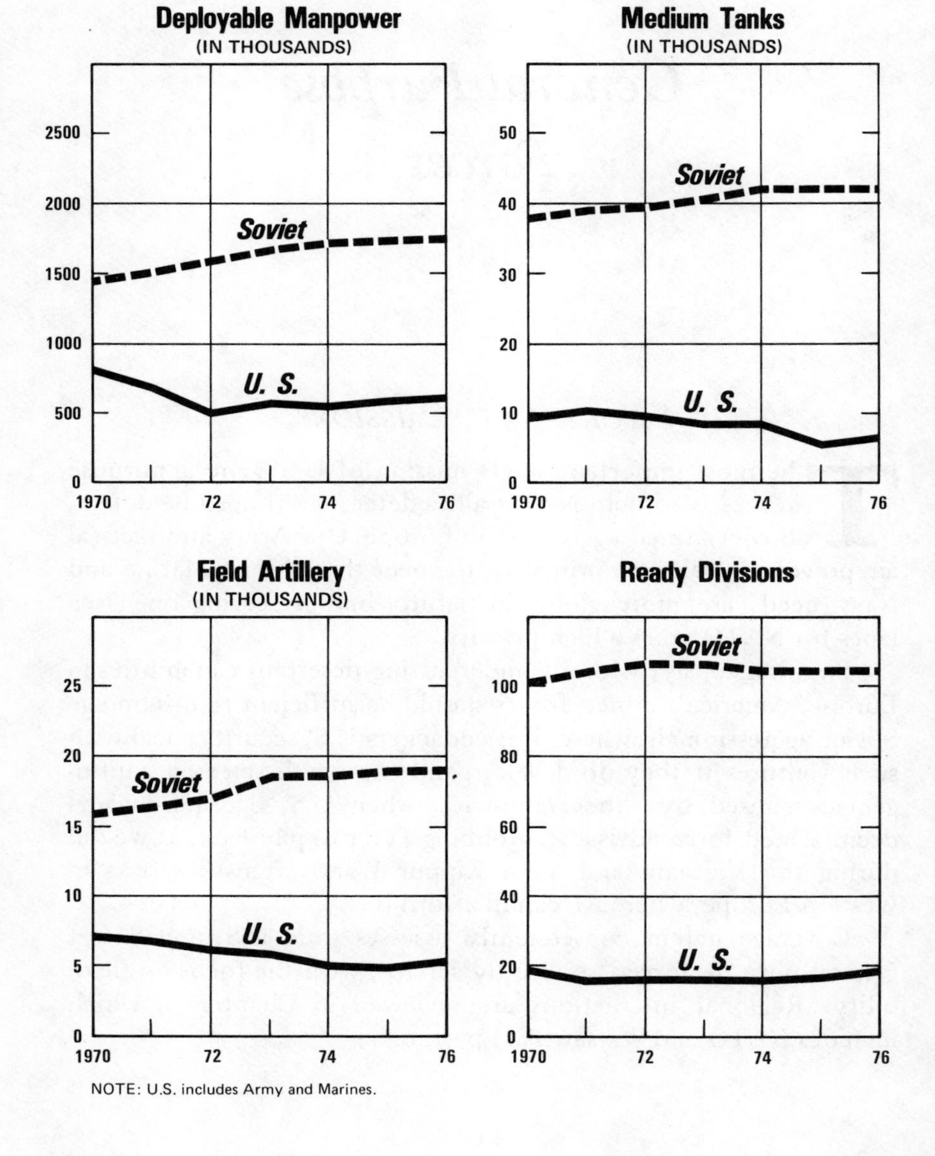

NOTE: U.S. includes Army and Marines.

GROUND FORCES

Preponderant ground combat power on both sides is [illegible] in armies. Marines and naval infantry provide supplemental strength, but are considered separately in this section, because their main missions and methods of operation are special.

Armies

A mammoth conscript army is the traditional source of Soviet general purpose strength. Other services are subsidiary, despite the emergence of a modern air force and navy. The much smaller U.S. Army currently consists of volunteers. Quantitative gaps that favor the Soviet Union are great in nearly every category.

COMPARATIVE MANPOWER

Armies everywhere are still manpower intensive, even in this mechanized age. Active deployable personnel strengths displayed in Figure 12 thus are very significant.[1]

U.S. Army rolls, drastically reduced by retrenchment after the Vietnam War, bottomed out at 420,000 in 1972, then began to recover by swapping overhead for sinew within a constant ceiling.[2] Still, Soviet increases during this decade cause the net U.S. loss to total almost half a million men. Soviet personnel, less command/support, now outnumber our own by more than three to one (1,725,000 to 505,000).

COMPARATIVE FIREPOWER

Firepower statistics in Figure 13 speak for themselves. Soviet quantitative superiority stands in stark relief. Qualities are competitive with (sometimes superior to) U.S. counterparts. A new main battle tank, a fine armored fighting vehicle (the BMP), a family of field army air defense weapons, and several artillery pieces, two self-propelled, are being deployed at rather rapid rates. Our Army's much-discussed main battle tank, mechanized infantry combat vehicle (MICV), and Patriot air defense system are still in gestation.

[1] Total army strengths are compared in Chapter 2. Not included in the discussion in this section are high command and general support forces, which amount to approximately a quarter of a million men in the U.S. Army, and nearly three-quarters of a million in the Soviet Union.

[2] George S. Brown, *United States Military Posture for FY 1976,* p. 68.

FIGURE 12 GROUND FORCES, DEPLOYABLE MANPOWER
(In Thousands)

	1970	*1971*	*1972*	*1973*	*1974*	*1975*	*1976*
UNITED STATES							
Army	684	594	420	477	451	485	505
Marines	85	66	62	65	65	70	70
Total	769	660	482	542	516	555	575
SOVIET UNION							
Army	1,450	1,510	1,570	1,650	1,700	1,720	1,725
Naval Infantry	10	10	10	11	11	12	12
Total	1,460	1,520	1,580	1,661	1,711	1,732	1,737
U.S. STANDING							
Army	-766	-916	-1,150	-1,173	-1,249	-1,235	-1,220
Marines/Naval Infantry	+75	+56	+52	+54	+54	+58	+58
Total	-691	-860	-1,098	-1,119	-1,195	-1,177	-1,162

U.S. Army strengths include field commands and mission-oriented base operating support.
Marine air wings are excluded.
Soviet forces exclude command and general support forces.

FIGURE 13 GROUND FORCE FIREPOWER
Statistical Trends and System Characteristics

ARMY	*1970*	*1971*	*1972*	*1973*	*1974*	*1975*	*1976*
Heavy and Medium Tanks							
United States	9,520	10,180	9,435	8,430	8,400	5,195	6,265
Soviet Union	38,000	39,000	39,500	40,500	42,000	42,000	42,000
U.S. Standing	-28,480	-28,820	-30,065	-32,070	-33,600	-36,805	-35,735
Light Tanks							
United States	1,660	1,600	1,600	1,575	1,575	1,570	1,570
Soviet Union	3,000	3,000	3,000	3,000	3,000	3,000	3,000
U.S. Standing	-1,340	-1,400	-1,400	-1,425	-1,425	-1,430	-1,430
APC/AFV							
United States	11,875	13,000	11,860	11,775	10,510	10,480	11,245
Soviet Union	30,000	30,000	30,000	35,000	35,000	38,000	38,000
U.S. Standing	-18,125	-17,000	-18,140	-23,225	-24,490	-27,520	-26,755
Artillery							
United States	6,885	6,635	5,840	5,540	4,710	4,625	4,885
Soviet Union	16,000	16,500	17,000	18,500	18,500	19,000	19,000
U.S. Standing	-9,115	-9,365	-11,160	-12,960	-13,790	-14,375	-14,115
Anti-Tank Guided Missiles							
United States	2,710	12,080	21,840	33,235	45,015	63,235	72,555
Soviet Union	4,500	4,700	4,800	5,000	5,500	6,000	6,000
U.S. Standing	-1,790	+7,380	+17,040	+28,235	+39,515	+57,235	+66,555

NOTES: U.S. light tanks indicate Sheridan armored assault vehicles.
Armored Personnel Carriers (APC) and Armored Fighting Vehicles (AFV) include wheeled and tracked vehicles.
The United States has no counterparts for about 10,000 Soviet heavy mortars.

FIGURE 13 (continued)

U.S. MARINES	*1970*	*1971*	*1972*	*1973*	*1974*	*1975*	*1976*
Medium Tanks	175	175	175	175	175	175	175
LVTPs	330	330	525	525	525	525	525
Artillery	250	250	250	250	270	270	270
Anti-tank Missiles	0	0		0	0	0	70
SOVIET NAVAL INFANTRY							
Light Tanks	140	140	175	175	175	200	200
APC/AFV	500	500	600	600	600	750	750
Heavy Mortars	120	120	150	150	150	180	180
Anti-Tank Missiles	60	60	75	75	75	90	90
GRAND TOTALS							
Medium Tanks							
U.S.	9,695	10,355	9,610	8,605	8,575	5,370	6,440
U.S.S.R.	38,000	39,000	39,500	40,500	42,000	42,000	42,000
U.S. Status	-28,305	-28,645	-29,890	-31,895	-33,425	-36,630	-35,560
APC/AFV/LVTPs							
U.S.	12,205	13,330	12,330	12,245	10,980	10,950	11,715
U.S.S.R.	30,500	30,500	30,600	35,600	35,600	38,750	38,750
U.S. Status	-18,295	-17,170	-18,270	-23,355	-24,620	-27,800	-27,035
Artillery							
U.S.	7,135	6,885	6,090	5,790	4,980	4,895	5,155
U.S.S.R.	16,000	16,500	17,000	18,500	18,500	19,000	19,000
U.S. Status	-8,865	-9,615	-10,910	-12,710	-13,520	-14,105	-13,845
Anti-Tank Missiles							
U.S.	2,710	12,080	21,840	33,235	45,015	63,235	72,625
U.S.S.R.	4,560	4,760	4,875	5,075	5,575	6,090	6,090
U.S. Status	-1,850	+7,320	+16,965	+28,160	+39,440	+57,145	+66,535

FIGURE 13 (continued)

U.S. SYSTEM CHARACTERISTICS

ARMOR	*First Deployed*	*Combat Weight (Tons)*	*Road Speed (mph)*	*Range (Miles)*	*Primary Arm*		*Crew/ Passengers*	*CBR Protection*
					Type	*Effective Range (Meters)*		
Medium Tanks								
M-48A3	1963	52	30	310	90mm	900	4	No
M-60A2	1971	57	30	280	152mm	Classified	4	No
M-60A1, A3	1961, 1976	53	30	310	105mm	4,400	4	No
Sheridan	1966	17	43	375	152mm Shillelagh	Classified Classified	4	No
APC/LVTP								
M-113	1962	12	40	300	50 Cal MG	1000	1/8	No
LVTP-7	1972	25	40	300	50 Cal MG	1000	3/25	No

ARTILLERY	*Type*	*Transport*	*Max Effective Range (Meters)*	*Nuclear Capable*
175mm	Gun	SP	32,800	No
8-in M-110	How	SP	16,800	Yes
155mm M-109A1	How	SP	18,000	Yes
155mm M-114	How	Towed	14,600	Yes
105mm M-101	How	SP	11,500	No
105mm M-102	How	Towed	11,000	No

NOTES: LVTP speed in water is 7.5 knots; endurance at sea 15 hours.

ANTI-TANK GUIDED MISSILES	*Type*	*Caliber*	*Max Effective Range (Meters)*	*Guidance*	*Weight (lbs)*	*Crew*
Dragon	Medium	5-in	1500m	Wire	30.5	1
TOW	Heavy	5.8-in	3000m	Wire	228	4

FIGURE 13 (continued)

SOVIET SYSTEM CHARACTERISTICS

ARMOR	*First Deployed*	*Combat Weight (Tons)*	*Road Speed (mph)*	*Range Aux Tanks (Miles)*	*Primary Arm* *Type*	*Primary Arm* *Effective Range (Meters)*	*Crew/ Passengers*	*CBR Protection*
Medium Tanks								
T-72	1975	40	30		{122mm? {115mm?	2,000+ 1,500	3	Yes
T-62	1962	40	30	310	115mm	1,500	4	Yes
T-55	1961	40	30	375	100mm	1,000	4	Yes
Light Tanks								
PT-76	1952	14	28	260	76mm	1,000	3	No
APC/AFV								
BTR-152	1959	9	45	390	None		19	No
BTR-50P	1957	15	25	170	None		22	Yes
BTR-60P	1961	10	48	300	None		16	Yes
BMP	1967	14	40	240	73mm Sagger AT	1,000 3,000	11	Yes

ARTILLERY	*Type*	*Transport*	*Max Effective Range (Meters)*	*Nuclear Capable*
203.2mm	Gun/How	Towed	27,000	Yes
180mm	Gun	Towed	30,000	?
152mm	Gun/How	Towed	17,000	?
152mm	Gun	SP	16,500	?
130mm	Gun	Towed	27,000	No
122mm	How	SP	15,300	No
122mm	How/AT	Towed	13-18,000	No
100mm	How/AT	Towed	8,500	No

ANTI-TANK GUIDED MISSILES	*Type*	*Caliber*	*Max Effective Range (Meters)*	*Guidance*	*Weight (lbs)*	*Number Rails*
Sagger	Medium	5.5-in	3000	Wire	25	1 on BMP, 6 on BRDM
Swatter	Heavy		2500	Radio	45	Always 4

The prognosis, however, is incomplete, despite Soviet progress. U.S. precision-guided munitions and mobile anti-tank missiles are causing disputes in the U.S.S.R., where some writers contend that such technologies threaten certain aspects of Soviet armored doctrine.[3] Talk continues, but tactical changes thus far have been slight.

MAJOR MANEUVER UNITS

Any army's cutting edge comprises maneuver units, of which divisions most affect the U.S./Soviet military balance. (See Figure 14.)

Ready Divisions The U.S. Army has never exceeded 16 active divisions since 1970. A maximum of about 19 might be attained if All-Volunteer Force recruiting standards were relaxed or incentives raised.[4]

Four Army divisions lack one regular brigade. Others lack one or more active maneuver battalions. Division readiness is unavoidably reduced, even though reserve component "roundouts" receive priority treatment and train part time with parent units.[5]

A fourth of our active divisions are stationed in Western Europe as part of NATO's on-site deterrent. Three in CONUS are earmarked to reinforce that force in emergency.[6] One is in Korea. Perhaps six in the United States should serve as a rotation

[3] Phillip A. Karber, "The Soviet Anti-Tank Debate," *Survival* (May-June 1976), pp. 105-11.

[4] *U.S. Army Force Design: Alternatives for Fiscal Years 1977-1981* (Washington, D.C.: Congressional Budget Office, July 16, 1976), pp. 45-49.

[5] The U.S. Army began its "roundout" program in 1972, when one reserve brigade and four separate battalions were designated to replace components missing from *under-strength* active divisions. Four brigades and 11 separate battalions (6 tank, 4 mechanized, and 1 infantry) now serve that purpose. All such elements must be ready to deploy with parent divisions on demand. Four other independent reserve brigades participate in an "augmentation" program intended to increase the combat power of designated *full-strength* divisions. These forces must also be ready to deploy with associated divisions on short notice, but requirements to do so depend on contingency plans. Data derived from Army staff officers in January, 1977.

[6] Four U.S. divisions are currently stationed in Germany. All others likely would be required as reinforcements if a major war occurred, but only three in CONUS (minus one brigade each already in Europe) are earmarked for early deployment. Heavy equipment for two of them is prepositioned. Assistant Secretary of Defense (Manpower and Reserve Affairs), *Manpower Requirements Report for FY 1976* (Washington, D.C.: February 1975), p. D-7; and *The Military Balance, 1976-1977* (London: International Institute for Strategic Studies, 1976), p. 6.

FIGURE 14 GROUND FORCE MANEUVER UNITS
Statistical Trends and Division Characteristics

DIVISIONS	*1970*		*1971*		*1972*		*1973*		*1974*		*1975*		*1976*	
United States	Active/ARNG, USMCR													
Armor	4	2	3	2	3	2	3	2	3	2	4	2	4	2
Mechanized	4	1	4	1	4	1	4	1	4	1	4	1	5	1
Infantry	5	5	3	5	3	5	3	5	3	5	4	5	5	5
Airmobile	2	0	1	0	1	0	1	0	1	0	1	0	1	0
Airborne	1	0	1	0	1	0	1	0	1	0	1	0	1	0
Tricap	0	0	1	0	1	0	1	0	1	0	0	0	0	0
Total	16	8	13	8	13	8	13	8	13	8	14	8	16	8
Marine	3	1	3	1	3	1	3	1	3	1	3	1	3	1
Grand Total	19	9	16	9	16	9	16	8	16	9	17	9	19	9
Soviet Union	Category I-II/Category III													
Tank	43	3	44	3	44	3	44	3	44	3	44	3	44	3
Motor Rifle	51	53	53	53	55	55	55	59	53	61	53	61	53	61
Airborne	7	0	7	0	7	0	7	0	7	0	7	1	7	1
Total	101	56	104	56	106	58	106	62	104	64	104	65	104	65
U.S. Standing	-85	-48	-91	-48	-93	-50	-93	-54	-91	-56	-90	-57	-88	-57
Grand Total														
U.S.	28		25		25		25		25		26		28	
U.S.S.R.	157		160		164		164		168		169		169	
U.S. Standing	-129		-135		-139		-139		-143		-143		-145	
BRIGADES														
United States	Active/Reserve Component													
Div Roundout	0	0	0	0	0	1	0	2	0	1	0	3	0	4
Div Augment	0	0	0	0	0	0	0	0	0	4	0	4	0	4
Separate	7	21	7	21	5	20	4	19	3	17	4	15	5	15
Total	7	21	7	21	5	21	4	21	3	22	4	22	5	23
Grand Total	28		28		26		25		25		26		28	
Soviet Union	0		0		0		0		0		0		0	
U.S. Standing	+28		+28		+26		+25		+25		+26		+28	

FIGURE 14 (continued)

REGIMENTS							
United States			National Guard				
Armored Cav	5 4	4 4	3 4	3 4	3 4	3 3	3 3
Soviet Union				Active			
Naval Inf	5	5	5	5	5	5	5

DIVISION CHARACTERISTICS

	Personnel Strength	*Medium Tanks*	*Armored Carriers*	*Artillery Pieces*	*Anti-tank Missiles*	*Maneuver Battalions*			
						I	*T*	*M*	
United States									
Army									
Armor	16,500	324	450	66	376	0	6	5	11
Mechanized	15,875	270	490	66	422	0	4	6	10
Infantry	15,875	54	120	76	366	8	1	1	10
Airborne	15,150	0	0	54	417	9	1	0	10
Airmobile	17,950	0	0	54	372	9	0	0	9
Marine	19,830	70	187 LVTPs	102	72	9	1	0	10
Soviet Union									
Tank	9,500	325	150	80	105	0	10	3	13
Motor Rifle	12,000	255	375	110	135	0	6	9	15
Airborne	8,000	0	100	54	145	9	0	0	9

NOTES: U.S. Marine division depicted includes personnel and weapons attached from force troops.
U.S. Light Anti-tank Weapons (LAWs) and Soviet rocket-propelled grenades (RPG-7s) are excluded. Soviet anti-tank missiles include about 100 on BMP armored carriers in each type division.
For maneuver battalions, I = infantry; T = tank; M = mechanized.

base for troops returning from overseas.[7] Thus, only two divisions (one in CONUS, one in Hawaii) are free for contingency purposes without spreading the force very thin or federalizing parts of the National Guard.[8]

In contrast, 55 Soviet Category I divisions are kept at 75 to 100 percent of top personnel strength, with complete equipment. Another 49 in Category II maintain average manning levels of about 60 percent. All officers, non-commissioned officers (NCOs), and key specialists are on tap to train as teams. Material shortages are minor. Experienced fillers thus can bring 104 divisions close to full strength very quickly.[9]

Soviet capabilities in the recent past have been buttressed considerably by beefing up manpower and firepower. Each tank division has 1,000 more men than in 1970, mainly mechanized infantry. Motor rifle divisions each have been bolstered by 2,000 troops and 67 tanks. Both types are buttressed with additional artillery, some of it self-propelled. Personnel strengths stay small compared with U.S. counterparts, but striking power is potent.[10]

Nevertheless, the reservoir of Soviet ready divisions is reduced by restrictions much like our own. Twenty-five Category I divisions abut the Czech and East German borders. Maybe 46 more, including Category II, are sited in Hungary, Poland, and European Russia as Warsaw Pact reinforcements. Another mix of 20 or more

[7] No rotation base would be needed if a general war erupted in Europe. *All* U.S. divisions were committed overseas during World War II. Most men returning to CONUS were separated from the Service, attended schools, or were assigned to staffs.

In peacetime, and during limited conflicts like those in Korea and Vietnam, U.S. divisions stay overseas, while personnel serve specified tours and then return to CONUS. Most men must receive assignments related to their skills, or readiness levels lapse. A satisfactory balance can be maintained if about 60 percent of the force stays in the United States, according to Army staff officers. When fewer than 40 percent serve as a rotation base, combat effectiveness suffers seriously.

Four U.S. divisions, for example, served as a rotation base for 15 others overseas in the late 1960s. Insufficient slots existed, especially for specialists. Training everywhere suffered. The four divisions, which played parts in NATO plans, were combat ineffective for a protracted period, because equipment was stripped and personnel passed through too rapidly.

[8] Five separate brigades and three armored cavalry regiments (ACRs) afford extra U.S. strength, but are not the equivalent of two or three divisions, because they lack staying power and the capacity for large-scale combined arms action. Three brigades serve special purposes in Alaska, Berlin, and Panama. Two ACRs and equipment for the third (now in CONUS) are positioned in Germany. Only two brigades remain in general reserve.

[9] Data derived from personal conversations with DIA analysts in January, 1977.

[10] *Ibid.* See also Erickson, *Soviet-Warsaw Pact Force Levels,* pp. 31-39.

man the Chinese frontier.[11] Some guards divisions around major cities, like Moscow and Leningrad, never move far from home stations. Perceived requirements, of course, could change, but fewer than 12 out of 104 ready divisions are currently uncommitted.

First-Line Reserves Since most U.S. and Soviet ready divisions are tied to continuing tasks, first-line reserves fulfill a crucial function.[12] Eight Army National Guard (ARNG) divisions comprise the U.S. complement.[13] That number has stayed constant since 1968. Fifteen separate brigades and three armored cavalry regiments not affiliated with active division roundout or augmentation programs complete the list of major maneuver units in the U.S. reserve.

ARNG infantry divisions, as a general rule, would require ten weeks of intensive preparation before being fully ready to fight, although deployment might take place a month earlier in emergency. Post-mobilization training for armored and mechanized divisions would take almost four months, if tank gunners and signal troops in particular were allowed time to attain proficiency. Minimum combat standards could be achieved in nine or ten weeks.[14] Any National Guard division, however, could replace Regular Army forces in CONUS soon after entering federal service.

Nothing in the Soviet inventory is really comparable to our National Guard. Twenty-some-odd so-called "mobilization divisions," not carried on active order of battle lists, could fill in 30 days, but manning levels now are minimal (200 to 300 men), stocks are in dead storage, and training would take several months.[15]

[11] *Ibid.*

[12] The generic term "reserves," as used in this study, refers to all reserve components, including the National Guard.

[13] Twelve divisions in the U.S. Army Reserve are regional replacement training centers, rather than maneuver units.

[14] Readiness Condition 1 (REDCON-1 or C-1) requires U.S. units to be "fully capable of performing missions for which organized or designed." Such units, with 95 percent of authorized manpower and 90 percent of materiel, "may be deployed to a combat theater immediately" after mobilization. C-3 units, which are marginally ready, have deficiencies that severely constrain capabilities, but could be committed in combat "under conditions of grave emergency." Army Regulation 220-1, Unit Readiness Reporting, March 17, 1975, pp. A-1, 2.

Times it would take for ARNG divisions to reach C-1 or C-3 status were furnished by Army staff officers in January, 1977.

[15] Unclassified data derived from personal conversations with DIA analysts in January, 1977.

Sixty-five Category III divisions come closer to corresponding with U.S. reserve components, although all have substantial active elements. The best are at about one-third strength. The poorest are simply cadres that total 10 percent. Combat equipment is almost complete, if elderly items count. Severe transport shortages would be solved in crises by taking trucks from the civil economy.[16]

Most such divisions, being stationed in densely populated regions with many reservists, could fill in about three days. Delays would be longer for those that rely on distant replacements. Redeployment to relieve Category I and II divisions in static sectors could commence as soon as units were up to strength, but considerable cramming would be required to create cohesive divisions.[17] Total elapsed times would be directly proportionate to training standards and percentages on active duty.

The Soviet aggregate, eight times larger than our National Guard, allows the Kremlin latitude not available to U.S. leaders. Perhaps 20 to 25 Category III divisions stand guard along the Sino-Soviet frontier,[18] a few more in the far south. The remaining 40 to 45 are first-line reserves.

Marines and Naval Infantry

The main mission of U.S. Marines is to seize and secure hostile coasts by amphibious assault, but the Corps is capable of sustained operations ashore, independently or in concert with our Army, if assisted logistically.[19]

Active ground force components comprise 70,000 men in three divisions that are backed by extensive combined arms combat and service support. (See Figures 12 through 14.)

Three divisions are positioned to cope promptly with assorted global contingencies and, after assembly, could contribute significantly to U.S. deterrent/defense capabilities vis-à-vis the Soviet

[16] *Ibid.*

[17] *Ibid.* Some sources contend that Cat III divisions can fill in 24 hours.

[18] About half of all Soviet divisions on the Chinese border are Category III. The Kremlin apparently anticipates no early aggression by either side in that area.

[19] U.S. Marines are organized, trained, and equipped as an air-ground team whose divisions are never committed to combat without associated tactical air wings. This section arbitrarily separates the two components to facilitate U.S./Soviet comparisons.

Union. If those forces were committed, the Marine's reserve division/wing team would provide the sustaining base.[20]

Soviet naval infantry, with a total strength of 12,000, compose five regiments of fewer than 2,000 men each. As currently constituted, with minimum fire support and not much staying power, they are suitable mainly for small-scale raids and conventional operations against second-class opponents.[21]

Combined Flexibility

U.S. active ground combat power, pooled with that of allies, presently serves deterrent purposes in Northeast Asia and NATO Europe. Five divisions (two Army, three Marine) are free to contend concurrently with contingencies elsewhere, if the situation stays stable in those theaters. Additional "brush fires" adverse to American interests would burn beyond our control, unless we called up reserves. Such action, however, could dilute essential U.S. deterrent powers by reducing abilities to reinforce rapidly at either point of primary decision.

Soviet flexibility superficially seems more favorable. As matters now stand, Moscow has more than 50 divisions in strategic reserve, including 10 that are combat ready. None, however, have recently been deployed beyond Soviet borders, except in satellite states, and abilities to sustain large-scale forces on far foreign shores are subject to serious question.[22] Proxies, including Cubans, take their place.

[20] The 4th Marine Division, manned by 19,000 reservists in paid drill status, plans to reach full readiness 60 days after recall. Its units maintain only about 50 percent of their equipment allowance at home armories. Shortages must be made up in emergency from Prepositioned War Reserves at the same time active divisions are expected to tax the supply system. Individual Ready Reserves normally receive at least 30 days' notice before callup. If that requirement were waived, combat proficiency could be achieved more rapidly. U.S. Congress, Senate, *Hearings before the Armed Services Committee on FY 1977 authorization,* Part 3, Manpower, 94th Congress, 2d Session (Washington, D.C.: U.S. Government Printing Office, 1976), pp. 1846-50.

[21] DIA analysts identify a marine division structure in the Soviet Far East, but its elements still exercise separately.

[22] Several Soviet divisions were stationed in northern Iran during World War II. They stayed from August 25, 1941 until May 9, 1946, when U.S. and U.N. pressures prompted withdrawal.

The likelihood thus seems low that Soviet divisions will be used for distant initiatives in the short-range future. Immense impediments inhibit employment in hot spots, such as the Middle East.[23] Huge Soviet reserves consequently bolster Warsaw Pact capabilities, as assessed in Chapter 6, but bear less on the global balance.

[23] Difficulties are described in U.S. Congress, House, *Oil Fields as Military Objectives: A Feasibility Study*, prepared for the Special Subcommittee on Investigations of the Committee on International Relations by the Congressional Research Service, 94th Congress, 1st Session (Washington, D.C.: U.S. Government Printing Office, 1975), pp. 18-21.

NET ASSESSMENT APPRAISAL

The basic trends which Collins summarizes in Graph 6 provide a good overview of the overall strengths of each side. It is interesting to note, however, that the trends he describes for ground forces also exist in U.S. and Sovet general purpose air forces.

Further, the Soviets are modernizing their forces with new equipment more quickly than the U.S. Chart One shows that the Soviets lead the U.S. in virtually every aspect of production. Moreover, the U.S.S.R. deployed new major types of weapons in every category during the period 1974–76 and the U.S. did not.

COMPARATIVE FUTURE INITIATIVES

The overall impact of future U.S. and Soviet technological innovation on the general purpose forces balance is unclear. Although a Joint Staff comparison of the major developments on each side is provided in Table One, it is impossible to translate current trends into comparisons of system effectiveness, a clear picture of how such systems will be deployed, or estimates of how many will be procured.

CHART ONE

CHANGES IN QUANTITIES OF MILITARY EQUIPMENTS – U.S./U.S.S.R. (1966–1976)

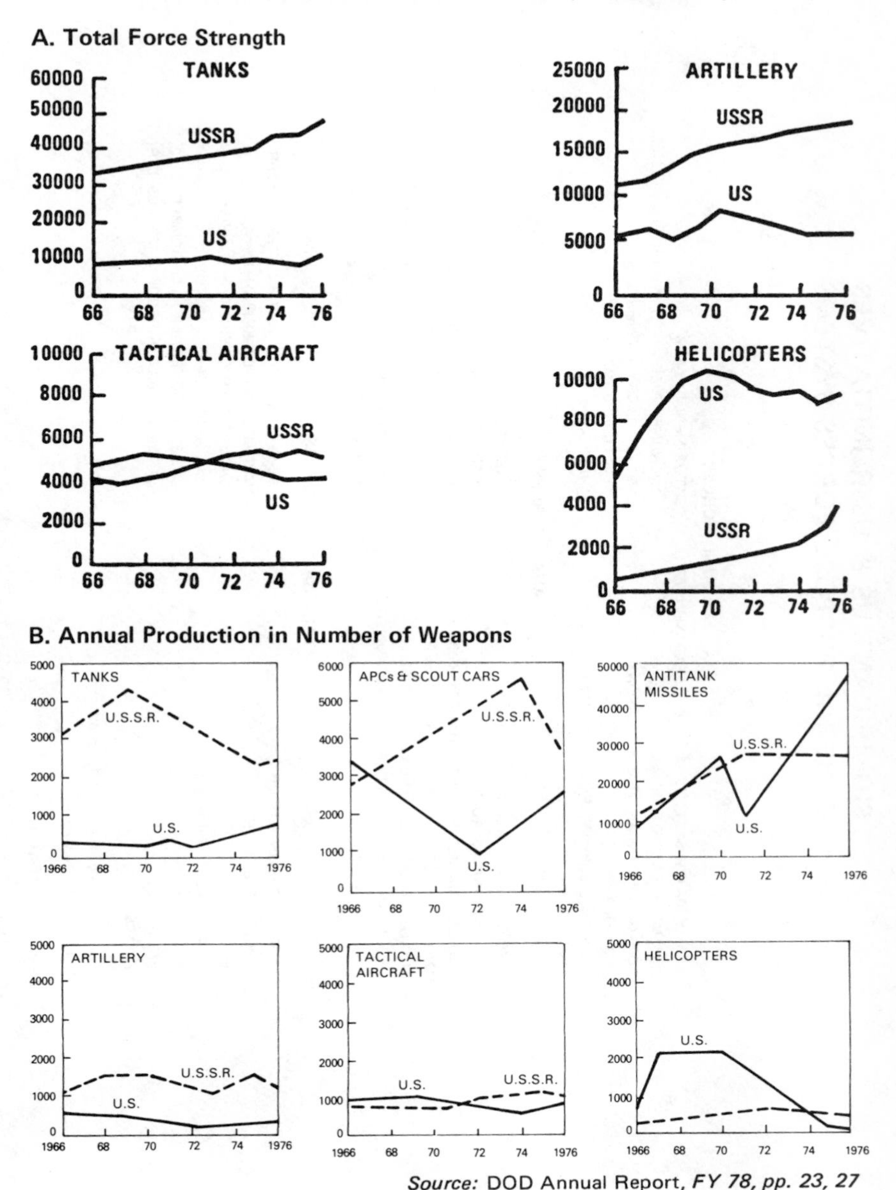

Source: DOD Annual Report, *FY 78, pp. 23, 27*

TABLE ONE

SIGNIFICANT US & USSR INITIATIVES GENERAL PURPOSE FORCES SYSTEMS

US	USSR
GROUND FORCES	
ARMY MAJOR SYSTEMS: AAH & UTTAS. XM-1 TANK & MICV. PATRIOT (SAM-D)	*T-72 MEDIUM TANK
*DRAGON & TOW ANTITANK WEAPONS	*NEW/IMPROVED ANTITANK GUIDED MISSILES
US ROLAND	*SP ARTILLERY (122mm/152mm)
GENERAL SUPPORT ROCKET SYSTEM (GSRS)	*TACTICAL SAMS (SA-8)
	*HIND A HELO
NAVAL FORCES	
**688 CLASS ATTACK SUB	*KIEV CLASS CARRIER
**LHA AMPHIB ASSAULT SHIP	*KARA CLASS CRUISER
GUIDED MISSILE FRIGATE (HARPOON EQUIPPED)	*KRIVAK CLASS DESTROYER
**NIMITZ CLASS CARRIER	*NEW CLASS LPD/LSD
GUIDED MISSILE DESTROYER (AEGIS)	*NEW, VERY LARGE REPLENISHMENT SHIP
NUCLEAR GUIDED MISSILE STRIKE CRUISER	*ALFA CLASS SSN
TOMAHAWK MISSILE	*TANGO CLASS SS
TACTICAL AIR FORCES	
*A-10 CLOSE AIR SUPPORT A/C	*SU-19 (FENCER A) VGW FIGHTER BOMBER
*F-15 (EAGLE) FIGHTER	*MIG-23 (FLOGGER B) FIGHTER
EF-111A	*SU-17 (FITTER C) FIGHTER BOMBER
F-14 (TOMCAT) FIGHTER	*YAK-36 (FORGER A) V/STOL
F-16 FIGHTER	*MIG-23 (FLOGGER) FIGHTER BOMBER
F/A-18	*MIG-25 (FOXBAT B) RECONNAISSANCE STRIKE
AWACS	

*Currently being deployed still in production

**Now in Series production

Source: Military Posture, FY 78

CHANGES IN U.S. AND SOVIET DIVISION STRENGTHS

Soviet divisions have grown steadily in strength and quality. There has been a striking shift in the Soviet combined arms balance. Tank divisions have acquired large amounts of supporting firepower, and motorized rifle divisions have increased sharply in overall firepower. This has been achieved with only a slight increase in manpower, but with major increases in armored mobility and armored fighting vehicles, conversion to self-propelled artillery, and increases in anti-tank weapons strength.

THE IMPACT OF CHINA ON THE BALANCE

At the same time, however, the Soviet Union must face a rapidly improving threat from the People's Republic of China. China is now experiencing enough stability to significantly improve its forces, and while she is still grossly inferior in quality and equipment to U.S. and Soviet forces, her absolute strength in weapons has already reached the level that she must be regarded as far more than a mass of infantry. This is shown in Chart Two.

CHART TWO

MAJOR WEAPONS AND EQUIPMENT – GROUND FORCES JAN. 1977

	US	USSR	PRC
TANKS	10,000	45,000	8-9,000
APC & FIGHTING VEH.	22,000	45-55,000	3-4,000
ARTILLERY	5,000	19,000	15-17,000
HEAVY MORTARS	3,000	7,000	5-6,000
HELICOPTERS	9,000	3,800*	3-400

*2300 IN FRONTAL AVIATION

Source: DOD, U.S. Military Posture, FY 78, p. 63

COMPARING STRATEGIC POSTURE OF THE U.S. and U.S.S.R.

The data on the Chinese balance illustrate the most serious problem in interpreting the trends in the balance of total U.S. and

Soviet ground forces. The entire Soviet army can never be committed against the entire U.S. Army, and the strategic position of the U.S.S.R. is fundamentally different from that of the United States.

The United States still enjoys the great strategic advantage of having no continental enemies. It can posture its ground forces as a partner in a broad European Alliance rather than as an uncertain master. There is no need to maintain large forces for internal security purposes, for border security, or to ensure control of its own population.

There is no doubt that the strategic pressures the U.S.S.R. faces have, along with other causes, led to a Soviet superiority which threatens U.S. interests. At the same time, the vast size of these forces imposes almost incredible costs on the U.S.S.R.

LAND-ORIENTED TACTICAL AIR POWER

Dissimilar strategies, geographic circumstances, and technologic competence have caused U.S. and Soviet tactical air combat power to develop along different lines that left our rival disadvantaged.

Assorted U.S. assets, positioned in allied countries or on aircraft carriers, possess global capabilities that can be supplemented swiftly with responsive reserves. Assigned missions span a wide spectrum.[24]

The Soviet side, dedicated to home defense in past decades, is still made up mainly of special-purpose aircraft designs merely modified since the 1950s and early 1960s. A modern force is emerging, but transition at current rates will continue into the mid-1980s.

High-Performance Combat Forces

The degree to which tactical air forces help deter or defeat opponents ashore depends on abilities to attain air superiority over

[24] For general discussions of U.S. tactical air power see William D. White, *U.S. Tactical Air Power: Missions, Forces, and Costs* (Washington, D.C.: The Brookings Institution, 1974), 121 pp.; and *Planning U.S. General Purpose Forces: The Tactical Air Forces* (Washington, D.C.: Congressional Budget Office, January 1977), 51 pp.

key contested areas, cut off enemy supplies/reinforcements, and furnish close support for ground forces. U.S. resources are structured to accomplish those tasks under a range of conditions. The Soviets are confined.

LAND-BASED FIGHTERS AND MEDIUM BOMBERS

Forward-based fighters of the U.S. Air Force provide America's primary land-based air power overseas. Rapid-reaction reinforcements in the United States add credibility to deterrent and combat capabilities.[25] Ready forces, assisted by in-flight refueling, can arrive almost anywhere in one to three days after notification, armed with conventional or nuclear weapons.[26]

Soviet counterparts (except for medium bombers) are found in Frontal Aviation. The largest concentration is focused on Eastern Europe, a lesser one on the Chinese frontier. The remainder are dispersed among military districts.[27]

Statistical strengths have stabilized at a level twice our own, counting Marine Corps tac air (Graph 7 and Figure 15),[28] but the changing complexion is at least as significant.

Intermediate- and medium-range ballistic missiles with no non-nuclear option were the principal weapons for deep interdiction purposes when this decade opened, because Badger and Blinder bombers (TU-16s, TU-22s) have poor penetration prospects. Short-range, fair-weather Frontal Aviation, lightly armed and with little

[25] About a third of all USAF fighter/attack aircraft in operational units are assigned to U.S. Air Forces in Europe (USAFE). A smaller slice is deployed with Pacific Air Forces (PACAF). Slightly more than half are held in CONUS reserve by Tactical Air Command (TAC). U.S. Congress, Senate, *Seminars: Service Chiefs on Defense Mission and Priorities. Hearings Before the Task Force on Defense of the Committee on the Budget,* Part III, Air Force (Washington, D.C.: U.S. Government Printing Office, 1976), pp. 70-72.

[26] F-111s from Mountain Home AFB, Idaho, reportedly took just 15 hours to arrive in Korea ready for combat during the 1976 DMZ incident in which two Americans were killed. "TAC's Dixon Watches Soviet Threat, Stresses Training Realism," *Aerospace Daily* (September 28, 1976), p. 128.

Air Force Reserve and National Guard units must be ready to move 72 hours after mobilization. "Virtually all . . . can reach Europe within a month or less," according to the Congressional Budget Office. *Planning U.S. General Purpose Forces: The Tactical Air Forces,* pp. ix, 27.

[27] Some Soviet fighter aircraft defended the Suez Canal a few years ago, but none are now stationed on foreign soil, except in satellite states.

[28] Soviet strategic air defense forces supplement tactical air power for air superiority purposes along that country's borders.

Graph 7

TACTICAL AIR COMBAT FORCES
Statistical Summary

(Note Different Scales)

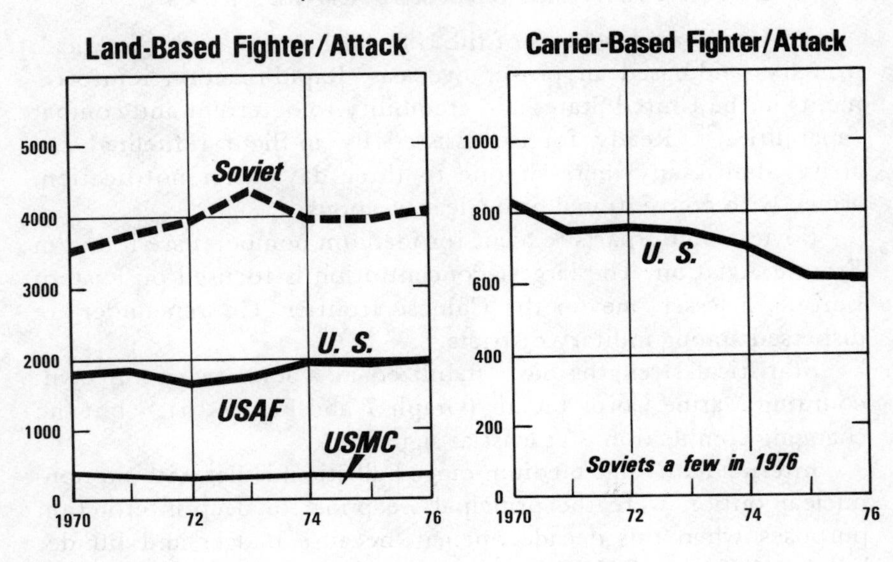

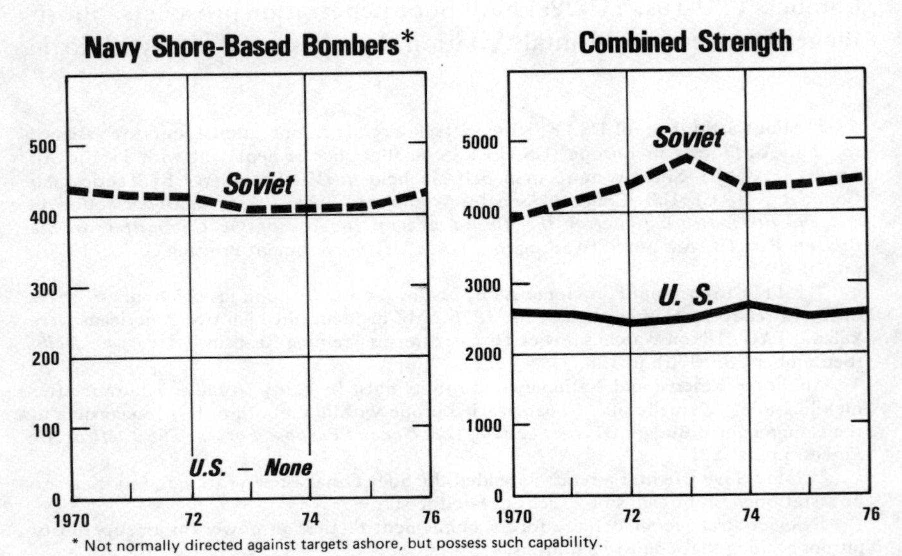

* Not normally directed against targets ashore, but possess such capability.

FIGHTER/BOMBERS	*1970*	*1971*	*1972*	*1973*	*1974*	*1975*	*1976*
United States							
Air Force							
A-7	4	80	144	144	253	224	210
F-4	968	971	933	965	1056	1055	1091
F-100	282	209	9	0	0	0	0
F-105	145	104	66	31	55	38	37
F-111	26	158	211	283	311	333	312
Total	1425	1522	1363	1423	1675	1650	1650
Marine							
A-4	94	101	67	89	85	65	76
A-6	76	47	45	49	49	56	57
AV-8	0	6	13	32	59	53	53
F-4	184	154	144	130	128	130	132
Total	354	308	269	300	321	304	318
Grand Total	1779	1830	1632	1723	1996	1899	1968
Soviet Union							
Bombers							
Backfire	0	0	0	0	0	25	60
TU-16	500	500	500	500	500	475	450
TU-22	175	200	200	200	200	170	170
Total	675	700	700	700	700	670	680
Fighters							
MIG-17	800	800	800	900	900	900	600
MIG-19	100	200	100	50	0	0	0
MIG-21	1400	1500	1600	1700	1500	1600	1700
MIG-23	0	0	100	200	300	300	500
SU-7	500	500	500	500	400	400	400
SU-17	0	0	50	100	100	100	100
SU-19	0	0	0	0	0	0	50
YAK-28	50	50	100	200	50	0	0
Total	2850	3050	3250	3650	3250	3300	3350
Grand Total	3525	3750	3950	4350	3950	3970	4030
U.S. Standing	-1746	-1920	-2318	-2627	-1954	-2016	-2062

FIGURE 15 (*continued*)

HELICOPTER GUNSHIPS	*1970*	*1971*	*1972*	*1973*	*1974*	*1975*	*1976*
United States	635	520	535	730	715	685	690
Soviet Union	0	0	0	100	125	200	300
U.S. Standing	+635	+520	+535	+630	+590	+485	+390

AIRCRAFT CHARACTERISTICS

	First Deployed	*Combat Radius (Miles)*	*Max Speed (Mach)*	*Paylod (lbs)*	*Typical Weapons*		*Nuclear Capable*	*All Weather*
					Guns	*Missiles*		
United States								
A-6	1963	750	0.9	10,000	None	18 Mk-82	Yes	Yes
A-7D	1966	550	0.9	7,200	1 20mm	12 Mk-82	Yes	Yes
AV-8	1969	200	0.9	2,500	2 30mm	AIM-9	No	Yes
F-4	1963	550	2.2	16,000	1 20mm	4 AIM-7E 11 Mk-117	Yes	Yes
F-100	1954	450	1.3	9,000	4 20mm	2 AIM-9 or AGM-12	Yes	No
F-105	1959	625	2.1	10,200	1 20mm	4 AGM-45 or 2 AGM-78	Yes	Yes
F-111	1967	745	2.2	14,500	1 20mm	24 Mk-82	Yes	Yes
Soviet Union								
Backfire	1975	2500	2.5	20,000		AS-4, AS-6	Yes	Yes
MIG-17	1953	360	0.9	1,100	3 23mm	4 Alkali	No	No
MIG-19	1955	425	1.0	1,100	3 30mm	4 Alkali	No	No
MIG-21	1956	550	2.1	2,000	1 23mm	4 Atoll	Yes	No
MIG-23	1971	550	2.5	2,800	1 23mm	4 AS-7	Yes	Yes
SU-7	1960	200	1.6	5,500	2 30mm	None	No	No
SU-17	1972	300	1.7	5,500	2 30mm	AS-7	Yes	Yes
SU-19	1974	300	2.0	11,000	1 23mm		Yes	Yes
TU-16	1955	2000	0.8	20,000	7 23mm	2 AS-5	Yes	No
TU-22	1962	700	1.4	12,000	1 23mm	1 AS-4	Yes	No
YAK-28	1961	[illegible]	[illegible]	[illegible]	[illegible]	[illegible]	[illegible]	[illegible]

MISSILE CHARACTERISTICS

	First Deployed	*Guidance*	*Range (Miles)*	*Speed (Mach)*	*Warhead*	
					Type	*Weight, Yield*
United States						
Air-to-Air						
AIM-7E		Radar	14	3.5+	HE	60 lbs
AIM-7F	1976	Radar	28	3.5+	HE	60 lbs
AIM-9B	1958	Infrared	2	2.5	HE	25 lbs
Air-to-Ground						
AGM-6						
AGM-12B	1959	Radio	7	1.8	HE	250 lbs
AGM-12C	1959	Radio	10	2.0	HE	1000 lbs
AGM-45	1964	PH	10	2.0	HE	145 lbs
AGM-65	1969	TV	Classified	Classified	HE	27 lbs
AGM-78	1968	PH	15.5	2.0	HE	1356 lbs
Soviet Union						
Air-to-Air						
Alkali	1953	Radar	3.7-5	1-2	HE	
Anab	1961	Radar, IR	5-6.2		HE	
Atoll	1956	Infrared	3-4		HE	
Advanced Atoll		Radar	3-4		HE	
Air-to-Ground						
AS-2	1961	Radar	130	1.2	HE	
AS-4	1962		450		Nuke	
AS-5	1967		200		HE	
AS-6		Radio	135		Nuke	200 KT
AS-7		Radio				

NOTES: See Figure 17 for U.S. carrier-based aircraft.
All aircraft are Unit Equipment (UE), excluding recon and special-purpose versions.
Backfire bombers are the same aircraft shown on Figure 8.
Combat radii correspond with payloads shown under average conditions.
Payloads are merely representative. External fuel tanks are included where applicable.
F-15 aircraft were all in Training Squadrons in 1976.
IR missile guidance is infrared. PH is passive homing.
U.S. armed helicopters are AH-1s. Soviet counterparts are MI-24s.

lift capacity, was better fitted for local air defense than for offensive strikes over enemy soil or close support for field armies.

Backfire bombers and multimission fighters with standoff firepower and better avionics now augment Moscow's arsenal. All can deliver nuclear weapons as well as conventional ordnance under adverse weather conditions. SU-19 Fencers are the first Soviet airframes created specifically for ground attack. Advanced armaments and penetration aids improve performance of all new types, especially against point targets. Older mainstays, after modification, conform to broader missions. The combat radius and destructive power of some ground attack regiments, for example, has quadrupled since 1970.[29]

Conversion, however, is just commencing. The newest aircraft are not yet numerous, other than MIG-23s. Air intercept training retains its traditional emphasis on strict ground control, with little attention to free air combat for air supremacy purposes. Soviet fighters, which cannot refuel in flight, are difficult to redeploy long distances over land, and depend on ships for movement over seas.[30] Consequently, U.S. forces are qualitatively superior, and should stay so, given our long lead in tactical aviation technology and F-15, F-16/A-10 programs, which are about to bear fruit.[31]

AMPHIBIOUS FIGHTER/ATTACK FORCES

The Soviets have no air power comparable to U.S. Marine air wings, which include more than three hundred fighter/attack aircraft in active squadrons (Figure 15).[32] Crews specialize in air cover and close air support for Marine divisions during amphibious assaults and sustained operations, but are prepared to participate in overall air efforts as directed.[33]

Marine forces exhibit adaptability unequalled by sister Serv-

[29] J-5 comments on the draft of this study, March 4, 1977.

[30] *Ibid.*, pp. 127, 185.

[31] Edgar Ulsamer, "The Quiet Revolution in USAF's Capabilities," *Air Force Magazine* (September 1975), pp. 38-44.

[32] Marine aircraft are an integral part of Navy tactical air power, procured with Navy dollars, supplied and serviced by a Navy system. Pilots in large part use Navy training facilities. This section, however, considers the special contribution of Marine air wings to land-oriented air power.

[33] Title 10, United States Code, Chapter 503, Section 5013; and Department of Defense Directive 5100.1, *Collateral Functions.*

ices, being able to function effectively either ashore or afloat.[34] Portable catapults, optical landing aids, arresting gear, lights, and other accoutrements associated with expeditionary airfields reduce requirements to seize or construct strips for high performance fighters at an early stage. Marine V/STOL aircraft transferred ashore on amphibious ships can commence operations even before installation is complete.[35]

NAVAL AIRCRAFT CONTRIBUTIONS

The U.S. Navy, with more than six hundred carrier-based fighter/attack aircraft and collateral functions connected with land combat,[36] can supplement or supplant Air Force and Marine forces in many circumstances that call for air power. (See Figure 18 in the next section for statistics.)

The Soviet Navy cannot yet compete. A few YAK-36 Vertical Takeoff and Landing (VTOL) aircraft aboard the carrier *Kiev* constitute its sole high performance capability. They reputedly are better than British-built AV-8 Harriers flown by U.S. Marines, but are too few to tip the U.S./Soviet balance.[37]

Helicopters for Fire Support

U.S. Army helicopter gunships, many armed with anti-tank (AT) missiles, afford significant close air support that is immediately responsive to ground commanders. Crews are combat tested.

Soviet counterparts, controlled by Frontal Aviation, not the Army, are neither so numerous nor so well trained, but forces are building up fast. MI-24 (Hind) helicopters, with 57 mm rockets and Sagger AT missiles, are formidable fire-support systems that can also carry considerable cargo or fourteen fully armed troops.[38]

[34] "Bare base" kits would allow Navy fighter/attack aircraft to operate off primitive strips ashore, but supply, maintenance, and other support facilities now aboard carriers would be costly to duplicate. Carrier aircraft could, however, share established installations with other Services.

[35] Marine Corps comments on the draft of this study, March 3, 1977. V/STOL stands for Vertical and/or Short Takeoff and Landing.

[36] Title 10, United States Code, Chapter 503, Section 5012; and Department of Defense Directive 5100.1, *Collateral Functions.*

[37] Edgar L. Prina, "Soviet Jet Called Superior to U.S. Version," *San Diego Union,* August 20, 1976, p. 21. Quotes Navy Secretary J. William Middendorf, II.

[38] Benjamin F. Schemmer, "Soviet Armed Helicopter Force Said to Double By Middle of 1977," *Armed Forces Journal* (December 1976), pp. 28-29; and "Hind-D Carries Added Weapons," *Aviation Week and Space Technology* (February 21, 1977), p. 21.

Combined Flexibility

America's tactical air combat assets in the aggregate afford flexibility not available to the Soviet Union, whose main strength still lies in mass.[39] Our side enjoys a clear qualitative edge in most respects, regardless of mission.[40] U.S. leadership is light years ahead of the Soviet system.[41]

Small size, however, creates an Achilles heel. The several U.S. Services are insufficient to cope with large-scale contingencies unless they assist each other. Even then, difficulties develop. Air Force-Navy-Marine collaboration, for example, was compulsory in Korea and Vietnam, to such a degree that deterrent/defense capabilities suffered in Central Europe. Projected U.S. procurement programs will do little to correct that shortcoming.

[39] This section is not a brief in favor of four U.S. air forces, which some critics castigate. (See, for example, "Four U.S. Tactical Air Forces," *The Defense Monitor* [Washington, D.C.: Center for Defense Information, October 1975], 8 pp.) It simply states superior U.S. flexibility as a fact.

[40] General David C. Jones, Air Force Chief of Staff, as cited by Ulsamer in "The Quiet Revolution in USAF's Capabilities," p. 39.

[41] David Binder, "Soviet Defector Depicts Grim Life at MIG-25 Base," *New York Times,* January 13, 1977, pp. 1, 12.

NET ASSESSMENT APPRAISAL

The strategic problems the U.S.S.R. faces in sizing its air forces are at least as great as in sizing its ground forces. The U.S.S.R. must defend far larger borders than the U.S. against potentially hostile states. It must plan to use air power to offset the mass of PRC ground forces. It must plan its tactical air forces to meet the air forces of Western Europe as well as the United States. And, it must provide a massive air defense against U.S. strategic bombers. It must accomplish all these tasks with a smaller economy than the U.S., and while trying to make the major improvements in every other branch of its forces which are described throughout this report.

COMPARING TACTICAL AIR FORCES

These differences in the U.S. and Soviet strategic posture make it difficult to compare U.S. and Soviet air forces. In Graph 7, Collins shows the different strengths of each side's tactical air forces by military service. This tends to make the U.S. seem strikingly inferior in some of the counts. In contrast, the Department of Defense usually compares all U.S. and Soviet tactical aircraft together, including some reserves. The DOD method may be more accurate as a picture of pure force size, and shows in gross terms, however, that U.S. and Soviet tactical air forces now have rough quantitative parity as shown in Chart Three.

CHART THREE

TACTICAL AIRCRAFT

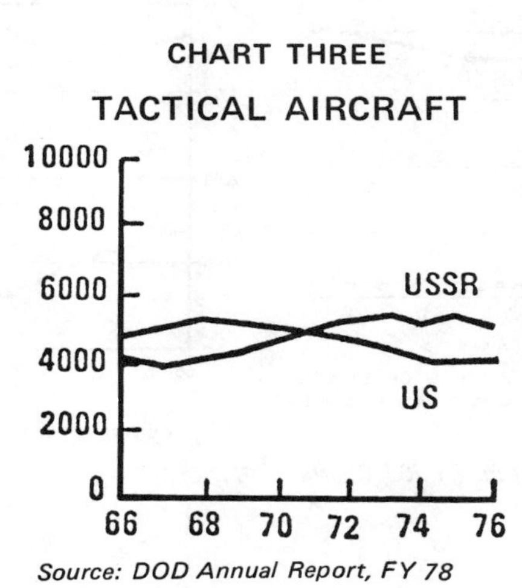

Source: DOD Annual Report, FY 78

THE IMPACT OF PVO STRANY ON THE BALANCE

Counts of tactical air forces are also complicated by the fact that the U.S.S.R. has large strategic air defense forces (PVO Strany), and the U.S. has very large helicopter, ASW, army aviation, and bomber forces. None of these forces are truly comparable, but none can validly be ignored. The U.S.S.R. has almost 75% as many tactical fighters in its strategic air defense forces as in its tactical aviation force, while the U.S. has only a few hundred such aircraft. Chart Four also shows that the Soviets have strong SAM defenses while the U.S. does not.

Some experts have argued that the Soviets could use part of their strategic air defense aircraft in tactical aviation missions in a war on NATO or in other contingencies, and there are reports that

CHART FOUR

US AND SOVIET TACTICAL FORCES COMMITTED TO STRATEGIC DEFENSE

US AND USSR STRATEGIC (HOME) DEFENSE
SURFACE TO AIR MISSILES ON LAUNCHERS

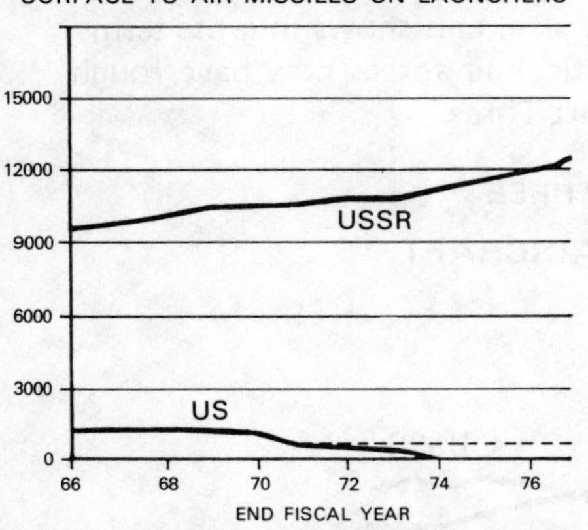

*DOTTED LINE INDICATES GENERAL PURPOSE FORCE MISSILES ASSIGNED TO CINC NORAD FOR USE IN STRATEGIC DEFENSE ORDER OF BATTLE

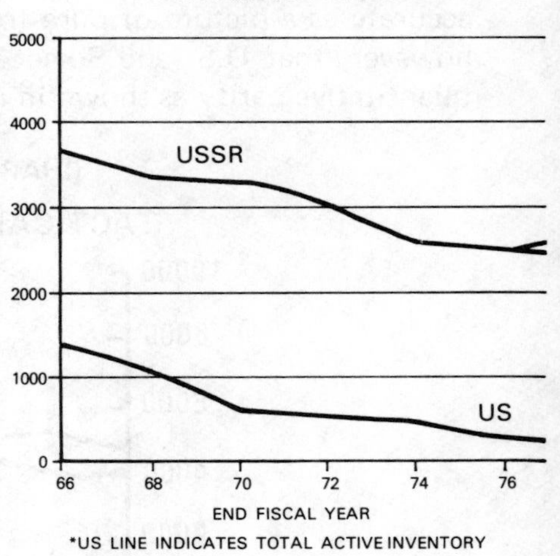

*US LINE INDICATES TOTAL ACTIVE INVENTORY

Source: OJCS, "Military Posture, FY 78," Charts 11, 12.

Soviet strategic air defense forces are reported to be trained for dual-capability. This, however, would pose problems for them. It would weaken their defenses when the risk of escalation is greatest; it would mean committing partially trained and often badly configured strategic air defense aircraft when other tactical aviation forces in the Soviet Union could just as easily be used; and it is unclear that most PVO Strany aircraft could be sheltered or properly handled by Soviet Tactical Aviation's AC&W ground control environment, and ground controlled intercept systems. Sheer numbers would not be useful unless their vulnerability can be minimized, and they could be managed and allocated effectively in battle.

THE ROLE OF OTHER MILITARY AIRCRAFT

In contrast, Chart Five shows that the U.S. has a massive, if rapidly declining, overall advantage in helicopters. This U.S. advantage in "other tactical" aircraft strength exists in many other areas of Army and Naval aviation, and generally is qualitative as well as quantitative.

Some experts do argue, however, that the new Soviet HIND attack helicopter may be giving the U.S.S.R. tactical superiority in anti-tank helicopter and infantry support capability, but this is unclear. Neither HIND numbers or weapons capability have been confirmed.

CHART FIVE

US AND SOVIET MILITARY HELICOPTER STRENGTH

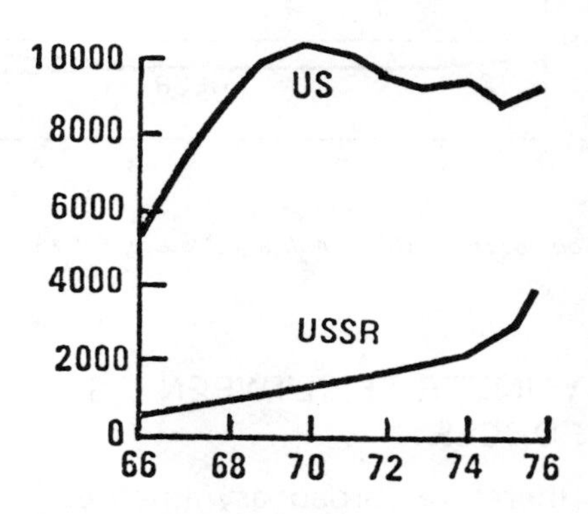

Source: DOD Annual Report, FY 78

THE ROLE OF STRATEGIC BOMBERS IN TACTICAL MISSIONS

The U.S. also has a massive advantage in heavy bomber conventional attack capability. This is sometimes forgotten in analyzing the European tactical air balance, but it is unlikely ever to be forgotten in Asia. This U.S. advantage is shown in Chart Six.

U.S. heavy bombers have almost incredible conventional payload capabilities. Unlike Soviet bombers, U.S. heavy bombers can be committed to tactical missions in many theaters in the world with reasonable survivability or exchange ratios, and without critically degrading time-sensitive strategic bomber sortie rates.

CHART SIX

US AND USSR
INTERCONTINENTAL BOMBERS
(INVENTORY)

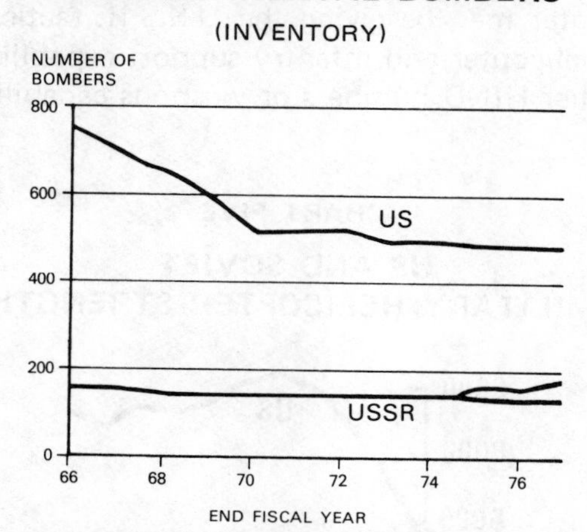

Source: DOD, OJCS, Military Posture, FY 78

OVERALL ASYMMETRIES BETWEEN U.S. AND SOVIET FORCES

There are, therefore, broad asymmetries in U.S. and Soviet tactical aviation capabilities. Fighter counts are of little meaning except when translated into specific theater or contingency capabilities. Even then, a much broader definition of tactical aircraft must be used to measure actual military strength.

MODERNIZATION AND COMBAT EFFECTIVENESS

Various experts have suggested the following differences may exist between U.S. and Soviet tactical air forces:

- U.S. air forces train realistically for combat. They train virtually all combat aircrews well using realistic mission profiles. The Soviets train only a few senior pilots well, and most Soviet pilots train rigidly to perform a few mission profiles.
- U.S. aircraft have superior human engineering. The Soviets as yet seem to have nothing approaching U.S. "heads up" displays, weapons delivery, and advanced navigation aids. Many Soviet aircraft have poor pilot visibility,

and are very difficult to fly at low altitudes. Soviet aircraft tend to burn easily and Soviet ejection seat design is poor.

- Soviet fighters may be more rugged and have longer operational availability between major service intervals, but many types cannot then be serviced in the field. Experts differ sharply over relative operational availability, and sortie rate or sortie generation capability.
- Most Soviet aircraft guns are primitive by U.S. standards. Soviet AAMs are poor, if improving, and Soviet PGMs are unproven and deployed on a comparatively limited number of aircraft types.
- The Soviets retain obsolete or obsolescent aircraft in their forces long after U.S. forces would withdraw them from U.S. force structures. Collins shows that 1,000 Mig 17 and SU-7 are still deployed in Soviet tactical aviation. Similar low performance aircraft are not counted in U.S. figures.
- Many types of Soviet aircraft are superior to older U.S. fighter "turn rates" and energy of maneuver at medium to high altitudes, but not if air combat forces the aircraft down to low altitudes. New U.S. fighters like the F-14, F-15, and F-16 will have effective superiority at all altitudes to all currently deployed Soviet types.
- The U.S. has vast operational and tactical experience. The U.S.S.R. has virtually none.
- Soviet fighter intercepts and attack missions are still dependent upon rigid planning and ground control. The range-payload and weaponry/avionics of most Soviet fighters gives the U.S.S.R. no other alternative.

At the same time, there is no debate over the fact that Soviet tactical aviation training and exercises have grown more realistic in recent years, or that new Soviet attack-fighters have range-payloads equivalent to that of contemporary U.S. types. There is continuing debate over whether the U.S.S.R. can, in fact, upgrade its avionics, weaponry, and overall force structure to U.S. standards.

GENERAL PURPOSE NAVIES

Sea power is a necessity for the United States, but not for the Soviet Union.[42] Commerce, always a U.S. tradition, assumes a

[42] Alva M. Bowen, Jr., specialist in naval affairs for the Congressional Research Service, acted as technical advisor and data source during the preparation of this section.

For general coverage, see Norman Polmar, *Soviet Naval Power: Challenge for the 1970's,* rev. ed. (New York: Crane, Russak & Co., 1974), 129 pp.; and *Understanding Soviet Naval Developments* (Washington, D.C.: Office of the Chief of Naval Operations, April, 1975), 79 pp.

salient role as dwindling natural resources increase our dependence on other countries for critical supplies. Petroleum products are most publicized, accounting for about 45 percent of domestic oil consumption, but mineral shortages are also important.[43] Routes must therefore be secured for friendly merchant ships under adverse conditions.[44] Essential sea lines of communication (LOCs) must also be kept open in wartime to ensure the free flow of military forces and logistic support between America, its allies[45] and/or contested areas.

The Soviet Union, with far fewer requirements for foreign raw materials and intrinsic interests that center on the Eurasian land mass, has only recently begun to break out of its continental cocoon. Its Navy is still cast as a spoiler that emphasizes negative sea denial (rather than positive sea assertion) capabilities, despite Admiral Gorshkov's grand design.[46]

This section shows how differences in size, composition, concepts, and accoutrements influence abilities of U.S. and Soviet general purpose navies to accomplish assigned tasks.[47]

Comparative Missions

U.S. deterrent/defense capabilities must be sufficient to satisfy several interlocking, overlapping naval missions in peacetime and war.[48] Some Soviet aims are the same. Others are quite different.

[43] Donald H. Rumsfeld, *Annual Defense Department Report for FY 1978* (Washington, D.C.: Department of Defense, 1977), p. 17.

[44] For a summary of commodities imported and exported by the United States over thirty-one sea routes to foreign countries, see *Essential United States Foreign Trade Routes* (Washington, D.C.: U.S. Government Printing Office, June 1975), 79 pp.

[45] Current U.S. defense treaties, congressional resolutions, executive agreements, policy declarations, and communiques are summarized in Annex C to *United States/Soviet Military Balance,* pp. 55-56. Two changes should be noted. The Tonkin Gulf Resolution terminated in January, 1971. The Southeast Asia Treaty Organization (SEATO) disbanded in September, 1975.

[46] Fleet Admiral S. G. Gorshkov, principal molder of the modern Soviet Navy, spells out his concepts for naval superiority in *Sea Power of the State* (Moscow: Military Publishing House, 1976), translated by U.S. Naval Intelligence Support Center (July 27, 1976), 363 pp. As Deputy Minister of Defense and Commander-in-Chief of the Soviet Navy since 1956, he has converted principle into practice with unprecedented continuity of purpose. The U.S. Navy saw nine Service Secretaries and seven Chiefs of Naval Operations during that same twenty-year period.

[47] Naval forces for strategic nuclear purposes were surveyed in the section on ballistic missile submarines. For sealift, including amphibious assault forces, see Chapter 5, which concerns mobility.

[48] Naval missions are addressed by Admiral J. L. Holloway, III, Chief of Naval Operations, in the enclosure to a letter to the Chairman of the House Armed Services Committee on January 12, 1977, pp. 3-6. See also Admiral Stansfield Turner, Command-

PEACETIME NAVAL PRESENCE

Peacetime presence to influence perceptions serves political and military purposes. This country consistently stresses one purpose. The Soviets strongly stress both.[49] Men who fashion America's foreign policy perceive the U.S. Navy as a peripheral and part-time instrument to exploit political, economic, and social opportunities. Its rival is routinely used (with sporadic success) to reap or retain an international reservoir of good will, which Soviet leaders try to translate into political persuasion, basing privileges, and other practical products that can have a significant bearing on the U.S./Soviet balance.[50] Their outposts along the African littoral, for example, overlook our oil LOCs to ports in the Persian Gulf.[51]

Both sides periodically parade flotillas or fleets as deterrent threats in times of tension or crisis to impress each other with intent or resolve. Four such confrontations have taken place in this decade, twice in the Mediterranean during Arab-Israeli disputes, twice in the Indian Ocean.[52]

SEA CONTROL

Freedom of the seas was a self-satisfying U.S. interest from late 1944, after the battle for Leyte, until increased Soviet capabilities started causing serious concerns in the mid-1960s. Since then, sea control, the prime prerequisite for all positive naval operations, has been an imperative U.S. mission.[53]

Positive U.S. missions demand abilities to deter or defeat enemy aircraft, submarines (including those bearing ballistic missiles), and surface combatants that try to interfere with friendly activi-

er-in-Chief Allied Forces Southern Europe, "Designing a Modern Navy: A Workshop Discussion," in *Power at Sea, II, Super-powers and Navies,* Adelphi Papers 123 (London: International Institute for Strategic Studies, 1976), pp. 25-27; and "The Naval Balance: Not Just a Numbers Game," *Foreign Affairs* (January 1977), pp. 342-47.

[49] For basic considerations, see U.S. Congress, House, *Means of Measuring Naval Power: With Special Reference to U.S. and Soviet Activities in the Indian Ocean,* prepared for the Subcommittee on the Near East and South Asia of the Committee on Foreign Affairs by the Congressional Research Service (Washington, D.C.: U.S. Government Printing Office, 1974), pp. 1-5.

[50] Gorshkov, *Sea Power of the State,* pp. 1-7, 312-22.

[51] "Sophisticated Soviet Base Nearly Completed in Somalia," *Baltimore News American,* December 16, 1976, p. 12. The foreword to *Jane's Fighting Ships, 1976-77* cites Soviet ports/anchorages in Aden, Somalia, The Seychelles, and Guinea, p. 121.

[52] *Understanding Soviet Naval Developments,* pp. 6-7; and U.S. Congress, House, *Means of Measuring Naval Power,* pp. 8-9.

[53] Title 10, United States Code, Chapter 503, Section 5012 and Department of Defense Directive 5100.1.

ties along selected ocean avenues or in associated areas. Local superiority (not necessarily numerical) is essential at specified times and places.

Sea denial, the Soviet speciality, is simpler to satisfy, since the Kremlin can apply power at times, places, and under circumstances of its choosing to prevent the accomplishment of U.S. tasks.

POWER PROJECTION

Naval air and/or amphibious forces can assist in controlling seas by projecting power ashore to seize and secure, damage, destroy, or otherwise exert control over critical terrain features (such as Norway's North Cape or the Dardenelles), enemy installations, and ships in port. Power projection capabilities can also support national purposes not associated with sea control, as occurred during U.S. campaigns in Korea and Vietnam.[54] American stress on such missions has been evident for many years. Soviet interest cropped up in a rather small way only recently.[55]

Comparative Force Structures

Naval strategists on both sides recognize that diversified forces are required to fulfill the foregoing missions, as shown on Figure 16, but U.S. and Soviet navies are nonetheless structured asymmetrically in almost every respect (see Graph 8 and Figures 17 through 21).[56]

AIRCRAFT CARRIERS

America's air power afloat has been cut in half since 1965, when 25 carriers (not counting helo platforms for amphibious assault) were still in active service.[57] Flexibility was first-rate. This country truly had two-ocean offensive capabilities at that time.

Reductions to the current complement of 13 carriers (Figure 17) have caused drastic revisions in forward deployment patterns since the start of this decade. Just four are positioned permanently

[54] *Ibid.*

[55] Turner, "The Naval Balance: Not Just a Numbers Game," pp. 342-43.

[56] For a summary of ship characteristics, see *Understanding Soviet Naval Developments,* pp. 11-29, 53-71. Aircraft are discussed in Norman Polmar, "Soviet Naval Aviation," *Air Force Magazine* (March 1976), pp. 69-75. Greater detail is contained in *Jane's Fighting Ships* and *All the World's Aircraft, 1976-77.*

[57] The complement in 1965 included 15 attack carriers (CVA) and 9 ASW carriers (CVS). Roughly one-third were commonly forward-deployed, three with Sixth Fleet in the Mediterranean and five with Seventh Fleet in the Western Pacific.

Figure 16

Naval Missions Related to Force Requirements

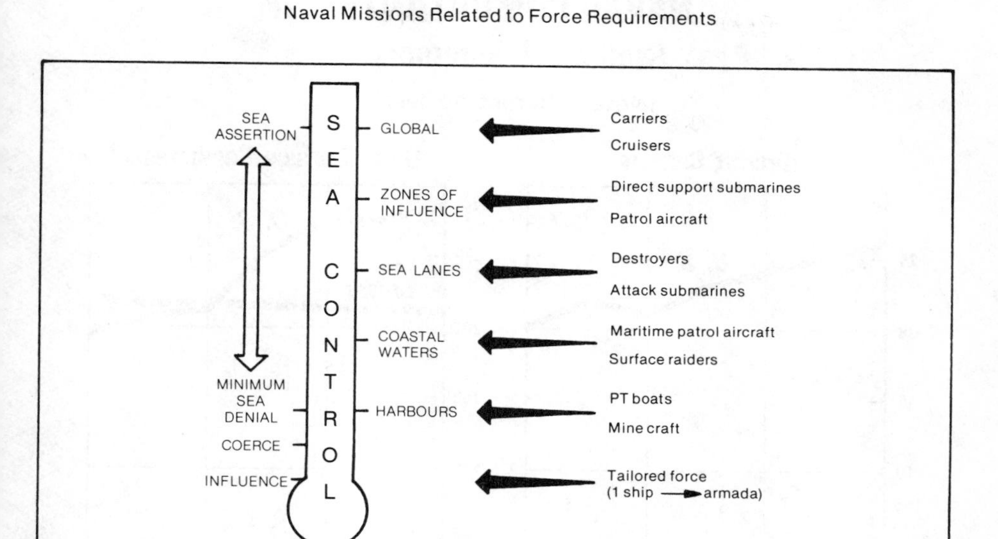

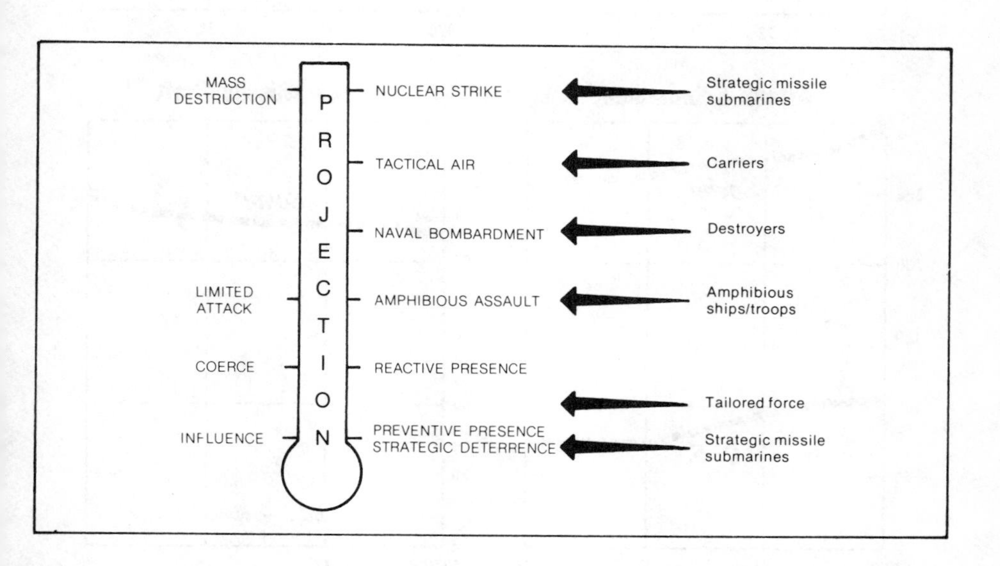

Source: Stansfield Turner, *Designing a Modern Navy: A Workshop Discussion,* Adelphi Papers 123 (London: International Institute for Strategic Studies 1976), p. 26.

Graph 8

NAVAL COMBATANTS
Statistical Summary

(Note Different Scales)

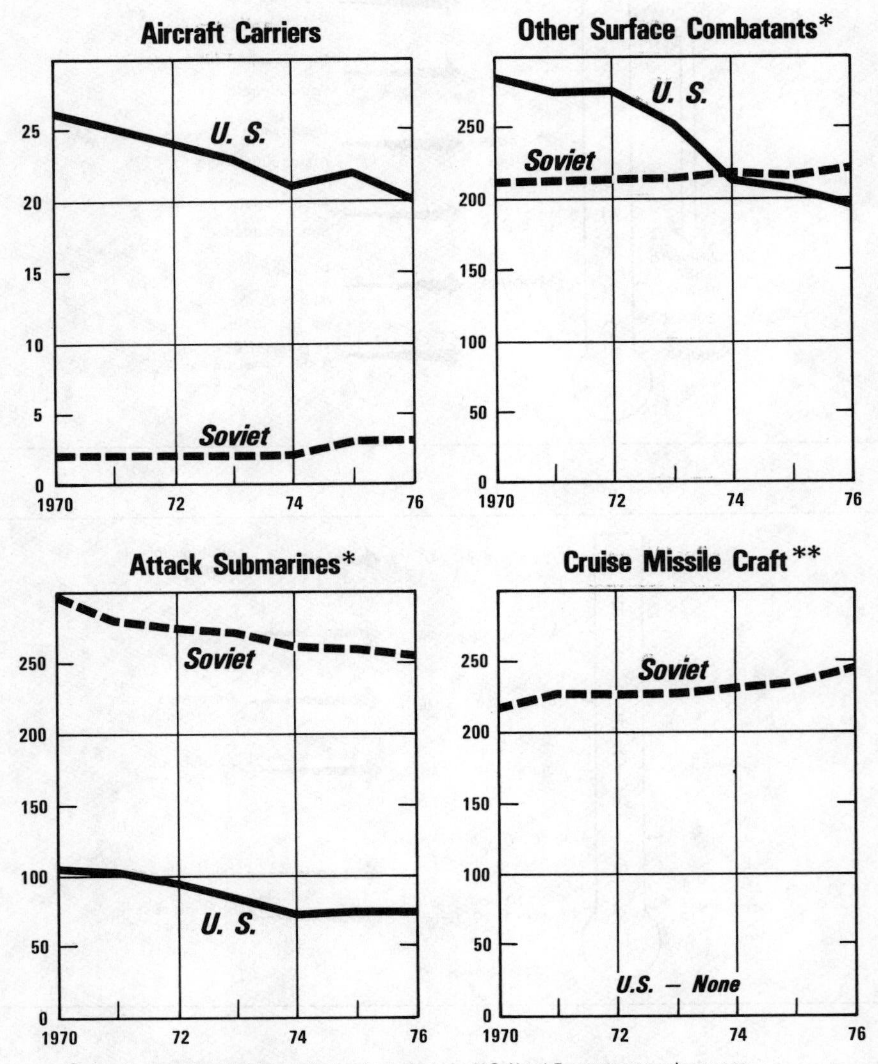

*Excludes coastal combatants and patrol boats; includes U.S. Naval Reserve ocean-going escorts.
**Includes cruise missile patrol boats.

FIGURE 17 CARRIER-BASED AIR POWER
Statistical Trends and System Characteristics

	1970	*1971*	*1972*	*1973*	*1974*	*1975*	*1976*
ATTACK CARRIERS							
United States							
Nuclear	1	1	1	1	1	2	2
Oil	14	13	13	13	13	13	11
Total	15	14	14	14	14	15	13
Soviet Union	0	0	0	0	0	1	1
U.S. Standing	+15	+14	+14	+14	+14	+14	+12
ASW CARRIERS							
United States	4	4	3	2	0	0	0
Soviet Union	2	2	2	2	2	2	2
U.S. Standing	+2	+2	+1	Par	–2	–2	–2
OTHER CARRIERS							
United States	7	7	7	7	7	7	8
Soviet Union	0	0	0	0	0	0	0
U.S. Standing	+7	+7	+7	+7	+7	+7	+8
GRAND TOTAL							
United States	26	25	24	23	21	22	20
Soviet Union	2	2	2	2	2	3	3
U.S. Standing	+24	+23	+22	+21	+19	+19	+17

NOTE: The Kiev is not strictly an attack carrier. The Moskva and Leningrad are not strictly ASW carriers. However, they correspond most closely with those categories.

FIGURE 17 (continued)

FIGHTER/ATTACK AIRCRAFT	*1970*	*1971*	*1972*	*1973*	*1974*	*1975*	*1976*
United States							
A-4	147	38	41	42	30	40	0
A-6	119	118	105	113	115	97	104
A-7	254	308	312	312	283	300	274
F-4	229	246	253	238	209	95	139
F-8	78	41	45	45	43	38	0
F-14	0	0	0	0	23	50	93
Total	827	751	756	750	703	620	610
Soviet Union							
Yak-36	0	0	0	0	0	0	15
U.S. Standing	+827	+751	+756	+750	+703	+620	+595
ASW AIRCRAFT							
United States							
S-2	91	83	73	68	52	31	5
S-3	0	0	0	0	0	28	73
SH-3 (Helo)	54	62	60	56	56	56	56
LAMPS (Helo)	0	0	11	17	37	43	56
Total	145	145	144	141	145	158	190
Soviet Union							
KA-25 (Helo)	80	95	95	145	160	170	180
U.S. Standing	+65	+50	+49	-4	-15	-12	+10

NOTE: UE aircraft only, excluding reconnaissance, training, and special-purpose versions.
Kiev class carriers can carry a mix of up to 30 fixed- and rotary-wing aircraft. All 30 could be YAK-36, but 10-15 have been counted thus far.

FIGURE 17 (continued)

AIRCRAFT CARRIERS	*OFFENSIVE WEAPONS SYSTEMS*							
	Fighter Aircraft	*Attack Aircraft*	*Recon Aircraft*	*ASW Aircraft*	*Average Age (Years)*	*Cruise Missiles*	*Other Weapons*	*Speed (Knots)*
United States								
Attack (CVA)	24 F-4/F-14	24-36 A-7, A-6	10 RA-SC	10 S-2/S-3	23	None	Sea Sparrow/ Terrier	35
Helicopter (LPH)	Carries a mix of about 20 CH-46, CH-53, UH-1, AH-1 Helos.				12	None	Sea Sparrow/ 4 3-in .50 cal	20
Soviet Union								
ASW (CVS)	Carries up to 30 V/TOL and/or ASW Helicopters				1	8 SS-N-3/12	Torpedoes, ASW Rockets, SS-N-1	30+
Helicopter (CHG)	0	0	0	20 Helo	10	None	Torpedoes, ASW Missiles	30

FIGURE 17 (continued)

AIRCRAFT CHARACTERISTICS

FIGHTER/ATTACK AIRCRAFT	*First Deployed*	*Combat Radius (Miles)*	*Max Speed (Mach)*	*Payload (lbs)*	*Typical Weapons: Guns*	*Typical Weapons: Missiles*	*Nuclear Capable*	*All Weather*
United States								
A-6	1963	750	0.9	10,000	None	18 Mk-82	Yes	Yes
A-7D	1966	550	0.9	7,200	1 20mm	12 Mk-82	Yes	Yes
F-4J	1961	475	2.2	15,500	None	8 AIM-7/9	Yes	Yes
F-14	1973	580	2.3	17,600	1 20mm	8 AIM-7/9 6 AIM-54	No	Yes
Soviet Union								
YAK-36	1976	300	Subsonic					

ASW AIRCRAFT	*First Deployed*	*Combat Radius (Miles)*	*Patrol Speed (mph)*	*ASW Weapons*	*Detection Devices*
United States					
S-2	1954	675	150	2 torpedoes or depth charges	MAD, radar, sonobuoys
S-3	1974	1,200+	200	4 torpedoes or other mix	MAD, radar, sonobuoys
SH-3 (Helo)	1961	100	135	2 torpedoes or other mix	MAD, radar, dipping sonar
LAMPS (Helo)	1973	60	150	1 torpedo	MAD, radar, sonobuoys
Soviet Union					
KA-25 (Helo)	1962	200	120	2 torpedoes or depth charges	

NOTE: Combat radii correspond with payloads shown under average conditions.
Payloads are merely representative. External fuel tanks are included wherever applicable.

overseas—two in the Western Pacific, two in the Mediterranean. One of the latter is on call for excursions into the Atlantic. Surge capabilities are slight.[58]

Still, 200 U.S. fighter/attack aircraft are available in key areas at all times. Soviet carrier air power is scant in contrast. The Kiev, with its few fighter aircraft and formidable missile armament, compares more closely with a strike cruiser than, say, with our Midway class carriers (in service since 1945), much less with the nuclear-powered Enterprise or Nimitz.[59] Moskva and Leningrad, called ASW cruisers, apparently are the first and last in their class.[60] The Soviets have no counterparts for seven U.S. helicopter carriers, which support amphibious assault forces and perform other useful functions.

OTHER MAJOR SURFACE COMBATANTS

Numbers of U.S. major surface combatants assigned to the Regular Navy have declined dramatically during this decade, while Soviet strength stayed steady (Figure 18). America's cruiser quantities are still roughly the same, but 90 destroyers were decommissioned, while only four were delivered.[61] (Statistical trends are summarized on Figure 18.)

Quality is even more telling. Soviet ships are somewhat smaller than U.S. counterparts,[62] and none are nuclear-powered (the United States now has seven, including aircraft carriers). They are generally faster,[63] however, and several major combatants mount

[58] Rumsfeld, *Annual Defense Department Report for FY 1977,* p. 160.

[59] For Kiev characteristics, including razor-sharp close-up photos, see "Soviet Navy Deploys New V/STOL," *Aviation Week and Space Technology* (August 2, 1976), pp. 14-17.

[60] Moskva and Leningrad, which lack fighter aircraft, apparently were designed to destroy ballistic missile submarines when short-range U.S. SLBMs required launch positions near the continental shelf. Land-based aircraft could cover the Soviet search. The situation changed when longer-range Polaris A-3 and Poseidon missiles replaced Polaris A-2s.

[61] U.S. and Soviet ship deliveries from 1965 through 1976 are contained in Michael McGwire, "Western and Soviet Naval Building Programmes 1965-1976," *Survival* (September/October 1976), pp. 204-09.

[62] *Kresta* cruisers displace 7,500 tons, slightly less than a U.S. Spruance class destroyer. Krivak destroyers, at 4,000 tons, are a little larger than our latest frigates.

[63] A speed advantage of one or two knots can be important in evading, engaging, or simply keeping station with enemy ships.

Developments, p. 23. For comparisons between Krivak and U.S. Perry class frigates (due to begin deployment in mid-1977), see Robert D. Heinl, Jr., "Navy Lagging in Surface Sea Power," *Detroit News,* August 22, 1976, p. 1; and a rebuttal by John B. Shewmaker, "U.S. Destroyer Seen Superior: Col. Heinl Overrated Soviet Warship?" *Detroit News,* September 18, 1976, Letter to the Editor.

FIGURE 18 NAVAL SURFACE COMBATANTS
(Less Aircraft Carriers)
Statistical Trends and System Characteristics

CRUISERS	*1970*	*1971*	*1972*	*1973*	*1974*	*1975*	*1976*
United States							
SSM	0	0	0	0	0	0	0
Other							
Nuclear	3	3	3	3	4	5	5
Oil	24	24	25	26	24	22	21
Total	27	27	28	29	28	27	26
Soviet Union							
SSM	9	10	12	15	17	19	21
Other	15	15	15	13	12	11	10
Total	24	25	27	28	29	30	31
U.S. Standing	+3	+2	+1	+1	−1	−3	−5
DESTROYERS							
United States							
Active							
SSM	0	0	0	0	0	0	0
Other	159	141	131	108	69	70	69
Total	159	141	131	108	69	70	69
Reserve	28	28	31	31	37	32	30
Grand Total	187	169	162	139	106	102	99
Soviet Union							
SSM	6	7	7	10	13	16	20
Other	71	70	70	67	67	65	64
Total	77	77	77	77	80	81	84
U.S. Standing	+110	+92	+85	+62	+26	+21	+15

NOTE: U.S. Naval Reserve ships shown are immediately available to augment active forces in emergency. Soviet SSM figures include surface-to-underwater missiles that probably have SSM capability.

FIGURE 18 (continued)

	1970	1971	1972	1973	1974	1975	1976
FRIGATES	Includes ships formerly called destroyer escorts and comparable craft over 1,200 tons.						
United States							
Active	47	57	66	67	64	64	64
Reserve	6	4	4	4	0	0	0
Total	53	61	70	71	64	64	64
Soviet Union	111	111	110	109	108	105	106
U.S. Standing	-58	-50	-40	-38	-44	-41	-42
SMALL COMBATANTS	SSMs are NANUCHKA. Others are GRISHA and POTI.						
United States	0	0	0	0	0	0	0
Soviet Union							
SSM	0	0	6	6	8	10	14
Other	70	75	77	80	85	88	88
Total	70	75	83	86	93	98	102
U.S. Standing	-70	-75	-83	-86	-93	-98	-102
SHORE PATROL	SSMs are Komar and Osa; Others are Fast Torpedo Boats, Sub-Chasers, hydrofoils, and the like.						
United States							
SSM	0	0	0	0	0	0	0
Other	16	17	16	14	14	14	8
Total	16	17	16	14	14	14	8
Soviet Union							
SSM	140	145	136	129	127	123	121
Other	665	555	480	460	430	394	391
Total	805	700	616	589	557	517	512
U.S. Standing	-789	-683	-600	-575	-543	-503	-504

FIGURE 18 (continued)

	SHIP CHARACTERISTICS						
UNITED STATES	*Current Number*	*First Deployed*	*Average Displacement (Tons)*	*AAA SAMs*	*ASW Weapons*	*Guns*	*Power Plant*
Cruisers							
Bainbridge (CGN)	1	1962	8,580	2 Terrier (Twin) 4 3-in	ASROC 2 Torpedo (Triplet)	None	Nuclear
Belknap (CG)	9	1964	7,930	2 3-in	ASROC (Twin) 2 Torpedo (Triplet)	1 5-in	Steam
California (CGN)	2	1974	10,150	2 Tartar	ASROC (Twin) 4 Torpedo	2 5-in	Nuclear
Leahy (CG)	9	1962	7,800	2 Terrier (Twin) 4 3-in	ASROC 2 Torpedo (Triplet)	None	Steam
Truxton (CGN)	1	1967	9,200	2 3-in	ASROC (Twin) 4 Torpedo	1 5-in	Nuclear
Virginia (CGN)	3	1976	11,000	2 Tartar (Twin)	2 Torpedo (Triplet)	2 5-in	Nuclear

FIGURE 18 (continued)

Destroyers							
Charles F. Adams (DDG)	23	1960	4,500	1 Tartar (Single or Twin)	ASROC 2 Torpedo (Triplet)	2 5-in	Steam
Coontz (DDG)	10	1960	5,800	1 Terrier (Twin)	ASROC 2 Torpedo (Triplet)	1 5-in	Steam
Forrest Sherman							
DD	14	1955	4,050	None	ASROC	3 5-in	Steam
DD	4	1967	4,150	1 Tartar	2 Torpedo (Triplet)	1 5-in	Steam
Gearing (DD) (modernized)	42	1945	3,500	None	ASROC 2 Torpedo (Triplet)	4 5-in	Steam
Spruance (DD)	8	1975	7,800	None	ASROC 2 Torpedo (Triplet) 1 ASW Helo	2 5-in	Gas Turbine

NOTES: Speeds average slightly over 30 knots.
Torpedoes are launch tubes only, not numbers of weapons.
Column 1 totals do not equal entire inventory, because several classes are not shown.

FIGURE 18 (continued)

SOVIET UNION	SHIP CHARACTERISTICS							
	NR	*First Deployed*	*Speed (Knots)*	*Average Displacement (Tons)*	*AAA SAMs*	*ASW Weapons*	*Anti-Surface Ship Weapons*	*Power Plant*
Cruisers								
Kara (CLGM)	4	1973	32	10,000	2 SA-N-3 2 SA-N-4 4 76mm 4 Gatling	ASW Rockets Depth Charges	8 SS-N-14 10 Torpedo	Gas Turbine
Kresta I (CG)	4	1967	32	7,500	2 SA-N-1 4 57 mm	MBU-2500A Depth Charges	4 SS-N-3 10 Torpedo	Steam
Kresta II (CG)	9	1970	32	7,500	2 SA-N-3 4 57 mm	ASW Rockets	8 SS-N-14 10 Torpedo	Steam
Kynda (CG)	4	1962	34	5,500	1 SA-N-1 4 76mm	2 MBU-3500A 6 Torpedo	8 SS-N-3 (Reloads)	Steam
Sverdlov (CL)	10	1952	32	17,500	32 37mm Some SA-N-4 Some Gatling	10 Torpedo	12 6-in 12 3.9-in Mines	Steam
Destroyers								
Kanin (DDG)	7	1968	34	4,500	1 SA-N-1 8 57mm	3 MBU-2500A	10 Torpedo	Steam
Kashin (DDG) (Converted)	5	1963	36+	4,500	2 SA-N-1 4 76mm	2 MBU-2500A Depth Charges	4 SS-N-2/11 5 Torpedo Mines	Gas Turbine
Kotlin (DD)	18	1954	35	3,885	16 45mm	MBU-2500A or Depth Charges	4 5.1-in 10 Torpedo Mines	Steam
Kotlin (DDG)	8	1962	35	3,885	1 SA-N-1 4 57mm 8 30mm	2 MBU-2500A	2 5.1-in 5 Torpedo	Steam
Krivak (DD)	15	1971	32	4,000	2 SA-N-4 4 76mm	2 MBU-2500A	4 SS-N-14 8 Torpedo	Gas Turbine

NOTES: Fourteen additional Kashin class destroyers lack any anti-ship missiles.
All Soviet SAMs have twin launchers.
Torpedoes are launch tubes only, not numbers of weapons.

FIGURE 18 (continued)

SOVIET UNION	*PATROL CRAFT CHARACTERISTICS*							
	NR	*First Deployed*	*Speed (Knots)*	*Average Displacement (Tons)*	*SAMs, AAA*	*Cruise Missiles*	*Torpedo Tubes Guns Other Weapons*	*Power Plant*
Patrol Craft								
Komar (PTG)	Few	1960	40	80·	2 25mm	2 SS-N-2	None	Diesel
Osa (PTG)	120	1960	35	220	4 30mm	4 SS-N-2/11	None	Diesel
Nanuchka (PGG)	14	1969	30+	800	1 SA-N-4 2 57mm	6 SS-N-9	None	Diesel

NOTE: CL = Light Cruiser; CG = Guided-Missile Cruiser; DD = Destroyer; DDG = Guided-Missile Destroyer; PTG = Guided Missile Patrol Boat; PPG = Patrol Combatant.

FIGURE 18 (continued)

	ANTI-SHIP MISSILE/ROCKET CHARACTERISTICS			
	First Deployed	*Range (Miles)*	*Warhead*	*Yield*
UNITED STATES				
ASROC (ASW)	1961	6	HE, Nuke	All
SOVIET UNION				
SS-N-14	1974	30	HE, Nuke	in
SS-N-12		Characteristics Classified		
SS-N-11		Characteristics Classified		Kiloton
SS-N-9		150	HE, Nuke	
SS-N-3	1960	150-250	HE, Nuke	Range
SS-N-2	1960	23	HE	
SS-N-1	1958	130	HE	

NOTES: ASROC carries a small torpedo or depth charge for use against submarines.
SS-N-14, a missile-borne ASW torpedo, probably has anti-surface ship capabilities.
SS-N-12 is replacing SS-N-3; SS-N-11 is replacing SS-N-2.
All Soviet SAMs probably have significant SSM capabilities.

a total of 85 cruise missiles created expressly to kill surface ships.[64] Perhaps more importantly, something like 164 SS-N-14 ASW missiles may also have anti-surface ship missions. Endurance and seaworthiness have increased, along with offensive combat capabilities. Several classes, such as Sverdlov, Kilden, and Kotlin, are reaching the end of their theoretical hull life, but many have been remodelled or reconstructed.[65]

COASTAL COMBATANTS

The Soviet Union has more coastal combat craft, including minesweepers, than the rest of the world combined.[66] Neither the U.S. Navy nor Coast Guard has anything to equal 14 Nanuchka class coastal combatants or 120-odd Osa patrol boats with cruise missiles for shoreline defense (Figure 18).

ATTACK SUBMARINES

The excellence of U.S. nuclear-powered attack submarines is widely acknowledged. They are not as fast as some Soviet boats (their Soviet Victors are the world's speediest), but are quieter and better equipped. The new Los Angeles class (SSN-688), now entering our inventory with wire-guided acoustic homing torpedoes and improved sonar systems, should strengthen the U.S. position.[67] Harpoon missiles will follow. Qualitative superiority, however, is insufficient when quantity is also essential.

ASW is the primary mission of U.S. attack submarines.[68] The 74 that remain after recent reductions therefore face distinct disadvantages trying to check three times their number in open oceans, even though most of their prey are diesel-powered and many are well past their prime (Figure 20). The balance would be better with the order of battle reversed.

Sixty-six Soviet submarines, which account for a third of the force, are fitted to fire anti-ship cruise missiles. Total tubes exceed 400. Papa and Charlie classes can fire from submerged positions.

[64] The U.S. Chief of Naval Operations considers Krivak class destroyers as "ton for ton the heaviest armed and most effective destroyer afloat." *Understanding Soviet Naval*

[65] *Jane's Fighting Ships, 1976-77,* foreword. Theoretical hull life is 20 to 35 years, depending on ship type and circumstances.

[66] *Understanding Soviet Naval Developments,* p. 25.

[67] Rumsfeld, *Annual Defense Department Report for FY 1977,* p. 174.

[68] *Ibid.* Submarines share ASW missions with several other fixed and mobile systems, but they play the paramount role.

Figure 19
Naval Choke Points Near the North Atlantic

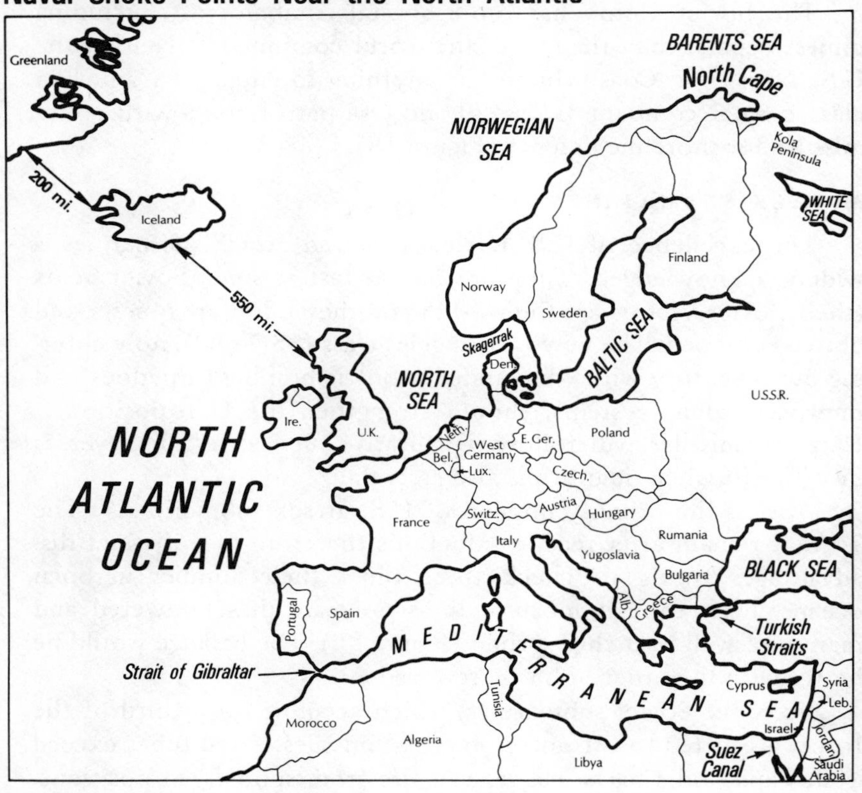

FIGURE 20 ATTACK SUBMARINES
Statistical Trends and System Characteristics

	1970	*1971*	*1972*	*1973*	*1974*	*1975*	*1976*
United States							
SSM	0	0	0	0	0	0	0
Other							
Nuclear	46	51	56	60	61	64	64
Diesel	59	50	38	24	12	11	10
Total	105	101	94	84	73	75	74
Soviet Union							
SSM							
Nuclear	35	38	40	41	42	42	43
Diesel	28	27	26	25	24	24	23
Total	63	65	66	66	66	66	66
Other							
Nuclear	24	26	28	31	34	37	38
Diesel	210	190	180	175	160	156	150
Total	234	216	208	206	194	193	188
Grand Total	297	281	274	272	260	259	254
U.S. Standing	-192	-180	-180	-188	-187	-184	-180

FIGURE 20 (continued)

ATTACK SUBMARINE CHARACTERISTICS

	Current Number	*First Deployed*	*SUBROC, Cruise Missiles*	*Sub-Surface Launch*	*Torpedo Tubes*	*Power Plant*
UNITED STATES						
688 Class (SSN)	1	1975	SUBROC	Yes	4	Nuclear
637 Class (SSN)	37	1966	SUBROC	Yes	4	Nuclear
594 Class (SSN)	13	1962	SUBROC	Yes	4	Nuclear
Skate Class (SSN)	5	1957	None		8	Nuclear
Skipjack Class (SSN)	5	1959	None		6	Nuclear
SOVIET UNION						
Cruise Missile						
Charlie (SSGN)	14	1968	8 SS-N-7	Yes	8	Nuclear
Echo II (SSGN)	29	1963	8 SS-N-3/12	No	10	Nuclear
Juliett (SSG)	16	1962	4 SS-N-3/12	No	6	Diesel
Papa (SSGN)	1	1973	8 SS-N-7	Yes	8	Nuclear
Attack						
Echo I (SSN)	5	1960	None		10	Nuclear
Foxtrot (SS)	60	1958	None		10	Diesel
November (SSN)	13	1958	None		8	Nuclear
Romeo (SS)	10	1961	None		8	Diesel
Tango	3	1973	None		6	Diesel
Victor (SSN)	19	1968	None		8	Nuclear
Whiskey (SS)	40	1951	None		6	Diesel
Zulu (SS)	15	1952	None		10	Diesel

NOTES: The few U.S. diesel-powered submarines, all commissioned in the 1940s and 1950s, are omitted. So are "one-of-a-kind" classes, like nuclear Lipscomb and Narwhal.

SS = Diesel Submarine; SSN = Nuclear Submarine; SSG = Diesel Cruise Missile Submarine; SSGN = Nuclear Cruise Missile Submarine.

Column 1 numbers do not equal entire inventory, because some classes are not shown.

FIGURE 20 (continued)

	ANTI-SHIP MISSILE/ROCKET CHARACTERISTICS			
	First Deployed	*Range (Miles)*	*Warhead*	*Yield*
UNITED STATES				
SUBROC (ASW)	1965	45	Nuke	All in Kiloton Range
SOVIET UNION				
SS-N-12	Characteristics Classified			
SS-N-7	1968	30	HE, Nuke	
SS-N-3	1962	150-250	HE, Nuke	

NOTE: SS-N-12 is replacing SS-N-3.

All classes carry torpedoes for close combat. The United States will have no cruise missile counterparts until Harpoon enters service.

LAND-BASED NAVAL AIR POWER

U.S. naval air power is mainly afloat. Soviet strength is almost all ashore (Figure 21). Land-based aircraft on both sides engage in ASW activities, active as well as passive, but only the Soviet Navy specializes in anti-surface ship strike forces. More than 300 aging Badger bombers, the basic component, can reach about 1,600 miles from home stations without refueling.[69] Their ability to pierce U.S. protective fighter shields would be poor if they carried gravity bombs, but cruise missiles can be launched at least 100 nautical miles from targets. Supersonic Backfires, whose naval numbers are increasing, open up new options.

B-52s armed with precision-guided munitions recently joined the competition, in accord with collateral functions of long standing,[70] which heretofore were finessed. Air Force tactical aircraft could augment SAC's shore-based strike capabilities, especially if forces receive Harpoon.[71]

Soviet Shortcomings

The "new" Soviet Navy suffers from several chronic shortcomings that it shares with the "old." Its chief handicap is the geographic strait jacket that makes timely mutual support almost impossible for four widely separated fleets based in the Baltic, Barents, and Black Seas, plus the Sea of Japan. All four risk being bottled up by bad weather and/or barriers at critical choke points. (Several show on Figure 19: The Greenland-Iceland-Faeroes-U.K. Gap, including the English Channel; the Skagerrak; Suez Canal; Turkish Straits; and Gibraltar.) The scarcity of all-weather ports is only partially overcome by extensive use of ice breakers and covered repair facilities.

[69] About 100 TU-16s are configured as tankers for aerial refueling. Soviet Long-Range Aviation units could augment them under some circumstances. (*Understanding Soviet Naval Developments,* pp. 27, 73. In-flight refueling for aircraft committed to surprise first strikes could take place with impunity, but slow-flying Badgers would be tempting targets during that process thereafter.

[70] Department of Defense Directive 5100.1 assigns the Air Force collateral functions in fields of sea interdiction, ASW, and aerial mine-laying.

[71] Robert N. Ginsburgh, "A New Look at Control of the Seas," *Strategic Review* (Winter 1976), pp. 86-89. See also Howard Silber, "B-52 Testing Its Ship-Sinking Ability," *Omaha World-Herald,* December 26, 1976, p. 1.

FIGURE 21 SHORE-BASED NAVAL AIRCRAFT
Statistical Trends and System Characteristics

ANTI-SURFACE SHIP BOMBERS	*1970*	*1971*	*1972*	*1973*	*1974*	*1975*	*1976*
United States	0	0	0	0	0	0	0
Soviet Union							
Backfire	0	0	0	0	0	0	30
IL-28	60	50	50	30	30	30	20
SU-17	0	0	0	0	0	0	Few
TU-16	320	320	320	320	320	320	320
TU-22	60	60	60	60	60	60	60
Total	440	430	430	410	410	410	430+
U.S. Standing	-440	-430	-430	-410	-410	-410	-430+
ASW AIRCRAFT							
United States							
P-3	210	210	213	214	202	199	203
Soviet Union							
Bear-F	0	0	0	0	0	15	20
BE-6	30	10	0	0	0	0	0
BE-12	60	75	75	100	100	100	100
IL-38	20	20	40	40	55	55	55
MI-4 (Helo)	130	130	130	130	115	105	70
Total	240	235	245	270	270	275	245
U.S. Standing	-30	-25	-32	-56	-68	-76	-42
GRAND TOTAL							
United States	210	210	213	214	202	199	203
Soviet Union	680	665	675	680	680	685	675
U.S. Standing	-470	-455	-462	-466	-478	-486	-472

NOTE: UE aircraft only, excluding reconnaissance, training, and special-purpose versions.

FIGURE 21 (continued)

ANTI-SURFACE SHIP BOMBERS	*AIRCRAFT CHARACTERISTICS*					
	First Deployed	*Crew*	*Nr Engines*	*Patrol Radius (Miles)*	*Detection Devices*	*Anti-Ship Weapons*
Soviet Union						
Bear-F	1975	5	4	3900		
Backfire	1975	2-4	2	2500		ASMs
IL-28	1950	3	2	685		Bombs or torpedoes
SU-17	1972	1	1	375		Bombs; rockets; ASM
TU-16	1955	7	2	1600		Bombs or ASM
TU-22	1962	3-4	2	700		Bombs
ASW AIRCRAFT						
United States						
P-3	1962	10-12	4	1200	MAD, radar, sonobuoy	Torpedoes or mines
Soviet Union						
BE-12	1961	4	2	1250	MAD, radar, sonobuoys	Bombs, mines, depth charges, torpedoes (various mixes)
IL-38	1968	12	4	2250	MAD, radar, sonobuoys	Bombs, mines, depth charges, torpedoes (various mixes)
MI-4 (Helo)	1953		1	75+	MAD, radar, sonar	None

NOTES: MAD stands for Magnetic Anomaly Detector.
BE-12 (and its predecessor BE-6) is a flying boat.

Soviet surface craft all must contend with lack of air cover when they sweep far from friendly shores. Land-based bombers for area defense and on-board SAM clusters for short-range and point defense are poor substitutes for defenses-in-depth that include carrier-based fighters.

Soviet naval forces are also short on stamina, except for late-model ships, such as Kiev, Kara, and Krivak. Small surface combatants, lacking large fuel capacities or nuclear power, have limited ranges. Restricted space for rations, ammunition, and other stores prohibit prolonged operations without resupply. Merchant tankers routinely refuel Soviet ships at sea, and trawlers serve some logistic purposes, but underway replenishment procedures, although improving, are still substandard compared with U.S. skills. Lengthy, large-scale operations would be next to impossible in sea areas remote from friendly port facilities.[72]

Finally, most conscripts quit after three years' service. Problems attendant to training 100,000 recruits every year (a fifth of the total force) almost beggar imagination in this age of technical specialization,[73] despite pre-induction preparations.

Soviet Strengths

Abilities of the Soviet Union to satisfy positive sea control and naval power projection missions against U.S. opposition will generally be restricted to regions along its periphery until limitations just outlined are alleviated. The capacity for coordinated attacks on U.S. men-of-war and merchant shipping, however, menaces American missions.[74]

THREATS TO THE U.S. SURFACE NAVY

A recent National Security Council study of U.S. strategy and

[72] U.S. Congress, House, *Means of Measuring Naval Power,* pp. 9-13; *Understanding Soviet Naval Developments,* pp. 17, 18.

[73] Robert Merry, "Jane's Editor Sings Russ Navy Blues," *Chicago Tribune,* March 8, 1976, p. 1.

[74] The Soviet Navy has conducted two world-wide exercises, one in 1970, the other in 1975. The most recent, code named "Okean-75," demonstrated dramatic progress during the intervening five years. Eight task forces, totalling more than 200 surface ships and submarines, plus land-based naval aviation, deployed in diverse ocean areas. Satellite surveillance, computerized data flow, and almost instantaneous communications coordinated functions that included convoy tracking and simultaneous cruise missile launches. See, for example, Bruce W. Watson and Margurite A. Walton, "Okean-75," *U.S. Naval Institute Proceedings* (July 1976), reprinted in *Congressional Record,* August 24, 1976, p. 514339-41; and "Soviets Seen Operating Two Types of Ocean Surveillance Satellite," *Aerospace Daily,* June 2, 1976, p. 169.

naval missions reportedly revealed three major Soviet threats to the U.S. surface navy. All three concern anti-ship missiles.[75]

Anti-Ship Missiles and Strategy The Soviets currently have a stable of assorted over-the-horizon anti-ship cruise missiles. Fifteen different sorts of surface warships, submarines, and aircraft can launch at least one kind.

Most of the long-range missiles (more than 100 miles) are jet propelled. Many short-range models are solid-fuel rocket powered. Speeds vary from 600 knots to several times faster than sound. Some missiles, with small visual and radar cross-sections, confound anti-aircraft gun crews by skimming across the sea's surface. Others, with steep trajectories such as Shaddock (SS-N-3), dive straight onto targets. In-flight corrections and terminal homing are the rule. The weight of some warheads exceeds a ton.[76]

Soviet strategy seems designed to seize and secure initiative with a single killing salvo. Missile-carrying surface ships, submarines, and aircraft, moving without any semblance of tactical formation, could trigger surprise, preemptive strikes on central signal from many directions, and perhaps from point-blank range.[77] U.S. carriers, cruisers, and support ships comprise high-contrast targets for Soviet missile seekers. To infrared sensors, they seem hot against cool sea backgrounds; to radars, they are large reflectors; to radiometric sensors, they are massive metal structures.[78]

Tactics close to shore tend to be somewhat different. Small, missile-bearing boats (nicknamed Wasp, Gnat, and Mosquito[79]) are difficult to distinguish in the coastal clutter of shallow-draft civilian craft and other reflectors. Short ranges and awesome weapons power could overwhelm warships caught unaware. Low-flying Soviet aircraft with cruise missiles complicate U.S. defensive problems. The impact on American power projection missions, particularly amphibious assaults, clearly could be profound.[80]

[75] "The Three Major Threats to U.S. Fleet," *Defense/Space Daily,* October 15, 1976, p. 246.

[76] William J. Ruhe, "Cruise Missile: The Ship Killer," *U.S. Naval Institute Proceedings* (June 1976), pp. 46, 47-48; and "Navy Faces Grave Cruise Missile Threat," *Aviation Week and Space Technology* (January 17, 1975), p. 101.

[77] Ruhe, "Cruise Missile: The Ship Killer," p. 47.

[78] "Navy Faces Grave Missile Threat," p. 101.

[79] Osa stands for Wasp; Nanuchka for Gnat; Komar for Mosquito.

[80] Ruhe, "Cruise Missile: The Ship Killer," pp. 48, 49, 52.

U.S. Countermeasures U.S. sea control tactics traditionally try to destroy enemy weapons before they endanger our ships, concentrating force on a few closely bunched combatant craft and other defendable targets. Surprise assaults by Soviet cruise missiles, launched at close range, could make that approach obsolete.

Active defenses alone appear inadequate. America's current ship-launched SAMs would be essentially ineffective against concerted attacks. "The time from detection to target engagement is [still] excessive and coordination among missile batteries on different ships . . . is poor. These difficulties are compounded by [SAM] system vulnerability to electronic countermeasures."[81] Even Phoenix-armed F-14s, which can engage six targets simultaneously, are subject to easy saturation if large-scale attacks box the compass.[82]

Diverting, rather than destroying, enemy missiles in flight therefore assumes increased importance. Authorities, however, generally agree that any navy which relies solely on decoys, jammers, chaff, and other electronic countermeasures for defense is doomed to take heavy losses when counter-countermeasures come into play.[83]

Successful defense likely will depend on SAMs and interceptor aircraft systems in combination with ECM, strategy, tactics, and doctrine. An appropriate package is not yet available.

THREATS TO MERCHANT SHIPPING

Soviet submarine threats to friendly merchant shipping are potent and pervasive.

[81] *Ibid.,* quoting the NSC study cited in Note 75. "There is indeed some doubt as to whether the Terrier system could effectively defend against cruise missiles fired from a single 'Charlie'-class submarine," according to a Congressional Budget Office study entitled *Planning U.S. General Purpose Forces: The Navy* (Washington, D.C.: U.S. Government Printing Office, December 1976), p. 41. See also Rumsfeld, *Annual Defense Department Report for FY 1978,* pp. 111, 112.

[82] Turner, "The Naval Balance: Not Just a Numbers Game," p. 350.

[83] Israeli patrol boats stymied Arabs armed with Styx missiles (SS-N-2s) during the Yom Kippur conflict in 1973, using chaff umbrellas and clever tactics. Roughly 50 missiles were fired without one hit, because the Arabs were still unequipped to wage electronic warfare. Surprises could be expected from sophisticated Soviets. *Aviation Week and Space Technology* (January 27, 1975), p. 121. See also Harry F. Eustace, "A U.S. View of Naval EW" and G. S. Sunderam, "Electronic Warfare at Sea," both in *International Defense Review* (April 1976), pp. 4f, 217 ff.

The Soviet Submarine Challenge Cruise missiles, with ranges from 25 to 250 miles, supplement new Soviet families of homing, acoustic, and wire-guided torpedoes. Submarines that serve as launch platforms can swim farther, faster, and deeper than their predecessors, while suppressing sound more effectively. Special features include inertial navigation, highly directional passive sonar, and receivers to warn of airborne and seaborne radar.[84]

The resultant menace is manifest. U-boat successes against Allied shipping were spectacular at the beginning of World War II, when the Nazi Navy boasted just 57 primitive boats. (Present Soviet holdings exceed four times that number.) By June 1942, 1,602 ships totalling 7,860,000 tons had gone to the bottom.[85]

America's ASW Response Successful ASW operations depend on abilities to find, fix, and finish enemy undersea raiders before they can wreak heavy damage. U.S. hunter-killer task forces are deficient on all three counts. Breakthroughs in the detection field are still in the blueprint stage, but beyond that, the size of America's specialized force is simply insufficient. ASW is mainly a time-consuming matter of attrition, in which numbers matter more and more as friendly losses mount. *At most,* we might account for 20 percent of all Soviet subs before the real carnage commenced among merchantmen.[86] Consequently, Soviet capacities to interfere with U.S. life lines at sea could prove to be low-cost, low-risk operations under certain circumstances, at least as long as a "Mexican standoff" persists at strategic nuclear levels.

Protection for petroleum tankers plying routes from the Persian Gulf to U.S. and European ports is a case in point. Convoys would reduce attrition, but U.S. escorts currently are inadequate even to shepherd ships along the 5,000-mile course to Capetown, if the Suez Canal were closed. Combat losses would cut effectiveness further. Land-based aircraft could provide part-time cover for un-

[84] George R. Lindsay, "Tactical Anti-Submarine Warfare: The Past and the Future," in *Powers at Sea I. The New Environment,* Adelphi Papers 122 (London: International Institute for Strategic Studies, 1976), p. 33; Merry, "Jane's Editor Sings Russ Navy's Blues," p. 1.

[85] Germany had 330 submarines in service by July, 1942. Lindsay, "Tactical Anti-Submarine Warfare," p. 32.

[86] Norman Polmar, "Thinking About Soviet ASW," *U.S. Naval Institute Proceedings* (May 1976), p. 110.

armed, unaccompanied tankers following random tracks across the Atlantic, provided appropriate base rights could be obtained in neutral or allied countries, but would be a poor substitute for on-the-scene ASW support.[87]

Similar problems are apparent in the Pacific, where our Navy reputedly could keep sea lanes open to Alaska and Hawaii, but would be hard pressed to control seas farther west against Soviet attack.[88]

MINE WARFARE

The U.S. Navy has long been skilled at mine warfare, which helped strangle Japanese shipping in World War II and sealed off Haiphong harbor thirty years later. Even so, Soviet mine*laying* capabilities are generally superior, although tactics are different.[89]

The Soviet edge in mine*sweeping* is even more clearly evident. U.S. ships of that sort on active service have all but disappeared. Sixty have been decommissioned since 1970, leaving only three. Twenty-two in the Naval Reserve would be hard pressed to clear important CONUS harbors expeditiously if extensive mine warfare occurred.[90] U.S. mine clearance capabilities in support of amphibious assault operations are also strained. Rotary-wing aircraft supposedly supplant our former surface force for that purpose, but many Soviet mines are laid at depths beyond their reach. Moreover, Marines and minesweeping helicopters would compete for space on aircraft carriers at times when that could affect operations adversely.

The Soviets, in contrast, maintain more than 360 ocean-going and coastal craft for minesweeping purposes. Their efforts might be easier than we would like if they were called on to break up U.S. barriers, because our clearance of Haiphong harbor and the Suez Canal were conducted in full view of Gorshkov's intelligence agents, who could copy techniques.

[87] U.S. Congress, House, *Oil Fields as Military Objectives,* pp. 19-20, 66-67.

[88] Admiral J. L. Holloway, III, Chief of Naval Operations, as quoted by Turner in "The Naval Balance: Not Just a Numbers Game," p. 351.

[89] Alva M. Bowen, Jr., specialist in naval affairs for the Congressional Research Service, was the principal source for the sub-section on naval mine warfare.

[90] Statistics were drawn from U.S. Navy computer printouts dated June 15, 1976.

Current U.S. Flexibility

One serious student of naval strategy summed up the U.S. situation most succinctly: as a seagoing power, the United States is entering an era of reduced options and reinforced risks.[91]

[91] Turner, "The Naval Balance: Not Just a Numbers Game," p. 339.

NET ASSESSMENT APPRAISAL

The quantitative trends in the naval balance that Collins describes in his tables and in Graph 8 are combined with Department of Defense reporting in Chart Seven. This chart confirms the statistical trends in Collins's analysis, and provides the following additional information:

- The PRC Navy provides no strategic counterweight of the kind it does in shaping the U.S. and Soviet ground force and air balance. Its naval forces are virtually non-existent.
- The declining U.S. advantage in carrier numbers is numerically counterbalanced by a massive Soviet advantage in stand-off weapons, and cruise missile craft and Soviet naval bombers.
- The U.S. numerical advantage in amphibious ships has disappeared. This does not reflect an actual Soviet parity in amphibious lift, since larger U.S. vessels have still much greater lift capacity. New types of Soviet ships may, however, be capable of bringing the U.S.S.R. to parity, and plans to improve aging U.S. amphibious vessels remain uncertain.
- The U.S. Navy has lost its quantitative advantage in other surface combatants.
- The Soviets retain their superiority in all submarine forces, and now have superiority in SLBMs. They are strikingly inferior, however, to submarine vulnerability, weaponry, and systems.

Chart Seven also shows that if ship numbers or numbers of new vessels determined world perceptions of the balance of power, the Soviets would have moved to parity or superiority in many critical aspects of naval strength. As Collins notes, however, Soviet and U.S. ships differ so much in size, weaponry, and missions configuration that a simple force count gives no real picture of military capability.

CHART SEVEN

Changes in Naval Force Levels— U.S./USSR (1966-1976)

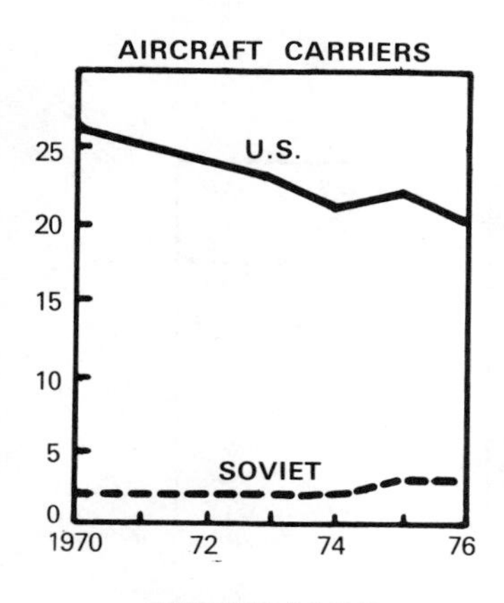

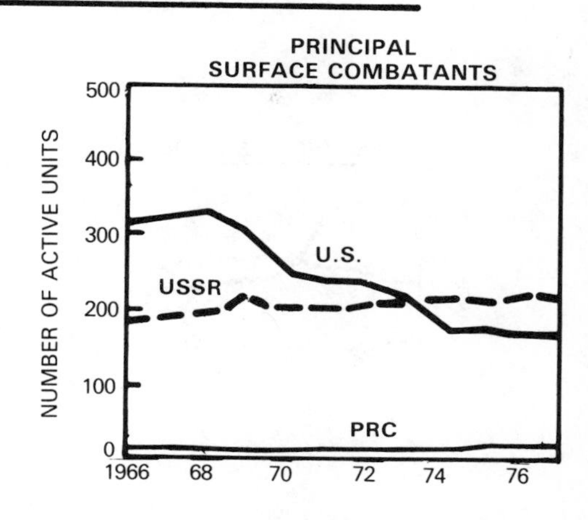

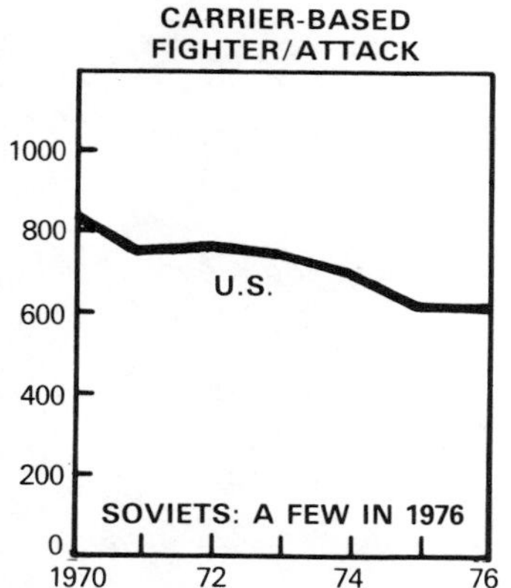

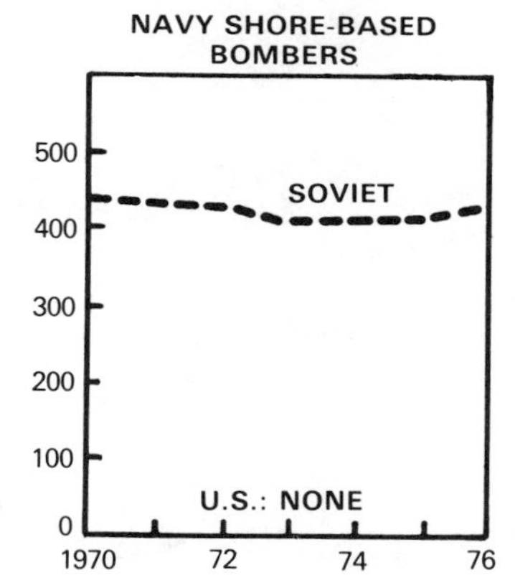

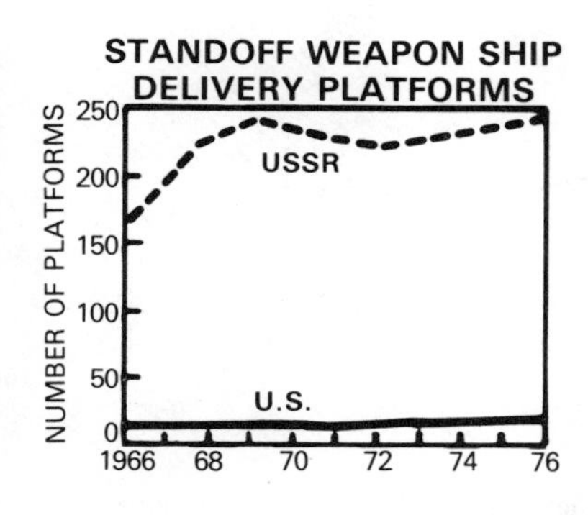

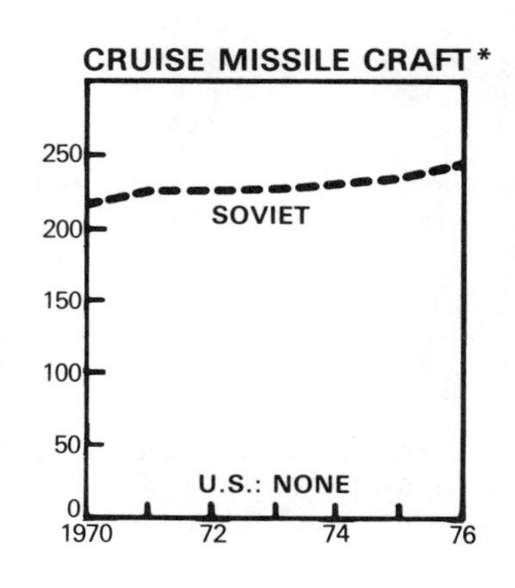

*Includes cruise missile patrol boats.

CHART SEVEN (Continued)

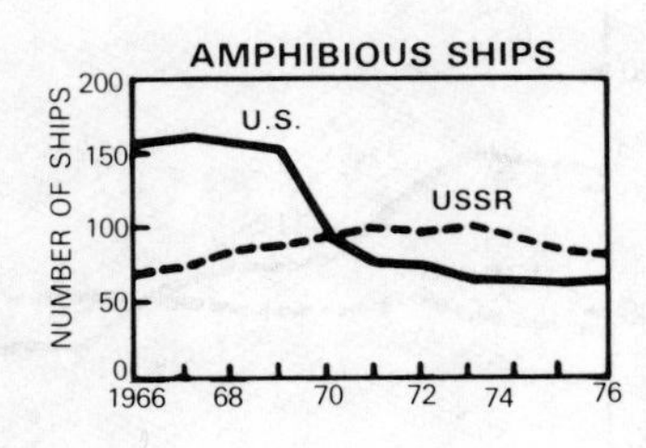

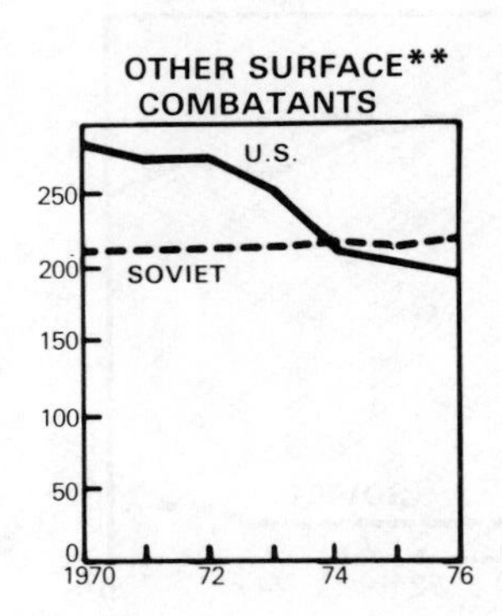

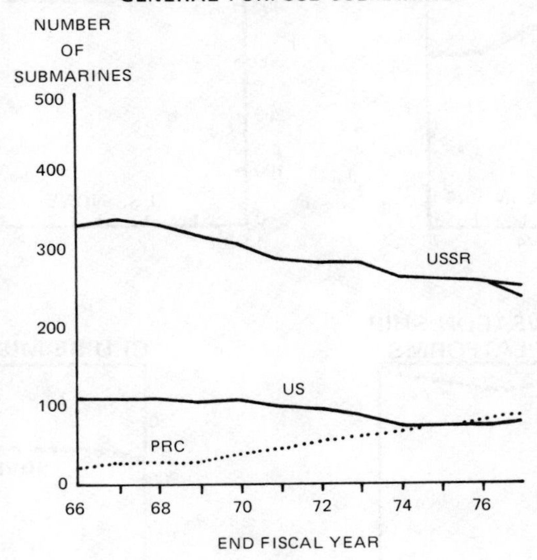

**Excludes coastal combatants and patrol boats; includes U.S. Naval Reserve ocean-going escorts.

CHART SEVEN (Continued)

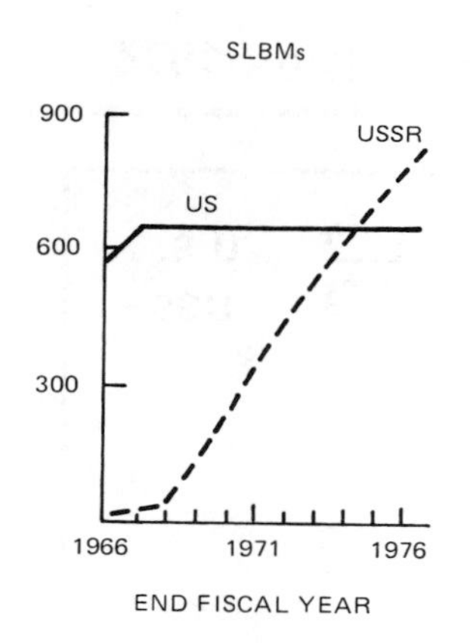

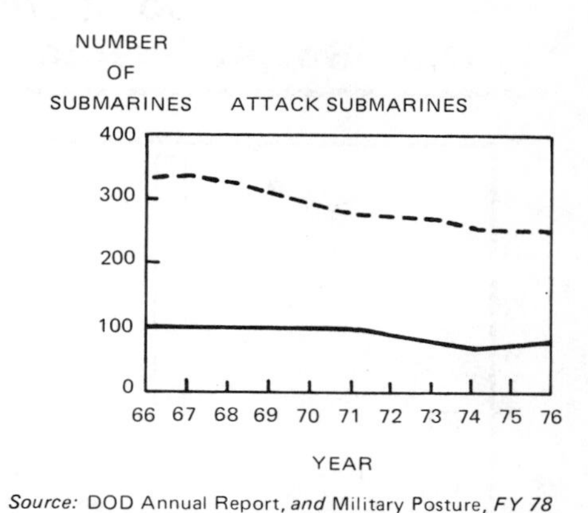

Source: DOD Annual Report, *and* Military Posture, *FY 78*

THE CHANGING STRUCTURE OF U.S. AND SOVIET NAVAL FORCES

The Soviet advantage in ship numbers is offset by an equal U.S. lead in total ship size or tonnage. In addition, both navies are cutting their total number of ships, but the U.S.S.R. is outbuilding the U.S. Navy at a rate of more than 3:1 in total combatant ship deliveries.

The United States, however, is still building much larger ships than the U.S.S.R. It delivered 249 ships during 1966-76, but these displaced roughly 2.1 million tons, or an average of over 8,000 tons per ship. In contrast, the Soviets built 765 ships but with a total displacement of only 1.5 million tons. This is an average of only 1,960 tons per ship, or less than 25% of the size of an average U.S. vessel.

UNDERSTANDING THE "NAVAL NUMBERS RACKET"

The problems in trying to assess naval force strength using simple "static" measures grow even worse as other measures of force strength are considered. For example, Chart Eight shows the average age of U.S. ships by class. It gives the impression that the U.S. Navy is generally older than the Soviet fleet:

CHART EIGHT

Comparison of U.S./Soviet Average Age of Ships 1974

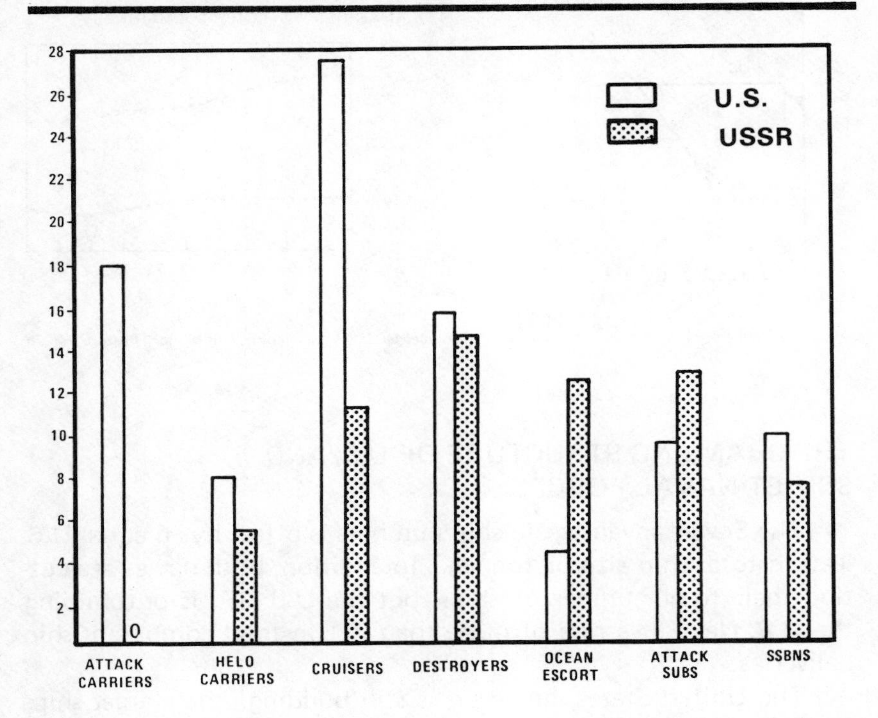

Source: Phillip A. Karber and John A. Lellenberg, "The State and Future of U.S. Naval Forces in the North Atlantic," NEW STRATEGIC FACTORS IN THE NORTH ATLANTIC, *IPC Science and Technology Press, Ltd., Guildford*

Chart Nine, however, again reverses the image. It indicates that the U.S. may have fallen behind the U.S.S.R. in naval construction in the mid-1960s, but that if it pursued its 1974 construction plans, it would surpass the U.S.S.R. by 1980.

This conclusion too, however, may be dubious. The U.S. Navy has rarely if ever completed its ambitious force improvement plans.

Unfortunately, just as there is a strategic forces "numbers game," there is a "naval numbers game." Not only do total navies

CHART NINE

Comparison of U.S./Soviet Ship Construction 1965-1980

A. SURFACE SHIPS

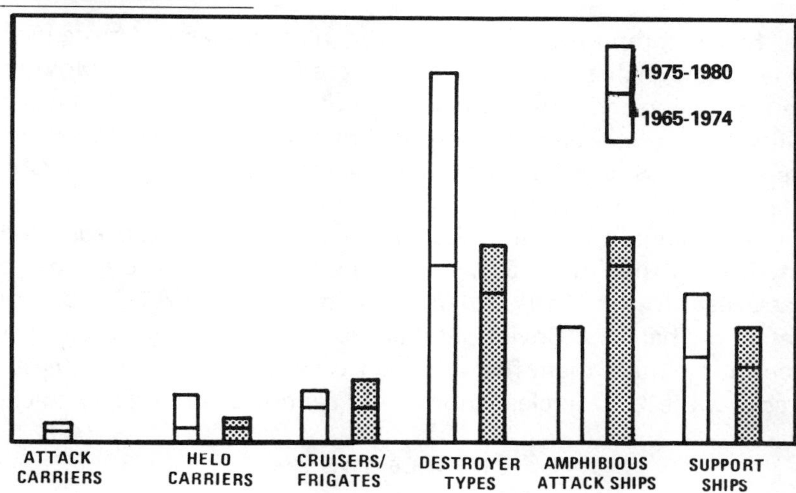

B. SUBMARINES

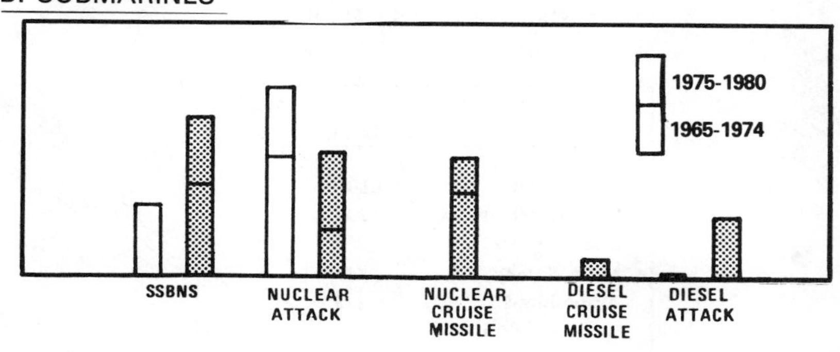

Source: Karber and Lellenberg, Ibid.

U.S.
USSR

not fight each other, but such broad "static" measures of the balance can be compared using criteria and time frames which can prove almost anything the analyst wishes. This is shown in Chart Ten which directly compares how ship numbers or ship tonnage can be used to give totally opposite pictures of the balance.

COMPARING U.S. AND SOVIET NAVAL FORCES BY REGION

Static comparisons can, however, be considerably more useful when they analyze naval strength presence by typical deployment area or region and mission. This may not measure war-fighting capability, but it does give an image of political or military "presence." Deployed U.S. and Soviet naval force strengths are shown in Chart Eleven, and show steady increase in Soviet activity.

As Collins notes, the U.S. has a significant advantage in the Atlantic in that the U.S.S.R. must exit its forces through narrow passages into the North Atlantic and must cross NATO submarine detection barriers. Soviet submarines are also noisy. Their speed does not protect them from the fact they are comparatively easy to track, while U.S. nuclear submarines cannot be effectively tracked

TABLE TWO

NATO & Warsaw Pact Fleets
Excludes U.S., U.S.S.R., France, Greece & U.K. FBMs

Type	NATO	Warsaw Pact
Submarines	105	17
Destroyer/Frigates	204	6
Coastal Patrol	253	364
Mine Warfare	243	178
Amphibious	243	33
Small oilers/tankers	130	25
Auxiliary Ships	509	84
Totals	1,698	707

Source: U.S. Navy, "European Communist Naval Order of Battle," 1 July, 1977 for Warsaw Pact, *Jane's Fighting Ships 1977-1978* for NATO.

CHART TEN

The Naval "Numbers Racket": U.S. and Soviet Total Naval Forces in 1974

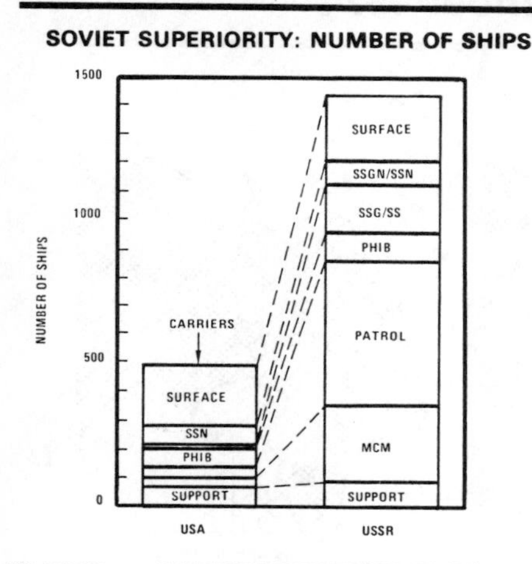

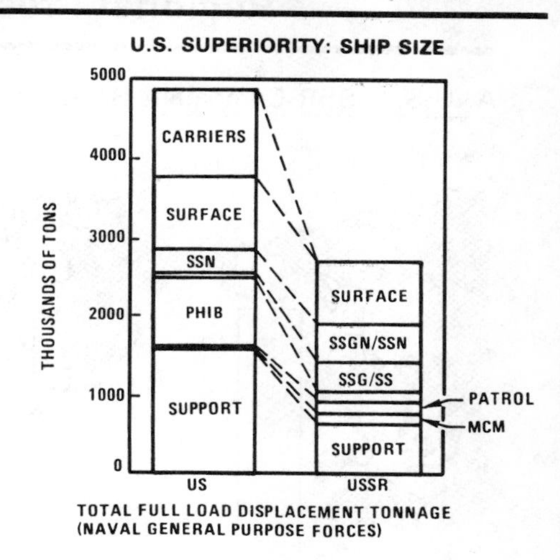

KEY:
SURFACE = MAJOR SURFACE COMBATANTS (OVER 1,000 TONS)
SSGN/SSN = NUCLEAR POWERED ATTACK SUBMARINES
SSG/SS = DIESEL POWERED ATTACK SUBMARINES
PHIB = AMPHIBIOUS LIFT SHIPS
PATROL = PATROL VESSELS (SURFACE COMBATANTS 1,000 TONS AND SMALLER, NOT INCLUDING INSHORE/RIVER CRAFT)
MCM = MINE COUNTERMEASURES VESSELS (SEA GOING VESSELS ONLY)
SUPPORT = MAJOR SUPPORT AND UNDERWAY REPLENISHMENT SHIPS

Source: Karber and Lellenberg, Ibid.

by Soviet attack submarines, and Soviet attack submarines have no large scale submarine tracking and detection systems through which they can vector in on U.S. submarines. Further, U.S. vessels have much greater endurance than their Soviet counterparts, and the U.S. has a considerable advantage in where it can base its land-based aircraft and in the strength and quality of its Allies.

Allied navies may, however, be a declining U.S. advantage. Most NATO navies no longer approach the United States in anti-submarine warfare weaponry and detection systems. They are not standardized, and some would find it difficult to operate effectively with U.S. naval forces in either surface or anti-submarine warfare.

QUALITATIVE COMPARISONS OF THE NAVAL BALANCE

The ultimate key to measuring the naval balance is quality and mission effectiveness, not ship numbers. Unfortunately, even less is known about the actual quality and mission performance capa-

CHART ELEVEN

Illustrative Comparisons of U.S. and Soviet Naval Strength by Deployment and Regional Presence: 1974

A. U.S./USSR Combatant Deployments

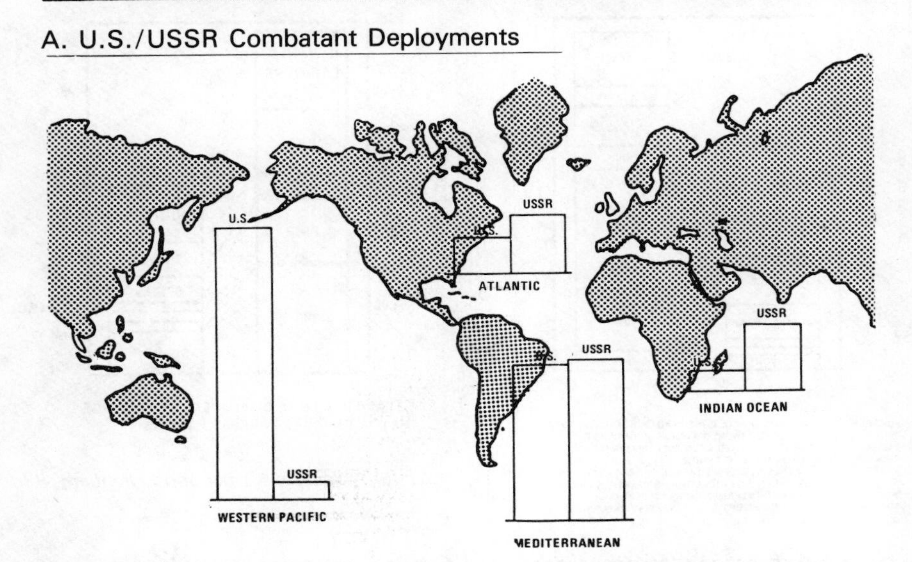

*Includes aircraft carriers, general purpose submarines, major surface combatants, minor surface combatants, amphibious ships, and mine warfareships.

B. Comparison of U.S./Soviet Atlantic Oriented Fleet Structure 1974

	U.S. 2nd Fleet Atlantic	USSR North Fleet
Carriers (Attack and Helo)	4	0
Surface Combatants	67	56
	71	56
Submarines		
Fleet Ballistic Missile	28	50
Submarines	28	50
Cruise Missiles	0	40
Attack	20	80
	48	170
Total	119	226

*Includes all United States ballistic missile submarines based in Charleston, Holy Loch, and Rota.

CHART ELEVEN (Continued)

C. U.S./SOVIET SHIP DAYS IN THE ATLANTIC

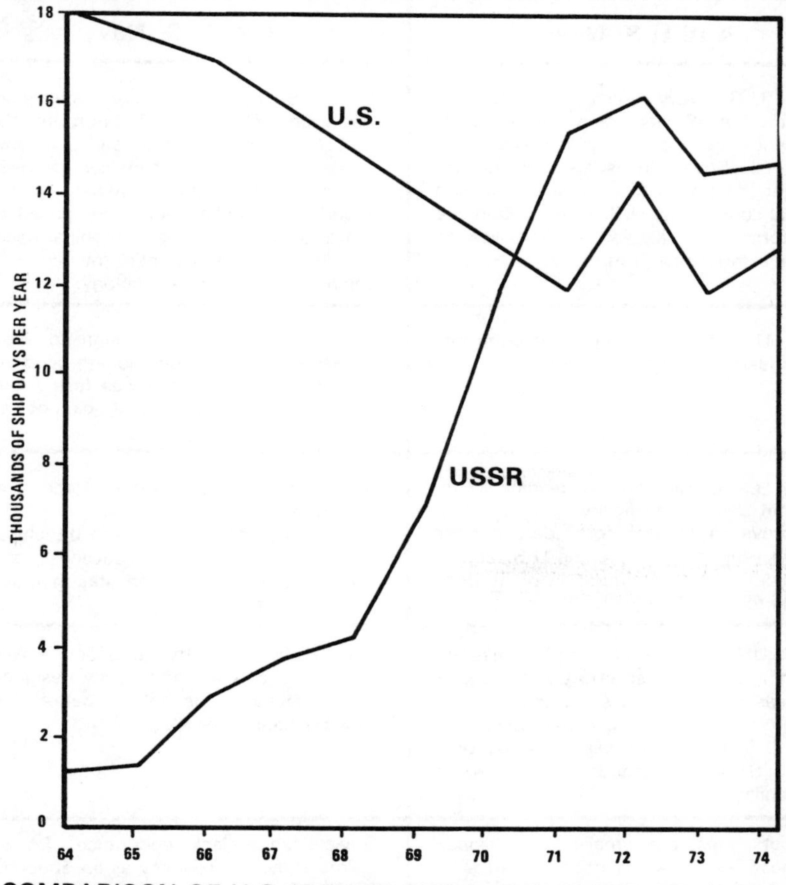

COMPARISON OF U.S./SOVIET SHIP DAYS IN THE ATLANTIC

Source: Phillip A. Karber

bility of U.S. and Soviet naval forces than is known about the precise details of the qualitative performance of U.S. and Soviet strategic nuclear forces. Further, navies usually act as task forces in most missions. These task forces involve complex mixes of surface to surface and ASW capability. Little, if any, unclassified data

CHART TWELVE

Opposing Views of Naval Quality

Anti U.S. Navy	Pro U.S. Navy
The U.S. Navy relies on carrier air against a Soviet navy with vastly superior surface to surface capability. A single Soviet first strike with missiles might destroy U.S. carriers and win naval supremacy. U.S. plans to upgrade its forces with missiles slip from year to year without real progress.	The U.S. has large ships which can easily be refitted with Harpoon and the SLCM when these are ready. It now relies on carrier air which can be used for power projection anywhere in the world. The Soviet fleet is composed of small specially configured ships using missiles of uncertain performance with rapidly obsolescent technology.
The U.S. fleet is aging, and confronts a modern Soviet fleet.	U.S. ships are large enough to allow massive refitting, conditioning, and reconfiguration. Soviet forces face future block obsolescence, and cannot be effectively refitted.
The U.S. carrier force is steadily declining in size and flexibility. U.S. plans to improve carrier task force defenses slip from year to year, and the U.S. cannot decide on the configuration of its post-1980 forces.	The present U.S. carrier force can generate far more sorties than the former force could with twice the ships. Each sortie has more effectiveness, and carriers can now stay on station much longer.
The USSR is conducting a massive training and exercise effort with steadily increasing effectiveness. Such training efforts are much larger than those of the U.S. Navy, and will eventually give the Soviets superior deep water capability.	The USSR is still trying to find a way to make a polyglot and poorly designed fleet effective. It is getting better, but lags far behind the U.S.
Soviet ships are smaller and leaner because Soviet sailors do not need U.S. luxury.	Soviets ships lack endurance. Soviet sailors show no reliability as an effective fighting force.
The new Soviet carrier fleet with missile capability is more modern and effective.	The Moscow class ASW carrier was built to try to kill U.S. SSBNs. Yet it lacks the ASW weapons and sensors to kill U.S. SSBNs, and the range to attack them at their current stations. It is unclear the Kiev or future Soviet carrier designs will be any better.
The Soviets have a vast lead in attack and general purpose forces submarines. It can saturate the steadily declining U.S. strength in ASW escorts.	All of which are so easy they are incapable of killing U.S. SSNs under most conditions and highly vulnerable. The U.S. has the advantage of submarine barriers.

CHART TWELVE (Continued)

Anti U.S. Navy	Pro U.S. Navy
The Soviet navy has a large lead in land based naval bombers.	Most of which are highly vulnerable, and individually less capable in many missions than U.S. carrier based fighters. U.S. B-52s have far more dual capability in naval missions than single purpose Soviet naval bombers.
The USSR has steadily stronger amphibious forces.	These forces have little real contingency capability, and no combat experience.
The Soviet navy is establishing an image of presence and power projection throughout the world.	It has yet served no successful purpose in advancing Soviet interests. U.S. Navy power projection has constantly done so for the United States.
The Soviets could block all maritime movement and paralyze the West.	Soviet fleet endurance is limited to two to three weeks at most. Careful use of Atlantic choke points and Soviet fleet bases would quickly attrit the Soviet navy.

are available on such capabilities and, as Collins notes, they can be radically affected by land-based air forces and by reconnaissance and sensor systems ranging from satellites to submerged sensors.

The U.S. Navy view of the actual ASW capabilities of each side in dynamic analyses or exercises have never "leaked" to the level where it is possible to conduct a meaningful unclassified discussion. Neither have U.S. Navy models of carrier, air, and cruise missile exchanges, of the naval defense of Atlantic shipping, or of Atlantic or Mediterranean sea control. This forces analysts working with unclassified information to speculate to an unusual degree. Consider, for example, the following opposing "expert" views of naval quality summarized in Chart Twelve.

There is little point in advocating any of the positions in Chart Twelve. These are critical issues, and there are many others, but they do not lend themselves to unclassified static analysis.

[Chapter 5]

Mobility Forces

MOBILITY MODES AND MISSIONS

The proper mix and amount of airlift, sealift, and land transportation depends on how much must be moved how far and how fast under specific conditions to serve particular purposes. Prepositioning selected stocks (such as armor, artillery, and ammunition) in or near prospective employment areas can reduce demands, but only up to some changeable point, beyond which such steps can be counterproductive.[1]

Intercontinental lift over open oceans is a U.S. essential. Russian requirements thus far have been more regional. Dissimilar demands consequently foster mobility force structures that are quite different in size as well as composition (see Graph 9).[2]

[1] Prepositioned stocks require protected facilities. Duplicate sets of equipment (one in use, one in storage) not only double costs, but some items are difficult to maintain, much less modernize at rates equal to those of items in active units. Land-based depots in particular are susceptible to preemptive destruction in some deployment areas. Relocating caches in emergency can *add* to, rather than *reduce,* demands on airlift and sealift. R. C. Rainville, *Strategic Mobility and the Nixon Doctrine* (Washington, D.C.: The National War College, 1971), pp. 60-61.

[2] U.S. Plans, programs, and alternatives are addressed in Rumsfeld, *Annual Defense Department Report for FY 1978,* pp. 228-37; George S. Brown, *United States Military Posture for FY 1978,* pp. 93-98; and *Mobility Forces: An Interim Report,* prepared by the Congressional Budget Office for the Subcommittee on Priorities and Economy in Government of the Joint Economic Committee (Washington, D.C., October 26, 1976), 13 pp. plus tables.

Graph 9

AIRLIFT AND SEALIFT
Statistical Summary

(Note Different Scales)

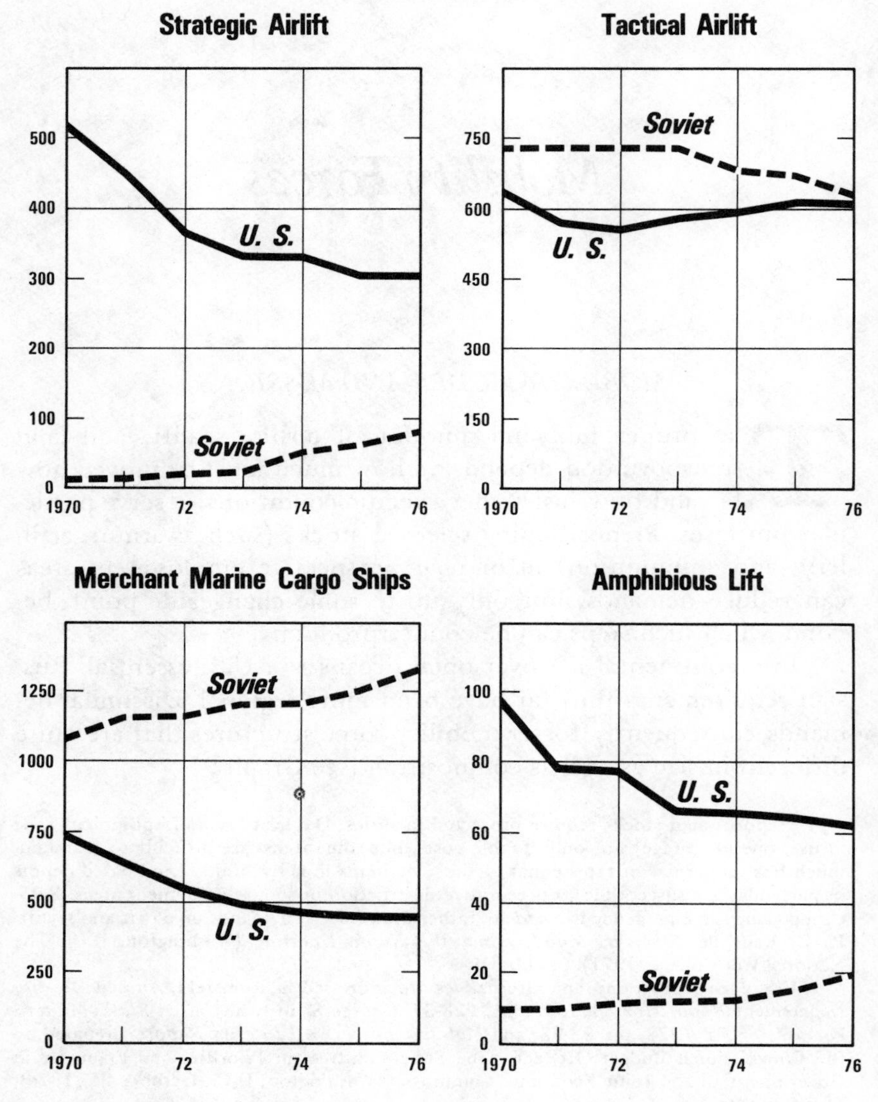

Note: U.S. airlift and merchant ships include reserve components.

MILITARY AIRLIFT

There is no clean dividing line between strategic and tactical airlift, both of which specialize in rapid redeployment to serve deterrent/defensive purposes, but primary missions are distinct.

Strategic Airlift

SOVIET POSTURE

AN-22 Cocks, with a cargo capacity second only to U.S. C-5s, can lift outsize items like T-62 tanks, Frog-3 rockets, and SA-4 SAMs on tracked launchers. Flashy turbofan Candids (IL-76) have some features similar to C-141s. Taken together, however, they total just 80 aircraft, a small fleet compared with the titanic armed force they are tasked to support (see Figure 22). Neither is fitted for in-flight refueling.[3]

Aeroflot, the Soviet counterpart of U.S. civil airlines, could increase cargo capacities about 25 percent, and triple spaces for passengers,[4] but its crews, along with those in military service, lack much experience in long-range operations over strange territory.

Those factors in combination rule out large-scale airlift operations in southern Africa, Latin America, and other remote locales, unless "stepping stones" are available.

U.S. POSTURE

America's strategic airlift assets, despite significant shortfalls, are without peer in the world.

Capabilities Major emphasis on modernization has paid off. Our present all-jet force, less than half the size of its propeller-powered predecessor in the late 1960s, can lift loads more than three times larger.[5] Seventy C-5s can accommodate outsize, oddly shaped cargo, such as heavy helicopters, tanks, and 20-ton cranes.[6] Aircraft and crews are qualified for in-flight refueling, which makes non-

[3] Terry L. Schott, "Soviet Air Transportation: Projection of Power," *U.S. Army Aviation Digest* (July 1976), p. 14.

[4] Brown, *United States Military Posture for FY 1978,* p. 94.

[5] U.S. Congress, House, Research and Development Subcommittee of the Committee on Armed Services, *The Posture of Military Airlift,* 94th Congress, 2d Session (Washington, D.C.: U.S. Government Printing Office, April 9, 1976), p. 28.

[6] Congressional Budget Office, *Mobility Forces,* p. 3.

FIGURE 22 MILITARY AIRLIFT
Statistical Trends and System Characteristics

STRATEGIC AIRLIFT	*1970*	*1971*	*1972*	*1973*	*1974*	*1975*	*1976*
United States							
Active							
C-5	2	28	49	70	70	70	70
C-133	38	14	0	0	0	0	0
C-141	234	234	234	234	234	234	234
Total	274	276	283	304	304	304	304
Reserve							
C-97	32	8	8	0	0	0	0
C-124	208	168	72	24	24	0	0
Total	240	176	80	24	24	0	0
Grand Total	514	452	363	328	328	304	304
Soviet Union							
AN-22	10	10	20	20	50	50	50
IL-76	0	0	0	0	0	10	30
Total	10	10	20	20	50	60	80
U.S. Standing	+504	+442	+343	+308	+278	+244	+224
TACTICAL AIRLIFT							
United States							
Active							
C-7	80	80	4	0	0	0	0
C-123	60	38	0	0	0	0	0
C-130	394	332	288	272	272	272	234
Total	534	450	292	272	272	272	234
Reserve							
C-7	0	0	32	48	48	48	48
C-119	48	0	0	0	0	0	0
C-123	8	8	40	72	72	72	64
C-130	44	114	188	188	198	222	262
Total	100	122	260	308	318	342	374
Grand Total	634	572	552	580	590	614	608
Soviet Union							
AN-12	730	730	730	730	680	670	630
U.S. Standing	–96	–158	–178	–150	–90	–56	–22

FIGURE 22 (continued)

GRAND TOTAL							
United States	1148	1024	915	908	918	918	912
Soviet Union	740	740	750	750	730	730	710
U.S. Standing	+408	+284	+165	+158	+188	+188	+202
UTILITY/CARGO HELICOPTERS							
United States							
Army							
CH-34	185	40	0	0	0	0	0
CH-37	25	0	0	0	0	0	0
CH-47	480	460	345	380	365	330	325
CH-54	70	55	50	50	50	50	50
UH-1	4030	3750	2810	2680	2945	2440	2565
Total	4790	4305	3205	3110	3360	2820	2940
Marine							
CH-46	224	164	133	140	91	118	141
CH-53	113	95	101	107	101	97	101
Total	337	259	234	247	192	215	242
Grand Total	5127	4564	3439	3357	3552	3035	3182
Soviet Union	800	900	950	1100	1325	1350	1475
U.S. Standing	+4327	+3664	+2489	+2257	+2227	+1685	+1707

FIGURE 22 (continued)

TRANSPORT AIRCRAFT CHARACTERISTICS

	Power Plant		*Cruising Speed (Knots)*	*Maximum Cargo Load (lbs)*	*Range With Max Load (Nautical Miles)*	*Minimum Runway Length (Feet)*		*Troops*	
	NR	*Type*				*T/O*	*Land*	*Pax*	*Para*
United States									
Active									
C-5	4	Jet	450	209,000	2,565	7700	4610	343	0
C-130E	4	Prop	280	42,000	2,000	2600	2700	91	64
C-141	4	Jet	425	63,600	2,835	6300	3840	131	123
Reserve									
C-7	2	Prop	140	6,000	100	1000	1000	31	25
C-123K	4	2 Jet 2 Prop	145	17,600	100	1325	1150	58	58
C-130A	4	Prop	290	25,000	1,075	1850	1850	89	64
C-130B	4	Prop	290	35,000	1,575	2400	2400	91	64
Soviet Union									
AN-12	4	Prop	320	44,000	750	2300	1640	28	80
AN-22	4	Prop	350	176,200	2,200	4260	2620	29	0
IL-76	4	Jet	430	88,000	2,700	UNK	UNK	122	0

NOTES: Coke (AN-24) and Curl (AN-26), both similar to Fokker F-27s, are scarcely suitable for most airlift purposes. Coot (IL-28), a medium transport, is assigned almost exclusively to Aeroflot, rather than Soviet Air Force units. All three are therefore excluded.

Ranges correspond with loads shown. Performance data predicated on wartime maximum gross takeoff weights, no wind, and maximum fuel reserves. Lighter loads allow longer ranges. Minimum runway lengths shown above apply to *average*, not *maximum*, gross takeoff weights. C-5 range is unrefueled.

Pax stands for passengers; para for paratroops.

C-130s, AN-12s, and AN-22s are turboprop powered.

stop performance possible at unrestricted range.[7] C-141s, proved over a seven-year period on the trans-Pacific "pipeline" to Southeast Asia, are still reliable mainstays that make up 75 percent of our strategic airlift stable. Utilization rates could be increased in emergency by mobilizing associate reserves that lost their last aircraft in 1975, but are collocated with active squadrons, and participate in operations.[8] C-130 E/H models, available for augmentation under certain circumstances, could increase strategic capabilities by about 8,500 tons in fifteen days.[9]

The U.S. Civil Reserve Air Fleet (CRAF), with roughly 135 long-range cargo and 94 passenger aircraft,[10] all modern jets, can be committed during crises in accord with contracts that connect commercial carriers with Military Airlift Command (MAC).[11]

Combined military/CRAF 30-day lift capabilities reportedly total 180,000 tons, 50,000 of which (28 percent) comprise C-5 sorties with outsize equipment.[12]

Constraints Absolute airlift abilities, however impressive in abstract terms, must be related to the real world and real requirements if they are to be meaningful. The time required to deploy Army divisions with unit integrity intact depends entirely on C-5s,

[7] C-5 aircraft have always been equipped with refueling receptacles, but crews were not qualified during the emergency airlift to Israel in 1973. Consequently, they not only were compelled to take on fuel enroute at politically sensitive Lajes Field in the Azores, but consumed 1.3 pounds of POL at Lod Airport in Israel for every pound of equipment delivered. *Ibid.*, p. 15.

Present U.S. aircraft could carry 38 percent more outsize cargo if refueled in flight, according to a recent MAC study.

[8] Brown, *United States Military Posture for FY 1978*, p. 94.

[9] Congressional Budget Office, *Mobility Forces*, p. 7, amended by Air Force comments on the draft of this study, March 7, 1977.

[10] Brown, *United States Military Posture for FY 1978*, p. 95.

[11] CRAF callups, by model and series, can be in three stages. The first two afford extensive voluntary civil augmentation airlift in times of limited crisis. Commander, MAC is authorized to activate Stage I. Stage II requires Defense Secretary approval. Stage III provisions, connected with grave national emergencies declared by Congress, the President, or (under specified conditions) the Office of Preparedness, permit requisitioning of the entire CRAF. To date, none of those plans have been implemented. Commitments during the Vietnam conflict comprised voluntary expansion of peacetime contracts. U.S. Congress, House, *The Posture of Military Airlift*, pp. 31-32.

[12] Air Force comments on the draft of this study, March 4, 1977. To put 30-day lift capacities in perspective, a single armored division with accompanying supplies (computed at 422 pounds per man day) weighs 55,900 tons. A mechanized division weighs 49,710. Infantry divisions, which make up most of our National Guard, weigh 29,370 each. Figures were furnished by Army DCSLOG planners on February 4, 1977.

the only aircraft (including those in CRAF) that can carry outsize cargo.[13] The current complement of 70 clearly is too few to implement NATO plans in prescribed time frames. Nearly five trips, for example, would be needed to move the medium tanks of one armored division, at the rate of two per load. One official study concludes that eight days would elapse before all outsize equipment could reach NATO airports after C-141s and CRAF delivered an inter-service package composed of 300,000 troops and 169,000 tons of cargo.[14]

C-141s, which "cube out" quickly,[15] fly many sorties with substantial lift capacity unused. Their inability to refuel in flight causes operational costs to soar and constrains mobility options.[16] Huge numbers are needed even to lift the combat elements of a light airborne division over long distances with a basic load of ammunition and five-day supplies of rations and fuel. A move from Fort Bragg, North Carolina to the Middle East would consume more than 700 sorties,[17] not counting comparable airlift required for associated Army support and forces from other services.[18]

CRAF equipment compensates less effectively than first glance suggests. Only 12 are wide-bodied freighters.[19] None of those are configured to carry outsize cargo.[20]

Finally, SAC's fleet of 615 KC-135 tankers[21] is sufficient to serve concurrently the peacetime U.S. B-52 alert force, transoceanic fighter redeployments, and strategic airlift operations only

[13] U.S. Congress, House, *The Posture of Military Airlift,* p. 18.

[14] Rainville, *Strategic Mobility and the Nixon Doctrine,* pp. 51-52, 73. The study cited is out of date, but C-5 holdings have stayed constant at 70 and similar delivery delays could be expected today.

[15] Large, irregular items such as vehicles often fill the cigar-shaped C-141 cargo compartment long before allowable cargo load (ACL) limits are reached in terms of pounds. That phenomenon, called "cubing out," has produced a prototype program for a "stretch" version of the C-141, which (if procured) would increase lift capacity about 30 percent.

[16] C-141s would have proved useless during the 1973 airlift to Israel if Portugal had refused landing rights at Lajes Field. The same tonnage could have been delivered with 57 fewer sorties while saving 5 million gallons of gas if in-flight refueling had been feasible. U.S. Congress, House, *The Posture of Military Airlift,* p. 15.

[17] A sortie is one round trip by one aircraft.

[18] Statistics were received by telephone from staff officers of the 82nd Airborne Division in April 1975.

[19] Thirteen more wide-bodied aircraft can be converted from passenger to cargo configuration. *MAC Monthly Reserve Air Fleet Summary.*

[20] Rainville, *Strategic Mobility and the Nixon Doctrine,* p. 49.

[21] Brown, *United States Military Posture for FY 1978,* p. 96.

during small-scale, medium-range contingencies. Competition could delay receipt of supplies and reinforcements under more stringent circumstances.[22]

Tactical Airlift

Tactical airlift is used primarily for *intra* (rather than *inter*) theater transportation. Various types of aircraft can quickly shift troops, supplies, and equipment by airlandings and/or aerial delivery.

SOVIET POSTURE

AN-12 Cubs, which account for almost 85 percent of all Soviet airlift aircraft, are inferior to U.S. C-130s in every respect. Still, they find space for most equipment assigned or attached to airborne divisions (see Figure 22 for statistics and characteristics). The same is true for IL-76s, which are just beginning to see service.[23]

Tactical transports are sufficient to airlift one Soviet airborne division about 1,000 miles with all combat equipment and three days' worth of accompanying supplies. Military aircraft could move assault elements of two divisions the same distance, provided heavy items were prepositioned. Augmentation from Aeroflot could triple the number of personnel.[24] Shorter hops increase capabilities, because turnaround times are reduced. Major elements thus could move rapidly to, say, the Middle East from departure airfields in the Caucasus, especially if committed in waves over several days.[25]

Reliable fighter support, however, could be a critical limiting factor in any objective area far from Soviet frontiers or satellite states. Forward basing conceivably could be found in friendly countries, such as Syria, but complexities would increase. The likelihood that large-scale Soviet airborne operations would occur under any conditions that exclude local air superiority thus seems slight.[26]

[22] Rumsfeld, *Annual Defense Department Report for FY 1978,* p. 230.

[23] Graham H. Turbiville, "Soviet Airborne Forces: Increasingly Powerful Factor in the Equation," *Army Magazine* (April 20, 1976), p. 22; Schott, "Soviet Air Transportation," pp. 14-16.

[24] DIA comments on the draft of this study, March 3, 1977.

[25] Erickson, *Soviet-Warsaw Pact Force Levels,* p. 51; Turbiville, "Soviet Airborne Forces," p. 27.

[26] Turbiville, *Soviet Airborne Forces,* p. 27.

U.S. POSTURE

America's tactical airlift, long the world's best, is still unexcelled, but in the absence of modernization measures shows clear signs of age.

Capabilities Tried, true, and time-tested C-130s, which make up most of the U.S. tactical airlift force, are ideally suited for multiple missions over medium ranges under combat conditions. They adapt equally well for airlandings and airborne assaults that involve minimal campaigns or mass movements.

Constraints MAC's active force is much smaller than in 1970 (534 aircraft then, 234 now).[27] Reserve components, equipped mainly with older model C-130s, are much larger (100 then, 374 now). Composite strengths have therefore stayed constant statistically, but combined capabilities have decreased. Reliance on reserves, once modest, is now marked.

Beyond that, C-130 cargo compartments are too tight for loads like self-propelled artillery and MICV,[28] which C-141s can lift only by slighting strategic airlift missions. We also are losing any capacity to conduct operations off crude strips less than 2,000 feet long. A few C-7 and C-123 aircraft, approaching the end of their service life, are still assigned to squadrons in reserve, but capabilities are absent in the active inventory.

Battlefield Mobility

Dual-purpose Soviet MI-24 Hind helicopters, which serve as weapons platforms as well as cargo/troop carriers, possess great possibilities.[29] Their numbers, however, now are few. Other makes are less impressive. Doctrine is still in development.

The combined U.S. Army and Marine fleet of 3,200 cargo/utility helicopters, cut by 60 percent since 1970, still possesses battlefield mobility much superior to that of the Soviets. That trend, however, could turn around if present stocks are not replaced with more capable models.

[27] U.S. tactical airlift assets were assigned to Tactical Air Command and theater commanders until March 31, 1975, when Military Airlift Command was made single manager.

[28] U.S. Congress, House, *The Posture of Military Airlift,* p. 7.

[29] Erickson, *Soviet-Warsaw Pact Force Levels,* p. 51; David A. Bramlett, "Soviet Airmobility: An Overview," *Military Review* (January 1977), pp. 16-18, 24-25.

MILITARY SEALIFT

Military sealift essentially serves two purposes: administrative movements and amphibious assault. Once again, asymmetries between U.S. and Soviet structures are clearly apparent (Figure 23).

Administrative Sealift

Airlift affords rapid response, but only sealift can carry mass tonnages over transoceanic distances to sustain forward deployed forces or move strategic materials in amounts essential for national security.

SOVIET STRATEGIC SEALIFT

The Soviet merchant marine, controlled by the Navy, currently includes 1,650 modern, highly automated ships whose characteristics have been shaped more by sea power concepts than purely commercial considerations. Most, being self-sustaining[30] and smaller than U.S. counterparts, are better able to operate in ports plagued by shallow harbors and skimpy facilities. Abilities were displayed to advantage during the Vietnam War, when Soviet merchantmen moved millions of tons 14,000 nautical miles around the Cape of Good Hope to Haiphong.[31]

Present trends tend toward Roll-on Roll-off (Ro/Ro) vessels, with ramps that allow wheeled and tracked vehicles to board and debark at will from open piers, with or without containers. Something like 20 are now in service. Finland is building two "Seabee" ships, based on U.S. technology. Barges, stowed topside and between decks, can be loaded and unloaded easily, and once in the water integrate easily with feeder systems that navigate inland waterways. The applicability to operate in out-of-the-way areas is apparent.[32]

[30] "Self-sustaining" in merchant marine terms connotes ships that can load and unload themselves, using the ship's booms, cranes, and other cargo-handling apparatus.

[31] Briefing by Larry Luckworth, U.S. Naval Intelligence Support Center, Washington, May, 1976, pp. 1-2 of text; *Understanding Soviet Naval Developments,* p. 39; Brown, *United States Military Posture for FY 1978,* p. 97. Sources refer to ship characteristics, not Soviet use.

[32] John D. Chase, "U.S. Merchant Marine—for Commerce and Defense," *U.S. Naval Institute Proceedings* (May 1976), pp. 134-35, 141; Rumsfeld, *Annual Defense Department Report for FY 1978,* p. 237; *Izvestia,* December 27, 1976.

FIGURE 23 MERCHANT MARINE
Statistical Trends and System Characteristics

	1970	*1971*	*1972*	*1973*	*1974*	*1975*	*1976*
United States							
MSC							
Nucleus							
Cargo	69	66	63	30	12	9	6
Tanker	25	24	17	16	20	19	21
Total	94	90	80	46	32	28	27
Charter							
Cargo	123	76	91	43	34	22	21
Tanker	36	31	34	23	22	13	9
Total	159	107	125	66	56	35	30
Grand Total	253	197	205	112	88	63	57
Remainder, Active Merchant Marine							
Cargo	351	300	205	249	257	272	268
Tanker	187	183	176	176	169	181	180
Total	538	483	381	425	426	453	448
Effective U.S. Controlled Fleet							
Cargo	34	22	20	18	14	11	10
Tanker	265	269	276	301	319	299	304
Total	299	291	296	319	333	310	314
Total Active							
Cargo	577	464	379	340	317	314	305
Tanker	513	507	503	516	530	512	514
Total	1090	971	882	856	847	826	819
United States							
NDRF							
Cargo	170	175	168	149	138	139	139
Tanker	20	27	28	26	24	28	19
Total	190	202	196	175	162	167	158

FIGURE 23 (continued)

RECAP							
Cargo							
Active	577	464	379	340	317	314	305
NDRF	170	175	168	149	138	139	139
Total	747	639	547	489	455	453	444
Tanker							
Active	513	507	503	516	530	512	514
NDRF	20	27	28	26	24	28	19
Total	533	534	531	542	554	540	533
All Active	1090	971	882	856	847	826	819
All NDRF	190	202	196	175	162	167	158
Total	1280	1173	1078	1031	1009	993	977
Soviet Union							
Cargo	1075	1150	1150	1200	1200	1250	1325
Tanker	325	300	300	300	300	300	325
Total	1400	1450	1450	1500	1500	1550	1650
U.S. Standing							
Cargo	-328	-511	-603	-711	-745	-797	-881
Tanker	+208	+234	+231	+242	+254	+240	+208
Total	-120	-277	-372	-469	-491	-557	-673

FIGURE 23 (continued)

UNITED STATES	CARGO SHIP CHARACTERISTICS							
	MARAD Design	*Ship Class*	*Capacity (M/T)**	*DWT At Design Draft*	*Design Speed*	*Boom/ Capacity*	*Container Capacity*	*Lighters/ Barges*
Break Bulk	VC2-S-AP2	Adelphi Victory	11325	7400	15	1-50T	None	None
	C2-S-37c	Sheldon Lykes	14300	9610	18	1-60T	None	None
	C4-S-57A	Challenger	16072	10290	20	4-10T 2-15T 1-70T	None	None
Partial Containership	C4-S-64A	Austral Pilot	17176	9800	20	4-15T 2-10T	203-20'	None
Self-Sustaining Cont.	C6-S-10c	President Polk	26700	17300	20	2-10T 4-22T 1-60T	380-20' 198-40'	None
Containership Non Self-Sustaining	C5-S-73b	C.V. Lightning	23800	11700	20	None	928-20	None
	C6-S-1N	American Ace	26100	17100	22	None	463-20' 234-40'	None
	-	Sea-Land Galloway (SL-7)	59300	20060	30	None	896-35' 200-40'	None
Lash	C8-S-81b	Lash Italia	32655	17990	21	1-30T 1-500T (Crane)	248-20'	49 with containers[1] 62 without containers
Sea Bee	C8-S-82A	Doctor Lykes	37187	25550	20	2000T Elevator	None	38 Barges[2]
RO/RO	-	Ponce	49200	11192	25	None	None	None
	C7-S-95	Maine	43390	11980	23	2-15T	None	None

*Measurement ton (M/T) equal to 40 cubic feet.
[1] Lash Lighter Capacity 475 M/T
[2] SeaBee Barge Capacity 1000 M/T
[3] Figure 23 excludes bulk cargo ships.
[4] Foreign flags are 1000 tons or more.
[5] MARAD stands for Maritime Administration

U.S. MERCHANT FLEETS

U.S. merchant shipping has definitely been on a dow
since World War II. The implications for national defe
inimical.

Military Sealift Command Six government-owned and 21 chartered dry cargo ships currently constitute the core assets immediately available to Military Sealift Command (MSC) for security purposes—166 less than those assigned at the start of this decade.[33] Thirty tankers make up the remainder.

Active Merchant Marine Capabilities so slight can be quickly exhausted. Consequently MSC leans ever more heavily on our active merchant marine, which is also shrinking.

Ships now in the inventory are tailored expressly for foreign trade, not military emergencies. The U.S. break-bulk tramp fleet has broken up.[34] Specialized container ships, which now predominate, capitalize on speed at sea and at pierside.[35] Loading and off-loading can take less than 24 hours, as opposed to several days for conventional freighters.

Unfortunately such vessels are ill-configured to carry vehicles, and the absence of onboard gantry cranes, which increase container capacity and decrease cost/maintenance problems, makes discharge difficult in undeveloped ports. Heavy-lift helicopters, balloon-supported aerial tramways (like those used by lumber companies), and equipment on self-sustaining ships currently serve as expedients to unload container ships in such circumstances, but efforts of that sort are expensive and inefficient.[36]

[33] MSC is the operating agency through which the Secretary of the Navy exercises his single manager responsibilities for all U.S. strategic sealift.

[34] Break-bulk ships, as opposed to those constructed to carry containers, transport undifferentiated dry cargo of various sizes and shapes. They suit military needs very well. "Tramps" are owned by independent operators, who haul cargoes of opportunity without regard for schedules or routes.

[35] Containers are rectangular steel boxes, eight feet square in cross-section. Standard lengths are 20, 24, 35, and 40 feet. Each can be carried on a trailer, railroad flat car, or in a barge. They stack vertically in container cells and on top of strong hatch covers, side-by-side, so that loading is simple and no space is wasted.

[36] Approximately 400 dry cargo ships were needed to sustain operations in Vietnam, because unloading and clearance facilities were inadequate. Consequently, scores of merchantmen constantly anchored offshore or quequed in Thai harbors waiting their turn, extending turnaround times. New container ships transferred cargo to old self-sustainers or LSTs at Cam Ranh Bay. Many of those useful craft, that saw yeoman service, have since been retired. Chase, "U.S. Merchant Marine—For Commerce and Defense," pp. 133-34; Brown, *United States Military Posture for FY 1978*, p. 98.

Even if the U.S. Merchant Marine were structured perfectly, potential problems would still exist, given existing legal limitations. Emergency callups are politically sensitive matters, and in prolonged conflict they could weaken this country's already poor competitive commercial position by diverting ships for defense.[37] Sharp distinctions consequently are made between major wars and minor contingencies. No emergency requisitioning, for example, ever occurred in the Vietnam War, although the President had such powers. As a direct result, civilian ships offered for charter were often second class in terms of requirements.[38]

National Defense Reserve Fleet Marginal abilities of the active Merchant Marine to satisfy national needs (with or without compulsory callups) reinforce reliance on the National Defense Reserve Fleet (NDRF), which moved 40 percent of all military cargo to Vietnam during peak periods.[39]

Capabilities, however, are minuscule, compared with those in the past. The "moth-balled" fleet of World War II cargo ships, which once numbered thousands, is now reduced to 139 that are worth reactivating. Law permits owners to trade in aging but effective Mariner class ships for equal value in clunkers they could sell for scrap, but few so far have taken advantage, so the NDRF rusts and rots.[40]

Those still seaworthy are small and slow,[41] but their self-sustaining characteristics and ability to accept outsize items like

[37] The Merchant Marine Act of 1936, as amended in 1970, prescribes a fleet "capable of serving as a naval and military auxiliary in time of war or national emergency," declared by Presidential proclamation. In such case, "it shall be lawful for the Secretary of Commerce to requisition or purchase any vessel or other watercraft owned by citizens of the United States, or under construction in the United States, or for any period during such emergency to requisition or charter the use of any such property." Title 46, Sections 1101 and 1242, United States Code.

[38] Lane C. Kendall, "Capable of Serving as a Naval and Military Auxiliary . . . ," *U.S. Naval Institute Proceedings* (May 1971), pp. 213-15.

[39] *The National Defense Reserve Fleet—Can it Respond to Future Contingencies?, Report to the Congress by the Comptroller General of the United States* (Washington, D.C.: General Accounting Office, October 6, 1976), p. 2.

[40] Mariner class ships are newer, larger, and faster than Victory ships, but are similarly configured.

Chase, "U.S. Merchant Marine—For Commerce and Defense," p. 140; Kendall, "Capable of Serving as a Naval and Military Auxiliary . . . ," p. 226.

[41] Cargo ship speed is an important factor for fast turnaround times. Steaming at a steady 15 knots, a Victory ship would take four days plus four hours longer than a 20-knot C-4 Challenger to travel 6,000 miles.

tanks, trucks, locomotives, rolling stock, and harbor craft makes them easily adaptable for military missions. Perhaps even more importantly, requirements for break-bulk ships will remain critical until commercial containers are approved for ammunition carriage sometime in the early 1980s.[42]

Current DOD plans call for supplementary sealift from NDRF storage sites to report ready for duty within 10 to 15 days after American servicemen and/or materiel are committed to contingency operations, but refurbishment in fact would take 30 days or more in many instances. Drydock schedules are cramped. Labor skills are short in some shipyards. The passage of time will only aggravate the availability of repair parts, which already are scarce. Union assurances that sufficient qualified crews could assemble in short order may prove optimistic. About eight months reputedly would transpire before all NDRF Victory ships could pass muster if reactivation proceeded on a regular schedule. Time could be cut to something like three months if crash programs were implemented.[43]

A program now is afoot to refit 30 NDRF ships fast enough to satisfy 5- to 10-day force generation requirements.[44] Five ships from this Ready Reserve Force (RRF) would join the MSC Nucleus Fleet. The remainder would be chartered. That program, however, will not reach fruition until 1981.[45]

Effective U.S. Controlled Fleet The Effective U.S. Controlled Fleet (EUSC) of 314 ships, owned by Americans but flying flags from Liberia, Panama, and Honduras, offers a fallback position of sorts. Written agreements list which ships might reasonably be available in emergency. How responsive they would actually be is subject to argument, but their military value is minimal in any case, since all but 10 are tankers or bulk cargo carriers best suited for hauling petroleum products and natural resources.[46]

[42] *The National Defense Reserve Fleet—Can it Respond to Future Contingencies?*, pp. 3-4; telephone conversation with MSC planners on February 9, 1977.

[43] *The National Defense Reserve Fleet—Can it Respond to Future Contingencies?*, pp. 4-13, 32.

[44] The present list of 30 ships will change in size, composition, and capability if MSC receives approval to swap a few Victory ships for more modern models.

[45] Brown, *United States Military Posture for FY 1978*, p. 97.

[46] Chase, "U.S. Merchant Marine—For Commerce and Defense," pp. 140-43.

Other Foreign Flags Other foreign-flag ships completely beyond U.S. control are more than ample to meet America's major contingency requirements. Dependability, however, could be poor if perceived interests of the countries concerned (including NATO allies) fail to coincide during crises with those of the United States. Even if *owners* show good will, there is no certainty that alien *crews* will agree to traffic in war zones, with or without a big bonus.[47]

Salient U.S. Shortcomings U.S. strategic sealift suffers from insufficient ships that can assemble in acceptable times and carry military-type cargo to points where facilities are undeveloped or destroyed. Army armored and mechanized divisions, with many more vehicles than ever before, would impose immense strains on merchant shipping, not just for initial deployment but to withstand combat losses, which could be heavy if history is any indicator. The trend toward fewer ships, constructed essentially to carry containers, thus is inversely proportionate to military demand. Even modest attrition from Soviet attacks could cripple our abilities to accomplish essential security missions.

Amphibious Sealift

Soviet amphibious sealift is extremely limited (Figure 24). Active U.S. assets are still comparatively strong, although they dropped from 162 ships to 62, after reaching their apogee in 1967.

The residue is satisfactory for battalion- and regimental-size landings, but lift requirements of a single Marine division/wing team would absorb all but four operational ships scattered from Manila to the Mediterranean. Lead times for assembly would be long, and combat losses irreplaceable.[48]

[47] *Ibid.,* p. 140; Kendall, "Capable of Serving as a Naval and Military Auxiliary . . . ," pp. 209, 216.

[48] One Marine division with its associated air wing normally embarks on a minimum of 49 amphibious ships, whose composition is as follows:

Command/control ship (LCC)	1	Landing ship dock (LSD)	9
Amphibious assault ship (LPH)	5	Amphibious cargo ship (LKA)	5
Amphibious transport (LPA)	3	Landing ship tank (LST)	16
Amphibious transport dock (LPD)	10	Total	49

Current assets are limited to 62 ships (see Figure 17 for LPH), of which 15 percent are normally in overhaul at any given time. Marine Corps comments on the draft of this study, March 3, 1977.

FIGURE 24 AMPHIBIOUS SEALIFT
Statistical Trends and System Characteristics

	1970	*1971*	*1972*	*1973*	*1974*	*1975*	*1976*
United States							
LCC	2	4	3	2	2	2	2
LKA	12	7	6	6	6	´6	5
LPA	4	3	3	2	2	2	0
LPD	12	14	15	14	14	14	14
LPSS	1	1	1	1	1	0	0
LSD	13	12	12	13	13	13	13
LST	46	30	30	21	20	20	20
Total	90	71	70	59	58	57	54
Soviet Union							
LST	10	10	12	12	12	15	20
U.S. Standing	+80	+61	+58	+47	+46	+42	+34

NOTES: LCC = Amphibious Command Ship
LKA= Amphibious Cargo Ship
LPA = Amphibious Transport
LPD = Amphibious Transport Dock
LPSS = Amphibious Transport Submarine
LSD = Landing Ship Dock
LST = Landing Ship Tank

LCM = Landing Craft, Mechanized
LCPL = Landing Craft, Personnel
LCU = Landing Craft, Utility
LCVP = Landing Craft, Vehicle, Personnel
LVTP = Landing Vehicle, Tracked, Personnel

Soviet LSMs (Landing Ships, Mechanized) correspond more closely with U.S. landing craft than with amphibious ships, and are so listed. They currently have 65.

FIGURE 24 (continued)

AMPHIBIOUS SHIP CHARACTERISTICS

	Speed (Knots)	*Troops*	*General Cargo (Cu Ft)*	*Vehicles (Sq Ft)*	*Ammo (Cu Ft)*	*Fuel Drums*	*Bulk Fuel (Gal)*	*Booms, Cranes, (Tons)*	*Boats*	*Helos*
UNITED STATES										
LKA Rankin Class	16	138	42,518	21,798	39,988	600	None	4 35T 2 10T 6 5T	6 LCM-6 6 LCVP 2 LCPL 1 LCM-8	None
LPA Paul Revere Cl	20+	1657	135,457	10,132	11,471	905	5900 AVGAS	2 60T 1 30T 3 10T 2 8T 2 5T	7 LCM-6 10 LCVP 5 LCPL	1 CH-53
LPD Austin Class	20+	925	2176	11,127	16,660	MOGAS: AVGAS: AV-LUB: JP-5:	22,335 97,328 4,500 224,572	1 30T 6 4T 2 1.5T	2 LCPL 2 LCVP	2 CH-53
LSD Thomaston Cl	20+	341	N/A	8,754	3000	AVGAS or MOGAS Diesel	 12,000 39,000	2 50T	Ship's Boats 2 LCVP 2 LCPL Sample Loads: 3 LCU; or 19 LCM-6; or 9 LCM-8; or 48 LVTP	1 CH-53

NOTES: Each class is different. Ships above are currently in widest use.
LPD can cargo ammo or general cargo, but not both.

FIGURE 24 (continued)

	LST CHARACTERISTICS							
	Speed (Knots)	*Troops*	*General Cargo (Tons)*	*Vehicles (Sample Loads)*	*Ammo (Cu Ft)*	*Bulk Fuel (Gal)*	*Boats*	*Helos*
UNITED STATES								
LST 1179 Class	20+	431	500 (beach) 2000 (over LST Installed Causeway)	25 LVT, 17 2½ T Trucks or 21 M-60 Tanks, 17 2½T Trucks	2,552	254,000 Diesel 7197 MOGAS 134,438 AVGAS	3 LCVP 1 LCPL	1 CH-53
SOVIET UNION								
LST Alligator	15	375		26 Tanks				

	LANDING CRAFT CHARACTERISTICS			
UNITED STATES				
LCM-6	9	80	34	3 ¼-T Trucks or 1 2½ T Truck/Trailer
LCM-8	12	150	60	1 M-60 Tank
LCU 1610 Series	8	400	180	3 M-60 Tanks
LCVP	8	36	8100	1 ¼-T Truck/Trailer
LVTP	7	28	None	None
SOVIET UNION				
LSM				
Polnocny A	18	200		6 Tanks
Polnocny C	18	30		5 Tanks

ROAD AND RAIL

Except for one gravel road that links CONUS with Alaska, the United States has no overland routes to any prospective areas of possible confrontation with the Soviet Union, which relies heavily on road and rail lines that lead to NATO territory, the Indian-subcontinent, China, and the Middle East.

Soviet networks generally compare poorly with those in this country. Not many highways are paved. Trucks are plentiful, but maintenance is poor. Railways are still restricted, even though traffic has doubled in the last decade. Trains enroute to and from central Europe (or China, for that matter), must change wheels at the border, because Russian broad gauge tracks are incompatible with those of satellite states.[49] The process takes about two hours for a 20-car train. Nevertheless, Soviet land lines of communication constitute impressive mobility means that are more important than airlift and sealift for many missions.

COMBINED FLEXIBILITY

Composite Soviet mobility forces are sufficient to influence a range of low-key contingencies in widely separated areas, such as Angola and the Arab states, but airlift/sealift shortages are still strong limiting factors for major military operations almost anywhere outside the home country or contiguous satellites.

Quick and efficient logistic support for allies is a U.S. airlift specialty. MAC's squadrons also afford means of reinforcing forward deployed forces rapidly or shifting sizable combat power anywhere in the world. Apparent flexibility, however, is conditioned by the dearth of sealift, which makes it almost impossible to sustain major efforts without allied assistance. That dangerous combination calls for caution under most conceivable circumstances, since aid is by no means assured.

[49] Russian broad gauge tracks are 5 feet wide. Standard gauge tracks in satellite states are 4 feet 8½ inches.

NET ASSESSMENT APPRAISAL

John M. Collins has provided an excellent overview of U.S. and Soviet mobility forces, but mobility forces are only one aspect of strategic mobility—the overall ability to project military presence or influence overseas—and strategic mobility must in turn be related to strategic requirements.

DETERMINING STRATEGIC REQUIREMENTS

For the last three decades the strategic objectives of the United States have tended to be variations on the theme of containment. They have sought to keep the U.S.S.R. limited in influence and military capability to its own territory and that of its Eastern European clients. In contrast, the basic strategic objectives of the U.S.S.R. have been to sustain its internal security and the control of Eastern Europe, while breaking out of American strategic containment. This struggle has been motivated by conflicting regional political interests, by a major conflict in political system and ideology, and by conflicting economic and military interests. It has, from the narrow viewpoint of U.S. and Soviet strategic planning, created the world shown in Chart One.

This chart illustrates the fundamental difference between U.S. and Soviet strategic interests. The U.S. is secure along its own borders while the U.S.S.R. is not, but the U.S. has the disadvantage that it must attempt to contain Soviet interests at great distance from its own territory and along a Soviet sphere of influence that stretches throughout the world. It must do so in an era where air power, political and economic influence, and military sales make physical or narrow containment impossible.

In contrast, the Soviet Union is confronted at its borders by hostile, neutral, or unstable states. It is threatened in a way that the United States never can be, and its very size deprives it of most of the advantages of interior lines of communication. It is strategically threatened by NATO at one end of a continent and by China at another. Its internal strategic mobility problem is probably as serious as the world problem in strategic mobility which confronts the U.S.

CHART ONE

U.S. and Soviet Strategic Posture and Requirements for Strategic Mobility: 1977

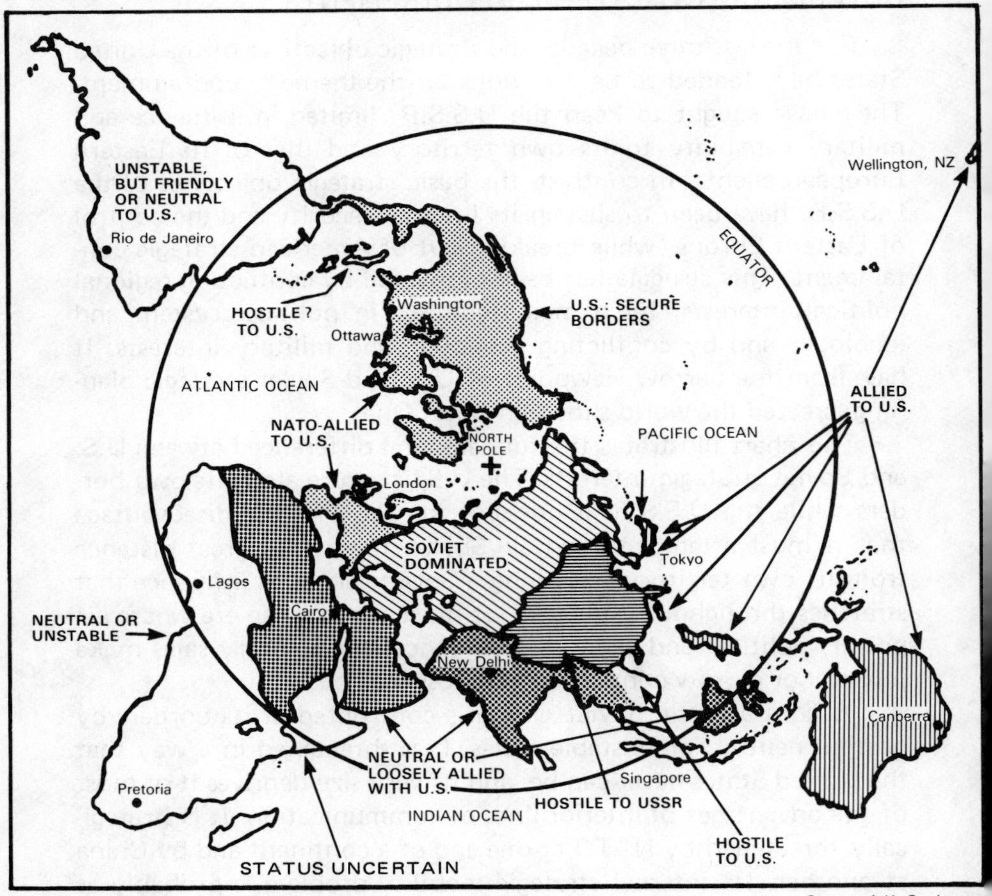

Source: A.H. Cordesman

TRADITIONAL MEASURES OF STRATEGIC REQUIREMENTS

These differences in strategic position lead to major differences in strategic requirements. The United States has sought to contain the U.S.S.R. by creating special relationships with the states along Soviet borders, and in the key areas where the U.S.S.R. might seek to break out of its containment. The treaty and mutual security agreements which have resulted create a wide range of requirements for U.S. strategic mobility.

However, three decades of strategic competition with the U.S. have allowed the U.S.S.R. to break out of the military straightjacket that the United States was initially able to impose after Soviet seizure of Eastern Europe. While Soviet overseas bases are often little more than minor service facilities, both sides have now created extensive basing structures. These are shown in Chart Two.

At the same time, the U.S. also has major economic interests which it must seek to defend. These are the result of its far greater dependence on imports, world-wide supplies, and foreign trade to sustain its economy.

The U.S. and its Allies also have a massive and growing dependence on oil imports. Arguably, the U.S.S.R. may also be forced to become a major oil importer by the early 1980s, but this is uncertain. At this point, only the West is absolutely dependent on the flow of oil shown in Chart Three.

In very condensed form, this overall structure shapes the traditional requirements for U.S. and Soviet strategic mobility and mobility forces. There is nothing stable about this structure for either the U.S. or U.S.S.R. Third world alignments and commitments have proved highly volatile, and it is far from clear that U.S. or Soviet ties to such nations have generally benefitted U.S. or Soviet interests. Foreign basing postures are also subject to sudden political changes. For example, the U.S. is now actively re-examining its future commitments to Korea, Taiwan, and the Philippines. Both power blocs are increasingly dependent on economic imports, and, in the future, the U.S.S.R. may be dependent on imports in many areas other than oil. It is a cliche in geopolitics that nations do not have "permanent allies, but only permanent interests." Yet, this is scarcely a truism. Even a brief review of the past three charts

CHART TWO
MARITIME BASES AND FACILITIES

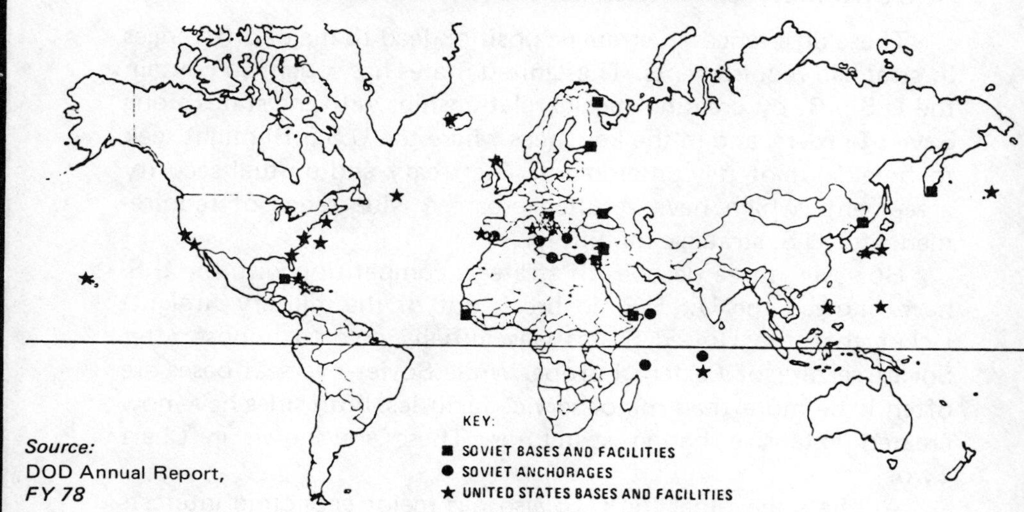
KEY:
SOVIET BASES AND FACILITIES
SOVIET ANCHORAGES
UNITED STATES BASES AND FACILITIES
Source:
DOD Annual Report,
FY 78

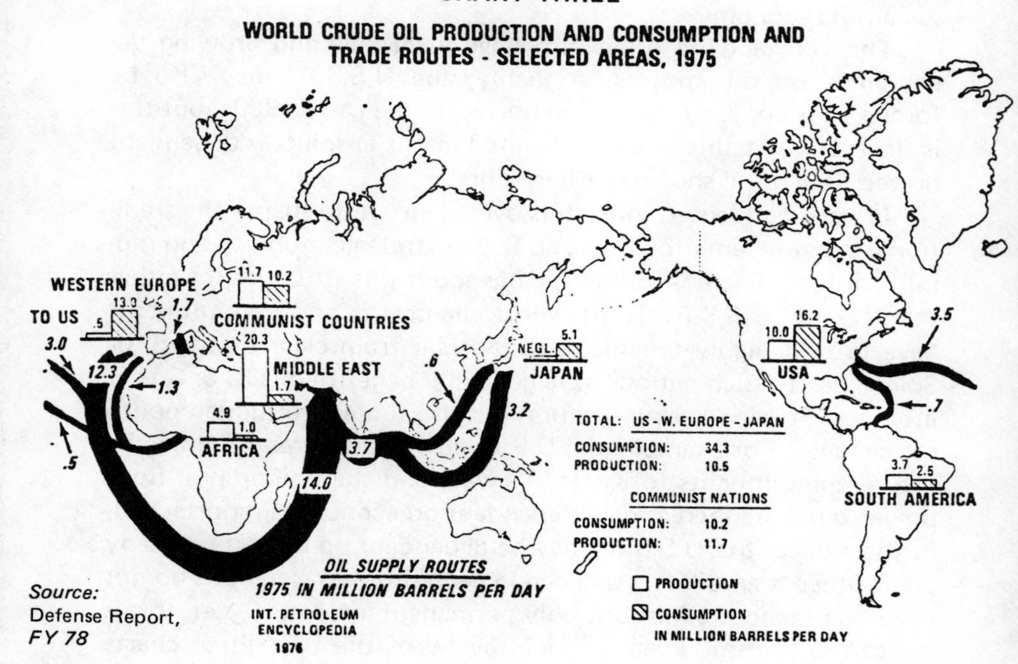
CHART THREE
WORLD CRUDE OIL PRODUCTION AND CONSUMPTION AND
TRADE ROUTES - SELECTED AREAS, 1975
WESTERN EUROPE
13.0
.5
TO US
1.7
11.7
10.2
COMMUNIST COUNTRIES
3.0
12.3
20.3
MIDDLE EAST
1.7
1.3
4.9
1.0
AFRICA
.5
.6
NEGL.
5.1
JAPAN
3.2
3.7
14.0
10.0
16.2
USA
3.5
.7
TOTAL: US - W. EUROPE - JAPAN
CONSUMPTION: 34.3
PRODUCTION: 10.5
COMMUNIST NATIONS
CONSUMPTION: 10.2
PRODUCTION: 11.7
3.7
2.5
SOUTH AMERICA
OIL SUPPLY ROUTES
1975 IN MILLION BARRELS PER DAY
INT. PETROLEUM
ENCYCLOPEDIA
1976
PRODUCTION
CONSUMPTION
IN MILLION BARRELS PER DAY
Source:
Defense Report,
FY 78

shows that "permanent" interests can change as well as allies, and have done so constantly during the last thirty years.

A few major consistencies have, however, emerged out of this complex and dynamic structure of global commitments. These are summarized in Chart Four.

The problem with the list in Chart Four is that neither the U.S. nor the Soviet Union is forced by its "permanent interests" into ultimate conflict or confrontation, but both powers have conflicting interests at lower levels of confrontation everywhere in the world. There are endless potential demands on both powers for strategic mobility and military presence, and these can shift without warning from decades of indifference to moments of crisis.

NON-TRADITIONAL REQUIREMENTS FOR STRATEGIC MOBILITY

It is the erratic nature of such confrontations that makes it so difficult for both the U.S. and U.S.S.R. to set requirements for alliances, strategic presence, or mobility forces. It is terribly easy for both powers to suddenly reach a crisis over highly marginal interests. It is also dangerous, however, for either power not to act. Inaction may be seen as weakness and exploited elsewhere, or lead to a disadvantage if the trivial escalates into the important. The relative balance of U.S. and Soviet strategic requirements is also so difficult to interpret in most specific cases that it is hard for both sides to reach a *modus vivendi* around inaction even when they try.

The winner of a U.S. and Soviet confrontation sometimes appears to be about as lucky as a lunatic who has struggled desperately to bury himself up to his neck in an ant hill; the loser appears to be another lunatic trying desperately to dig the first one out and replace him. Some of the current "anthills" might lead to U.S. and Soviet incidents or affect U.S. mobility forces; there is no way to predict their future importance.

Unfortunately this struggle is an unavoidable reality of current international politics, and the erratic confrontations between the U.S. and the U.S.S.R. are ultimately the result of conflicts between "permanent" interests. Basic changes must occur in the political postures of both states, and their allies, before "detente" can be turned into military stability.

This grim reality can be illustrated all too well by an analysis

CHART FOUR

Major U.S. and Soviet Requirements for Strategic Mobility: "Permanent" Requirements

U.S. "Permanent" Interest	Soviet "Permanent" Interest
• Defense of Europe	• Defense of the USSR
• Defense of Japan	• Continued control of Eastern Europe
• Survival of Israel	• Elimination of U.S. Containment
• Control of Atlantic, Mediterranean, and Pacific lines of Supply	• Elimination of NATO and China as hostile power structures
• Assured Supply of Oil and Critical Imports; Assured movement of Foreign Trade	• Advancement of Soviet Aligned Communism on World-Wide Basis
• Stable Relations with Canada and Mexico, and Alliance or Neutrality in Latin America	• Neutrality, or hostility to U.S., from Other Powers
• Stability and Neutrality or Alliance from Major Oil Producing States	
• Neutrality or Friendship from Other Powers	

Source: A.H. Cordesman

of the confrontations that have involved U.S. and Soviet military power during the last thirty years. A recent analysis by Barry M. Blechman and Stephen Kaplan of the Brookings Institute has provided an outstanding analysis of U.S. and Soviet involvement in use of military forces as a political instrument.[1] They compared

[1]Barry M. Blechman and Stephen S. Kaplan, "The Use of the Armed Forces as a Political Instrument," The Brookings Institution, December 31, 1976, Washington, D.C.

CHART FIVE

Patterns in Major Military Incidents Involving U.S. and Soviet Forces: 1946-1975

A. Distribution of Incidents by U.S. Administration

Administration	Average number of incidents per year in office
Truman (from January 1946)	5.0
Eisenhower	7.3
Kennedy	13.4
Johnson	9.7
Nixon	5.1
Ford (through October 1975)	4.3

B. Distribution of Incidents by Time Period and Region
Percentage of Total for Time Period

Time Period	Western Hemisphere	Europe	Middle East and North Africa	South Asia and Africa	Southeast and East Asia
1946-48	21	63	13	0	4
1949-55	13	29	8	4	46
1956-65	37	12	15	9	27
1966-75	13	15	32	4	36
1946-75	28	20	18	6	28

C. Soviet and Chinese Participation
Percentage of Total for Time Period

Time Period	Soviet Union	China
1946-48	63	4
1949-55	42	21
1956-65	28	14
1966-75	30	11
1946-75	34	14

CHART FIVE (Continued)

D. Principal Participants
Number of Incidents in Which each Participated

USSR	73	Israel	13
United Kingdom	35	National Liberation Front (South Vietnam)	13
China	30	Taiwan	12
Cuba	25	Turkey	12
North Vietnam	21	Organization of American States	11
France	17	Dominican Republic	10
Egypt	16	Greece	10
United Nations	16	Jordan	10
South Vietnam	14	Yugoslavia	10

Source: Adapted from Blechman and Kaplan

over 207 major incidents involving military forces between 1946 and 1975 and found the patterns summarized in Chart Five.

The figures in Chart Five give a clear picture of just how complex U.S. and Soviet requirements for strategic mobility can be, and how difficult they can be to relate to "permanent interests." This is also illustrated in Chart Six, which shows that regardless of the politics of the detente, there is no clear correlation between overall U.S. and Soviet relations and the number of potential demands on strategic mobility.

U.S. AND SOVIET MILITARY SALES, MILITARY AID, AND THE USE OF CLIENT OR PROXY FORCES

There is another trend which must also be considered in any net assessment of U.S. and Soviet strategic mobility. Both the United States and Soviet Union have been increasingly able to project military power without using their own forces.

CHART SIX

Interaction between U.S. and Soviet Relations and Military Incidents: 1945-1976

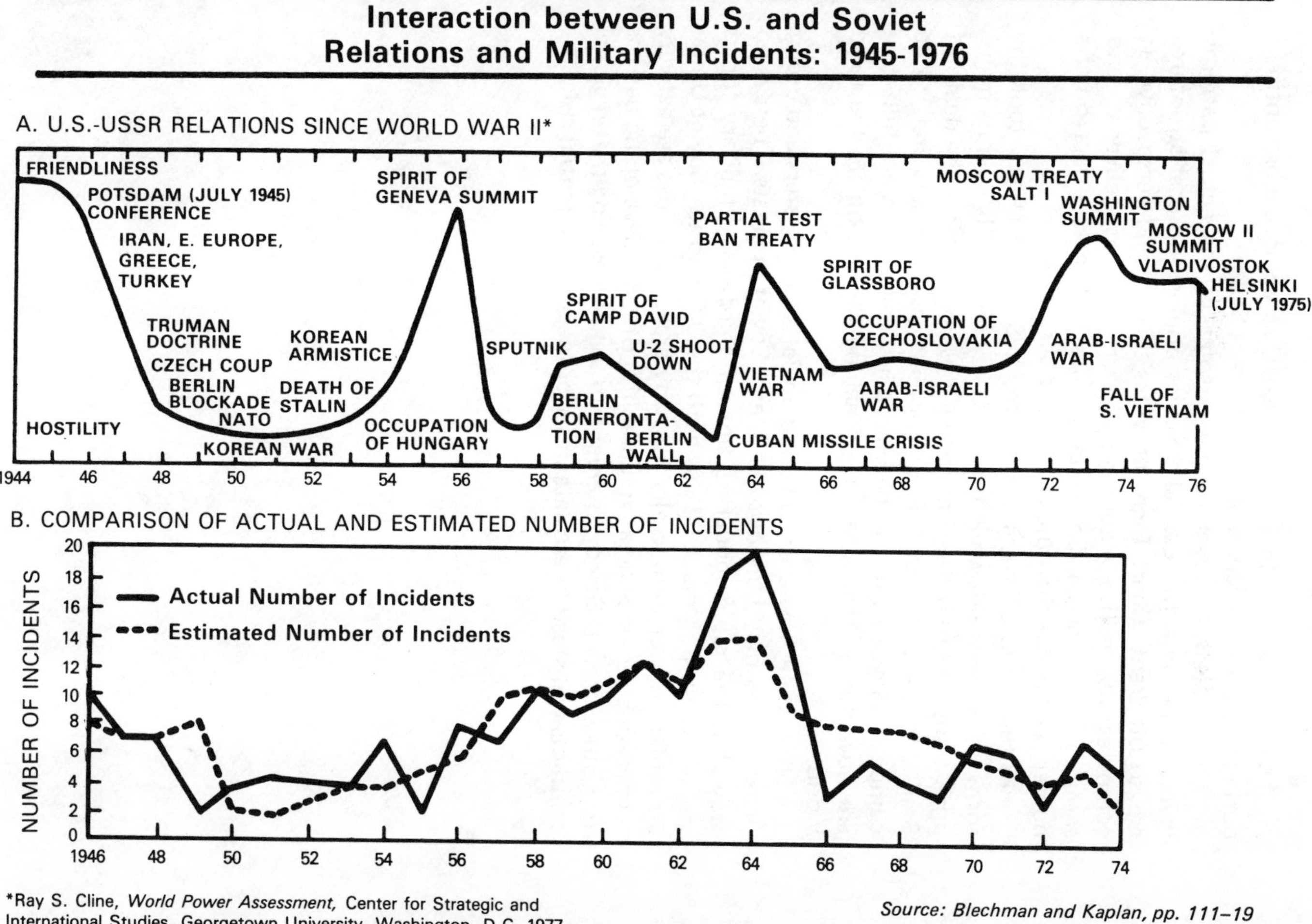

*Ray S. Cline, *World Power Assessment,* Center for Strategic and International Studies, Georgetown University, Washington, D.C. 1977

Source: *Blechman and Kaplan, pp. 111–19*

The vast growth in military sales has come almost entirely during the last ten years as shown in Chart Seven.

Chart Seven, however, disguises some of the actual patterns involved, because the U.S. and Soviet pricing and cost structures are so different. Chart Eight provides this missing perspective. It compares the military equipment each side has delivered, and shows that the actual Soviet effort has had more effect on the numerical balance than the U.S. effort has.

In the initial post war period, the U.S. effort largely took the form of military aid, while the U.S.S.R. consistently sold its major arms transfers from the time when these began after the death of Stalin. If the U.S. and U.S.S.R. sometimes have appeared to be competitors over who can be buried in a third world ant hill, they are now doing a splendid job of selling the ant hill to the third world.

Ultimately, however, U.S. and Soviet arms transfers and use of proxy forces may hurt both U.S. and Soviet interests. There is no way that the rapid modernization of third world military forces can avoid limiting the ability of both the U.S. and the Soviet Union to project power unilaterally into any area. Both sides are actively eliminating the once great vulnerability of most developing nations to both U.S. and Soviet intervention, while ensuring that regional conflicts will steadily escalate in seriousness and potential risk.

CHART SEVEN

Arms Transfer Agreements to Third World by Suppliers 1972-1976

Source: CIA

CHART EIGHT

Export of Major Weapons to Developing Regions by Major Suppliers, Cumulative 1971-75

Equipment Type	Total	United States	Soviet Union	France	United Kingdom
		(Numbers of Weapons)			
Land Armaments					
Tanks and Self-Propelled Guns	10325	3560	5220	440	1105
Artillery	3420	785	2550	80	5
Armored Personnel Carriers and Armored Cars	10435	5240	4190	780	225
Naval Craft					
Major Surface Combatants	84	63	6	—	15
Minor Surface Combatants	232	87	35	32	78
Submarines	33	22	6	2	3
Guided Missile Patrol Boats	37	—	33	4	—
Aircraft					
Combat Aircraft, Supersonic	2253	593	1385	275	—
Combat Aircraft, Subsonic	745	460	180	—	105
Other Aircraft	820	440	100	75	205
Helicopters	1150	460	380	265	45
Missiles					
Surface-to-Air Missiles	6630	1850	3950	240	590
Air-to-Air Missiles	2255	2155	—	50	50
Air-to-Surface Missiles	6260	6030	—	230	—

Source: WORLD MILITARY EXPENDITURES AND ARMS TRANSFERS, *1966-1976, U.S. ACDA, Table VI, page 81*

[Chapter 6]

NATO and the Warsaw Pact

FUNDAMENTAL FOCUS

Previous sections portray U.S. and Soviet armed forces as separate entities. Complete assessment of comparative strengths, however, must take allies into account, especially in the European area, where NATO and the Warsaw Pact are in contiguous confrontation.

Estimating who would win or lose if war broke out is beyond the scope of this study, which simply surveys current trends and assesses consequent problems, paying particular attention to the crucial center sector, which is the strategic center of gravity and point of decision.[1] If defense fails there, NATO can forget the rest of Europe.

NATO'S CENTER SECTOR

West Central Europe, with which we have strong political, economic, military, technological, and cultural ties, rates second only to North America in strategic importance among regions of the Free World. U.S. interests there may indeed be vital,[2] for if the

[1] NATO's center sector is herein construed to include Denmark, the Federal Republic of Germany, France, the Low Countries, Luxembourg, and the United Kingdom.

[2] The only vital national interest, by definition, is survival. States cease to exist if they fail to safeguard that essential. Serious threats to survival therefore compel stringent countermeasures.

Soviets were able to add that prodigious source of strength to their present holdings, the power balance might shift so far in their favor that this country could not compete.

Comparative Force Postures

NATO is quantitatively outclassed by the Warsaw Pact in almost every category, and is losing its qualitative edge in several respects that count.

QUANTITATIVE COMPARISONS

NATO is outnumbered, not just on prospective battlefields, but in backup (see Figure 25). U.S. and German troops along the Iron Curtain are the only NATO contingents strengthened since 1970. Contributions from most other countries have been cut.[3] France plans to prune its forward-based forces by almost 15 percent in the immediate future.[4] Financial straits could soon cause further reductions in Britain's Army of the Rhine.[5]

As it stands, Soviet forces alone substantially outnumber NATO in most instances. Twenty-five Category I divisions along the Iron Curtain in East Germany and Czechoslovakia compare favorably with 24 NATO counterparts.[6] Soviet tanks, artillery, and aircraft of all types (except ground attack) in those two countries exceed NATO's total (see Figure 25).

Soviet reinforcements could reach the current line of contact more rapidly and in much greater numbers than forces from the United States, which contains nearly all of NATO's uncommitted

[3] U.S. and German forces account for two-thirds of NATO's divisions (17 out of 24).

The United States has added two brigades to Seventh Army since 1970. A second wing of F-111 aircraft will soon be stationed in England. A fighter wing in Germany will convert to F-15s. Its present complement of F-4s will relocate in that country. European Basing Public Announcement, News Release by Office of the Assistant Secretary of Defense (Public Affairs), October 27, 1976.

For other changes, see *The Military Balance, 1970-71* (London: International Institute for Strategic Studies, 1970), pp. 22-30 and *The Military Balance, 1976-77,* pp. 18-25.

[4] U.S. Congress, Senate, Report of Senator Sam Nunn and Senator Dewey F. Bartlett to the Committee on Armed Services, *NATO and the New Soviet Threat,* 95th Congress, 1st Session (Washington, D.C.: U.S. Government Printing Office, 1977), pp. 10-11. France remains a member of NATO, but its armed forces are not under NATO control.

[5] Robert Keatley, "NATO Retrenchment Threat May Hint New Round of European Woes," *Wall Street Journal,* October 28, 1976, p. 8.

[6] Category III divisions airlifted into Czechoslovakia could quickly replace five Category I divisions deployed in that restless satellite since the abortive 1968 uprising.

FIGURE 25 NATO'S CENTER SECTOR
Statistical Summary

	1970			1976		
	UNITED STATES	*SOVIET UNION*	*U.S. STANDING*	*UNITED STATES*	*SOVIET UNION*	*U.S. STANDING*
Personnel	240,000	750,000	–510,000	271,000	840,000	–569,000
Divisions						
Committed						
Armor	2	14	–12	2	14	–12
Other	3	13	–10	3	13	–10
Total	5	27	–22	5	27	–22
Ready Reinforcements						
Armor	2	14	–12	2	14	–12
Other	9	7	+2	9	7	+2
Total	11	21	–10	11	21	–10
Sub Total	16	48	–32	16	48	–32
First-Line Reserves						
Armor	2	0	+2	2	0	+2
Other	10	9	+1	10	11	–1
Total	12	9	+3	12	11	+1
Total Divisions	28	57	–29	28	59	–31
Medium Tanks	2065	7900	–5835	2120	9100	–6980
Tactical Aircraft						
Light Bombers	0	200	–200	0	100	–100
Fighter/Attack	180	700	–520	250	800	–550
Interceptors	0	800	–800	0	950	–950
Total	180	1700	–1520	250	1850	–1600
MRBM/IBBM	0	650	–650	0	550	–550

FIGURE 25 (continued)

	1970			1976		
	NATO	*WARSAW PACT*	*NATO STANDING*	*NATO*	*WARSAW PACT*	*NATO STANDING*
Personnel	1,099,300	1,190,000	-90,700	1,045,200	1,216,000	-170,800
Divisions						
Committed						
Armor	8	24	-16	8	24	-16
Other	15	28	-13	16	27	-11
Total	23	52	-29	24	51	-27
Ready Reinforcements						
Armor	2	14	-12	2	14	-12
Other	10	7	+3	10	7	+3
Total	12	21	-9	12	21	-9
Sub Total	35	73	-38	36	72	-36
First-Line Reserves						
Armor	2	2	Par	2	2	Par
Other	11	13	-2	11	16	-5
Total	13	15	-2	13	18	-5
Total Divisions	48	88	-40	49	90	-41
Medium Tanks	6535	14,500	-7965	6615	16,000	-9385
Tactical Aircraft						
Light Bombers	15	300	-285	185	100	+85
Fighter/Attack	1400	1000	+400	1250	1200	+50
Interceptors	350	1100	-750	375	1200	-825
Total	1765	2400	-635	1810	2500	-690
MRBM/IRBM	0	650	-650	0	550	-550

FIGURE 25 (continued)

NOTES: U.S./NATO committed divisions include all active divisions in NATO's center sector. Soviet/Warsaw Pact counterparts are limited to divisions in East Germany, Czechoslovakia, and Poland. All are Category I.

U.S./NATO ready reinforcements include all other active U.S. Army divisions, less one in Korea; one U.S. Marine division; and one British division in the U.K. Soviet counterparts are restricted to Category I and II divisions in the Baltic, Belorussian, and Carpathian Military Districts. There are no satellite state divisions in this class.

U.S./NATO first-line reserves include one active U.S. Army division; three U.S. Marine divisions; and one Dutch division. Warsaw Pact forces are Category III divisions, including those in the Baltic, Belorussian, and Carpathian Military Districts of European Russia.

U.S., West German, and Soviet divisions have increased in size since 1970. Three German divisions, for example, had only two brigades each at that time. All 12 now have three brigades. The British Army has the same total number of brigades as in 1970, but has added a division headquarters.

Most of NATO's increased strength is in separate brigades and regiments, which do not show on these charts. Some studies include them as "division equivalents." The British Military Balance, 1976-77, for example, counts 29 NATO divisions, including 11 armored division equivalents.

U.S./NATO medium tank statistics include 535 U.S. prepositioned stocks (POMCUS), plus 130 in divisions that serve as maintenance float. Only tanks now in place are shown.

Aircraft statistics exclude U.S. dual-based forces in CONUS.

Personnel strengths are active forces only for U.S./NATO, but include Soviet Category III Divisions.

NATO and Warsaw Pact comparisons include the United States and Soviet Union.

France is excluded from all calculations. Its divisions and tactical aircraft, which are not now under NATO's control, would be difficult to reintegrate into the current command structure in emergency.

Special mention: Every U.S. Army and Marine division, active and reserve component, is shown on these charts. The Soviet Union has 109 others, some Categories I and II. Many of those would be available for service in Central Europe if a crisis arose.

strength.[7] Reserves in European Russia could relieve two Category I divisions in Poland and four more in Hungary. Forward-deployed forces could be further strengthened on short notice from eight armies composed of 32 divisions and 6,800 main battle tanks that now are maintained in the Baltic, Belorussian, and Carpathian Military Districts.[8]

Friendly forces consequently are poorly prepared to absorb attrition, which could be awesome during early stages of an all-out war.

QUALITATIVE COMPARISONS

The statistical tale is just the tip of the iceberg. Improvements on the Soviet side are undermining NATO's long-standing qualitative superiority, which was compensatory strength.

Soviet ground combat firepower, mobility, and staying power have been beefed up. The best weapons face NATO forces, not the Chinese frontier. T-72 tanks are starting to arrive in considerable numbers. BMP-76 armored carriers, despite drawbacks, permit Soviet infantry to fight while mounted, while NATO must fight on foot.[9] New anti-tank missiles, mainly mobile, merge with armor formations and thus add to combat power. A shift is under way from towed to self-propelled artillery, which eventually will enable many battalions to support advancing armor more adroitly. Soviet guns in general not only outnumber, but out-range NATO's, and have higher rates of fire. Engineer bridging capabilities, unequalled in the world, suggest that Western Europe's wide streams could be crossed quickly.[10] Logistical shortcomings, once the weakest link in the Soviet chain, are being reduced systematically.[11]

[7] See chart on page 99 of *The Military Balance, 1976-77*. Most European reserves are scheduled simply to bring existing formations up to full strength (same source, p. 100).

[8] Erickson, *Soviet-Warsaw Pact Force Levels,* p. 69-71. Soviet divisions in Hungary are shown on Figure 32 below, which concerns NATO's south flank.

[9] Magnesium armor on BMPs proved disadvantageous during the Yom Kippur conflict. Gas tanks on the rear door are also undesirable. Even so, these armored fighting vehicles, with a 76 mm gun and Sagger anti-tank missiles, are superior to NATO's current armored personnel carriers, which have no firing ports in the troop compartment and are armed with a single machinegun. (German forces are the sole exception.)

[10] Egyptian troops using Soviet engineer bridging in October, 1973, crossed the Suez Canal in great strength and in far faster time than Israeli intelligence previously indicated was possible.

[11] Coverage was compiled from Erickson, *Soviet-Warsaw Pact Force Levels,* pp. 74-75; Benjamin F. Schemmer, "Soviet Build-up on Central Front Poses New Threat to NATO," *Armed Forces Journal* (December 1976), pp. 30-33; U.S. Congress, Senate, *NATO and the New Soviet Threat,* pp. 4-5; *The Military Balance, 1976-77,* pp. 101-02.

Mobile air defense systems, being emended and extended in light of Middle East war experience, free many fighter-interceptors for air superiority missions. Some frontal aviation regiments have 25 percent more aircraft than they did in the recent past. Concrete shelters and increased dispersal assure their security more fully than in previous years. New types of tactical aircraft afford a fourfold improvement in payload and range that would allow them to strike critical targets deep in NATO territory without first redeploying from peacetime bases well behind present frontiers. Armed helicopters are helping to bolster close air support abilities, which got short shrift in the 1960s.[12]

NATO never has matched Moscow's medium- and intermediate-range ballistic missiles (MRBMs, IRBMs), which are being supplemented or supplanted by a mobile model (SS-20) with MIRVs.[13] Increased stocks of tactical nuclear weapons, such as Scaleboards, Frogs, and Scuds, cover targets as close as 10 and as far as 500 miles from firing points.[14] Chemical warfare capacities of all kinds are considerable. NATO has no defense against such threats.

No change by itself is crucial, but emerging Soviet capabilities, with great stress on offensive shock power, create a new strategic environment when considered in combination. NATO obviously has made some significant progress during the seven-year period surveyed by this study. However, it failed to keep pace, because member states continually compromised on requirements.[15] Consequently, the overall balance has not been so lopsided since the early 1950s, before NATO's bulwark was complete.

Integrating Structures

Several shortcomings are common to both coalitions, but unity of command coupled with central position affords strengths to the Communist side that NATO has never been able to equal.

[12] Erickson, *Soviet-Warsaw Pact Force Levels,* pp. 74, 75-76; Schemmer, "Soviet Build-up on Central Front Poses New Threat to NATO," pp. 31-32; U.S. Congress, Senate, *NATO and the New Soviet Threat,* pp. 5-6.

[13] SS-4 and SS-5 MRBMs/IRBMs, installed around 1960, have ranges of roughly 1,200 and 2,300 miles respectively. Each, armed with a single one-megaton warhead, is sufficiently accurate to hit within a mile or less of its target half the time. SS-20s, which carry three MIRVs each, reportedly have CEPs that approximate 440 yards over 2,500 miles. William Beecher, "Portable Red Missiles Housed in 'Garages,'" *Washington Star,* January 9, 1977, p. 9; *The Military Balance,* 1976-77, p. 73.

[14] Erickson, *Soviet-Warsaw Pact Force Levels,* p. 69.

[15] *The Military Balance, 1976-77,* p. 97.

SHARED SHORTCOMINGS

Political infidelity, poor motivation, or both are commonly cited as Warsaw Pact weaknesses, but in fact afflict NATO as well. France, for example, has not been a full partner for ten years, during which time the alliance has suffered all sorts of ills, many associated with lack of space for dispersion. Relations reportedly are improving, but the two French divisions in Germany still have no operational sectors.[16]

Some forces on each side are poorly trained and equipped. Not even East and West Germany, the best of the lot, are armed as adequately as their senior partners (German armor is a salient exception).

Soviet logistic shortcomings, corrected to some extent, are still evident. How well existing systems could sustain deep armored attacks, which consume huge quantities of ammunition and POL, is widely questioned. NATO's inflexible setup, predicated on separate supply and maintenance arrangements for each member country, is equally suspect. Stock levels show little consistency. Cross-servicing capacities for a hodge-podge of materiel are exceedingly limited. Shortages between authorized and actual inventories of critical items are common, even for U.S. units.[17]

Neither side has an operational peacetime chain of command. The Warsaw Pact is mainly an administrative organization. How (indeed whether) it would perform under combat conditions is subject to speculation.[18] The conversion of NATO's complicated command structure to wartime footing would be time-consuming and exasperating, especially in emergency. Indeed, the whole problem of command, control, and communications (C^3) seems to impinge on NATO's survival prospects at least as much as the physical balance of forces.[19]

[16] *Ibid.*, pp. 98, 102. This study excludes French forces from all calculations.

[17] *Ibid.*, p. 102; U.S. Congress, Senate, *NATO and the New Soviet Threat,* pp. 13-16, 18, 20; Les Aspin, "A Surprise Attack on NATO: Refocusing the Debate," remarks in the House, *Congressional Record* (February 7, 1977), pp. H912-13.

[18] Erickson, *Soviet-Warsaw Pact Force Levels,* p. 67.

[19] NATO presently has just one operational (as opposed to planning) headquarters, which is the nerve center for Allied Forces in Central Europe (AFCENT), at Boerfink, West Germany.

The DOD Director for Net Assessment expressed special concern for NATO's C^3 problems in comments on the draft of this study, March 1, 1977.

WARSAW PACT STRENGTHS

Soviet forces furnish a far greater share of Warsaw Pact strength than U.S. counterparts contribute to NATO.[20] Close integration is enhanced, because strategy, tactics, and command decisions derive directly from the Kremlin. Advantages are apparent, compared with the Atlantic Alliance, which relies on committee decisions in times of crisis before acting on compromise plans.[21]

The Soviets provide Warsaw Pact cohorts with standard arms and equipment that reduce costly duplication of R&D efforts, simplify logistic support, and foster flexibility. NATO, in sharp contrast, is afflicted with incompatible accoutrements and supplies. Neither ammunition nor repair parts are readily interchangeable between combat forces of different countries that share common causes and boundaries. Parochial national interests preclude early resolution of resultant problems.[22]

In compilation, the Soviet side, with interior lines and strategic initiative, displays military advantages denied NATO's loose coalition of fifteen nations.

Current Threats

Soviet power alone would pose serious potential threats to NATO's center sector, even if most satellite forces were pinned down for local security and air defense purposes.[23]

[20] The United States consistently contributes about 10 percent of NATO's ground forces, 20 percent of its naval forces, and a quarter of its tactical air forces. An additional 50,000 American specialists (such as subordinate elements of Defense Communications Agency), are stationed in Europe, but are not controlled by U.S. European Command (EUCOM).

[21] U.S. Congress, Senate, *NATO and the New Soviet Threat,* p. 10.

[22] U.S. Congress, House, Report to the Committee on International Relations by the Congressional Research Service, *NATO Standardization: Political, Economic, and Military Issues for Congress* (Washington, D.C.: March 29, 1977), 58 pp.

A view which suggests that standardization has several drawbacks is described in John K. Daniels, "NATO Standardization–The Other Side of the Coin," *National Defense* (January-February 1977), pp. 301-04.

[23] John Erickson indicates that satellite ground forces "earmarked" to supplement Soviet troops are substantially less than those on full order of battle lists. All 6 East German divisions apparently play parts in Soviet plans, but only 9 out of 15 Polish divisions and 6 out of 10 in Czechoslovakia seem to have combat missions. The political reliability of these select forces may be less shaky than popularly presumed, according to Erickson, who points out that *military* elites in Eastern Europe have been most consistently (sometimes irrationally) loyal to Moscow. *Soviet-Warsaw Pact Force Levels,* pp. 67, 79.

WARSAW PACT CAPABILITIES

The Soviets, in concert with selected allies, could exercise all or part of the following combat capabilities,[24] if they chose to run serious risks:

- Inflict catastrophic damage on the Continental United States with strategic nuclear weapons as a prelude to war in Europe.
- Invade Western Europe with little or no warning,[25] using air and ground forces now in East Germany and Czechoslovakia.
- Support conventional operations with tactical nuclear weapons targeted against NATO forces, airfields, ports, command/control centers, and supply installations.
- Challenge NATO for air superiority over Western Europe.
- Reinforce initial efforts rapidly with ready reserves in European Russia and Poland.
- Seriously inhibit reinforcement and resupply from the United States by interdicting trans-Atlantic air and sea lanes.
- Mobilize additional combat power.

SOVIET INTENTIONS

Capabilities just enumerated are tempered by Soviet intentions, which separate *possibilities* from *probable courses of action.*[26]

[24] Capabilities constitute the ability of countries or coalitions to execute specific courses of action at specific times and places. Fundamental components can be quantified and compared—so many tanks, ships, and planes with particular characteristics. Time, space, climate, terrain, organizational structures, and so on can also be calculated with reasonable reliability. Capabilities rarely are subject to rapid change. Technological breakthroughs, typified by the advent of atomic weapons, sometimes cause exceptions.

[25] One critic advances two uncommon arguments against a surprise Soviet attack, *Congressional Record* (February 7, 1977), p. H913. Neither is *necessarily* valid.

First, he contends, "It is hard to believe the U.S.S.R. would start a ground war without simultaneous attack at sea . . . . A sudden surge in [Soviet naval] deployments would tip us off." Exercises such as Okean-75, however, could camouflage intent, and sea control would be of reduced importance in any case if the Soviets decided to destroy NATO's ports.

In addition, he asks, "Would the Soviets not set in motion extensive and time-consuming [civil defense and other] preparatory measures before beginning a conflict that could rapidly escalate?" The Kremlin, however, could conclude that Assured Destruction threats are sufficient to deter the United States from defending NATO with strategic nuclear weapons, and therefore abstain from executing city evacuation plans.

[26] Intentions deal with the determination of countries or coalitions to use their capa-

History indicates that the Kremlin's hierarchy is essentially conservative, despite its revolutionary tradition. National character, communist doctrine, and unshakable convictions that time is generally on their side tend to repress impulses and reduce unwarrented risks. Political, economic, social, psychological, and technological competition have superseded naked force as policy tools since the Cuban missile crisis, although military power looms increasingly large as a possible option.

Bearing that backdrop in mind, premeditated Soviet attacks across the Iron Curtain, even for limited objectives, seem likely to occur only if Moscow entertains serious doubts concerning NATO's defense abilities and/or resolve. Even then, issues would have to be immediate and immense, unless Kremlin leaders believed *actual risks* were low in relation to *anticipated gains*. Whether those conditions will soon be satisfied is a matter of serious concern in the U.S. intelligence community and among net assessment specialists.

SOVIET MILITARY DOCTRINE

Soviet military doctrine suggests that the Warsaw Pact would have three main objectives if a major war should ensue: early destruction of NATO's defense forces, early occupation of NATO territory, and early isolation of Western Europe from its U.S. ally.[27] Unclassified analyses conclude that Soviet concepts for such operations stress surprise, shock, and quick exploitation.[28] Conventional and nuclear capabilities would be used in combinations suited to the occasion, without any scruples concerning collateral damage and casualties.[29] Employing nuclear arms is not

bilities in specific ways at specific times and places. Interests, objectives, policies, principles, and commitments all play important parts. National will is the integrating factor. Intentions are tricky to deal with, since they are subjective and changeable states of mind, but estimates of capabilities and intentions in combination are essential for decisionmakers who hope to design sound strategies.

[27] Thomas W. Wolfe, *Soviet Power and Europe, 1945-1970,* p. 456.

[28] Rumsfeld, *Annual Defense Department Report for FY 1977,* pp. 101-02; Eugene D. Betit, "Soviet Tactical Doctrine and Capabilities and NATO's Strategic Defense," *Strategic Review* (Fall 1976), pp. 95-107; John Erickson, "Trends in the Soviet Combined Arms Concept," *Strategic Review* (Winter 1977), pp. 38-53.

[29] The nature of many Soviet nuclear delivery systems, which stress missiles with large yields and low accuracy, raises grave doubts that the U.S.S.R. could contain collateral damage and casualties, even if it tried.

considered escalatory, since Soviet strategists contend that *political aims,* not *weapons systems,* establish the scope of war.[30]

NATO's Counter Strategy

NATO's common sense of purpose and associated policies form the framework within which strategic concepts must be shaped to counter Soviet threats.[31]

COMMON INTERESTS

Most U.S. interests in Europe coincide with those of our NATO allies, but emphases differ. Europe's survival and independence, for example, would be *directly* endangered by Soviet aggression. America's would not. Some choices that are seemingly open to us are not open to the rest of NATO. That condition has complicated the formulation of an agreed NATO strategy since the mid-1960s, when burgeoning Soviet nuclear strike forces caused West Europeans to question whether the United States would risk general nuclear war to satisfy interests that are not immediately vital. If Moscow ever entertained serious doubts concerning U.S. conventional commitments, NATO's credibility could be shattered.

DETERRENT/DEFENSE OBJECTIVES

To satisfy its security interests despite potential threats, NATO seeks to deter all forms of Warsaw Pact aggression, from encroachment to general war, and to defend NATO territory without serious loss or damage should dissuasion fail (Figure 26).[32]

Strategists in Western Europe understandably stress deterrence even more than their U.S. counterparts. Extensive hostilities on NATO soil would be "limited" from the U.S. standpoint, but could be lethal to our partners. Should war occur, *our* overriding objective would be to obviate damage to the United States. *Theirs* would be to safeguard Free Europe. Those schisms in defense priorities shape opposing schools of thought, whose views differ re-

[30] Erickson, *Soviet-Warsaw Pact Force Levels,* p. 69.

[31] Sections on NATO strategy depend primarily on John M. Collins, *Grand Strategy: Principles and Practices* (Annapolis, Maryland: U.S. Naval Institute Press, 1973), pp. 129-40.

[32] The three defense objectives in Figure 26 would apply equally if general war or encroachment occurred.

FIGURE 26 NATO'S DETERRENCE AND DEFENSE OBJECTIVES

Deterrent Objectives	*Defense Objectives*
Prevent General Nuclear War	Stabilize the Situation Expeditiously
Prevent Local Nuclear War	Repel Invaders
Prevent Conventional War	Limit Damage to NATO
Prevent Encroachment	

garding what stance would best ensure deterrence, and where the war should be fought if battle were unavoidable.

SUPPORTING POLICIES

Fundamental policies that shape NATO's military planning are summarized in Figure 27.

Obvious contradictions between policies and objectives, between various policies, between official policies and member state proclivities, and between NATO policies and Soviet military doctrine all cause compromises and increasing controversy.

STRATEGIC CONCEPTS

Policies constitute separate guidelines. Strategy is the concept of operations that ties them together.

The Switch to Flexible Response NATO's deterrent and defense posture originally was predicated on threats of massive retaliation against the U.S.S.R. in the event that the Warsaw Pact provoked a war in Western Europe. That simple, relatively low-cost strategy sufficed as long as U.S. nuclear capabilities were markedly superior to Moscow's. As the Soviets strengthened their position, massive retaliation gradually lost credibility as a deterrent. Worse yet, if deterrence foundered, massive retaliation guaranteed a general nuclear war which NATO could not "win" in any sense of achieving a favorable outcome.

A sweeping strategic reappraisal therefore culminated in the mid-1960s. Predominantly conventional defenses soon were deemed too expensive. Predominantly tactical nuclear defenses were deemed too unpredictable. Neither of those tacks could cope with a wide range of contingencies. After prolonged debate, a consensus

FIGURE 27 NATO'S DETERRENT AND DEFENSE POLICIES

Deterrence/Defense	*Burden-Sharing*
Limited War Second Strike Containment (not Rollback) Flexible Response Forward Defense High Nuclear Threshold Minimum Civilian Casualties Minimum Collateral Damage Central Control Non-provocative Posture Comprehensive Capabilities Lowest Credible Force Levels Heavy Reliance on: CONUS Reserves Mobilization	Fundamental Philosophy: An attack against one member is an attack against all, whether it occurs on the flanks or in the center sector. U.S. Provides: Primary Nuclear Capability Most Sea Power Substantial Air Power Substantial Land Power Europe Provides: Most Land Power Limited Nuclear Capability Limited Sea Power Substantial Air Power Installations and Facilities*

*United States pays construction costs in many cases.

Adapted from John M. Collins, *Grand Strategy: Principles and Practices,* (Annapolis, Maryland: U.S. Naval Institute Press, 1973), p. 131.

eventually prevailed in NATO councils that the low-option, low-credibility, high-risk strategy of massive retaliation was imprudent. In December, 1967, the Alliance therefore embraced a complex, costly strategy called flexible response, which could contribute credibly to deterrence and would afford multiple war-fighting options if a conflict erupted.[33] (See Figure 28 for a comparison of NATO's past and present strategies.)

CURRENT STRATEGIC SUMMARY

America's strategic retaliatory forces, with their Assured Destruction capability, provide the primary deterrent to general nuclear war between NATO and the Soviet Union (but do *not* similarly discourage Soviet use of tactical nuclear weapons, whose utility will shortly be shown).

NATO's strategy for limited war within its center sector contemplates a strong forward defense to repel invaders immediately or contain them as near the Iron Curtain as possible. That concept

[33] Robert L. Pfaltzgraff, Jr., *The Atlantic Community: A Complex Imbalance* (New York: Van Nostrand Reinhold Co., 1969), pp. 37-69.

FIGURE 28 PAST AND PRESENT NATO STRATEGIES COMPARED

	Massive Retaliation	*Flexible Response*
Type War		
Global; General	X	X
Regional; Limited		X
Main Theater of Operations		
U.S. - U.S.S.R.	X	
Western Europe		X
Main Objectives		
Deterrence	X	X
Defense		X
Options if Deterrence Fails		
Sustained Defense		X
Available Forces Only		X
Reinforcement		X
Conventional Only		X
Tactical Nuclear Assistance		X
"Tripwire" Defense	X	
Strategic Bombardment	X	X
Special Requirements		
U.S. Nuclear Superiority	X	
U.S. Nuclear Sufficiency		X
Local Air Supremacy		X
Sea Control		X
Strategic Mobility		X
Mobilization		X
Force Requirements		
Specialized	X	
Comprehensive		X

NOTES: General War is the last resort option of flexible response.
"Tripwire" forces are largely symbolic. Defensive *capabilities* may be considerable, but the *intent* is to trigger a massive response if the contingent is attacked.
Adapted from John M. Collins, *Grand Strategy: Principles and Practices* (Annapolis, Maryland: U.S. Naval Institute Press, 1973), p. 133.

demands sufficient versatility to cope with aggression at the most appropriate level on the conflict scale, and to escalate under full control if necessary. Nuclear weapons are held in reserve, ready for use whenever and wherever decision-makers decree. To execute its strategy successfully, the Alliance must gain and maintain air supremacy over Western Europe and control selected seas. Should NATO's standing forces prove insufficient, stiffening would come from ready reinforcements and strategic reserves.

In essence, NATO strives to deny the Soviets any hope of success unless they attack in such strength that compelling U.S. interests would be compromised and the risk of rapid escalation would be excessive.

NATO's Pressing Problems

Problems related to prompt and effective implementation of NATO's strategy are becoming so severe that some authorities are calling for a complete reappraisal of force requirements. The following coverage stresses NATO's weak spots.

CONVENTIONAL DEFENSE

The way in which NATO deploys its military power in peacetime is critical. Forces concentrated for conventional combat could expect unprecedented casualties if the enemy launched a nuclear war. Forces dispersed to escape the effects of nuclear weapons would be poorly prepared for classic defense. Compromise solutions are ill-suited for either environment.[34]

NATO presently is disposed for conventional combat, presuming that the Soviets would withhold nuclear weapons during the opening stage. Any surprise attack would be met, and repulsed if possible, by forces presently in place. If those elements were unable to stem the tide alone, they would strive to buy time for NATO to reinforce, make calculated decisions concerning escalation, or negotiate a solution.

Geographic Considerations Three strategically significant avenues of approach are available to the Warsaw Pact. The northernmost dead-ends at Hamburg, a shallow but lucrative goal. The most dangerous invasion routes traverse the broad North German Plain, part of a 1,000-mile corridor that cuts through NATO's center sector in transit from Russia to France. The third thoroughfare, in the south, follows the Fulda Gap through rugged uplands from Thuringia to the Rhine (see Figure 29).

NATO's Present Dispositions NATO's much-criticized dispositions athwart those three avenues result from historical accidents rather

[34] Carl H. Amme, Jr., *NATO Without France* (Stanford, Calif.: The Hoover Institution on War, Revolution and Peace, 1967), pp. 117-21.

Figure 29

Map of NATO's Center Sector

than strategic design. In large part, they parallel British, French, and American occupation zones at the end of World War II. The *Bundeswehr* shares responsibility with forces from Britain and the Low Countries for the critical North German Plain, but the United States, on the southern flank, still guards the most easily defended terrain. Amending maldeployments, by shifting U.S. ranks north or holding them in mobile reserve, might make military sense, but the cost of moving would be immense, and the diplomatic difficulties could prove discouraging.[35]

The Concept of Forward Defense The prescription for forward defense originally was a *political expedient* to ensure wholehearted participation by West Germany, which has persistently rejected any proposition that arbitrarily cedes German ground.[36] The objective, therefore, has always been to block major attacks and stabilize the situation quickly.

That task is imposing. The present line of contact would be difficult to defend, particularly along the flat northern plain, but forward defense has been a *military necessity* since 1967, when de Gaulle evicted NATO from France. The first sharp Soviet surge would sever friendly supply lines, which presently radiate from Bremerhaven, Rotterdam, and Antwerp, then run closely behind and parallel to the prospective front. Airfields also would be overrun.

NATO can no longer defend in depth, even if forward positions proved pregnable. Its forces formerly could fence with the foe all the way to the Pyrenees if necessary, along established lines of supply and communication. At West Germany's waist, the theater now is barely 130 miles wide, about the same distance that separates Washington from Philadelphia. Maneuver room for armies is at a premium. NATO forces and facilities are fearfully congested. Every lucrative military target, including command and control centers, airbases, ports, and supply depots, is within reach of Soviet IRBMs and MRBMs. An enemy breakthrough would compel NATO to retreat across Belgium toward Dunkerque or south toward the Alpine wall. Even if France invited NATO back in emer-

[35] U.S. Congress, Senate, *NATO and the New Soviet Threat,* p. 12.

[36] Carl H. Amme, Jr., "National Strategies Within the Alliance: West Germany," *NATO's Fifteen Nations* (August-September 1972), p. 82.

gency, many handicaps would remain, since facilities there have deteriorated or been dismantled.[37]

NATO's freedom of choice obviously would be constricted under present circumstances, and decision times compressed. How long the Atlantic Alliance could hold along the Iron Curtain would depend on a host of variables, including—but not restricted to—the nature of the conflict (nuclear or non-nuclear); the scale of Soviet attack (comprehensive or limited objectives); the amount of warning (hours, days, or weeks); the capabilities of opposing forces; NATO's will; and the weather. If strong enemy elements cracked through the crust, our main line of resistance could be enveloped, unless friendly forces quickly regrouped behind the unfordable Rhine, the first major defensible terrain feature to the rear.

TACTICAL NUCLEAR DEFENSE

If conventional defenses crumble, NATO plans to use tactical nuclear weapons, after consultation among its members. The time, place, and circumstances under which the Alliance would "go nuclear" have deliberately been left vague to complicate enemy planning.

Rationale For a High Nuclear Threshold Early resort to nuclear weapons theoretically could improve NATO's ability to sustain a strong forward defense, but a high threshold (crossed only after pressures became unbearable) would be salutary for several reasons.

Severe civilian casualties and collateral damage would be unavoidable if tactical nuclear weapons were exploded in large numbers. Limited target acquisition capabilities make it technically impossible to deliver ordnance infallibly onto stationary targets, let alone military forces on the move. Moreover, in a war for survival, the temptation to engage "suspected" targets would be high. Numerous deaths from accidental fallout probably would follow, even if both sides agreed to abstain from surface and subsurface detonations. Neutron weapons, available to NATO but not the Warsaw Pact, would do little to alleviate such problems.

Controls would be tenuous at best. Nuclear weapons could

[37] France has not undertaken any agreement to realign herself militarily with NATO. The use of French forces and territory in time of crisis would be subject to political decision. NATO planners therefore treat that possibility as one of many contingencies.

be administered very selectively—for defensive purposes only, on NATO territory only, against military targets only, using air bursts only or atomic land mines only, and low yields only—but none of those restrictions would be as readily distinguishable by the enemy as the "firebreak" between nuclear and conventional combat.

Since the first side to disregard arbitrary restraints might accrue a decisive advantage, the pressures to escalate would be enormous. Surprise Soviet ballistic missile strikes on key installations at the onset of a war could in fact confront NATO with the shocking choice between surrender and suicide, by blasting essential installations and blocking the arrival of reinforcements and resupply. The absence of missile defenses therefore constitutes a potentially fatal flaw for NATO, but not for the Soviet Union, which faces no similar threat.

Manpower Requirements Manpower requirements for tactical nuclear warfare might *exceed* those for conventional combat. NATO's forward defense forces have to be strong enough to make the enemy mass. Otherwise, Soviet assault troops would present few profitable targets. However, friendly formations would also suffer from nuclear attack, and attrition rates would be high. Eventual ascendance thus might be attained by the side with the greatest reserves of matériel and trained manpower.[38]

NATO unfortunately has few readily accessible reserves. All major ground combat forces are "on line." On-the-spot fighter squadrons are insufficient to perform assigned tasks. In exigency, the early augmentation of elements now in place therefore would be imperative. Their arrival, however, would depend on adequate warning, which might indeed be available, but is definitely not assured.

SPECIAL REQUIREMENTS

NATO's strategy of flexible response depends on several capabilities that were of reduced moment when massive retaliation was in vogue: air superiority, sea control, and strategic mobility.

[38] Alain C. Enthoven and K. Wayne Smith, *How Much is Enough?* (New York: Harper and Row, 1971), p. 125.

Air Superiority Freedom of action on the ground demands dominance in the air. NATO's aerospace defenses nevertheless are dangerously thin, when taken in context with Soviet threats. Revetments at air bases reduce dangers somewhat, but U.S. Hawk and Hercules batteries are short of missiles, and surprise attacks by nuclear-tipped IRBMs/MRBMs could neutralize friendly air power in parking areas, except for those on alert.[39]

Once aloft, NATO's tactical air forces confront masses of Soviet interceptors, SAMs, and air defense artillery, which shield static point targets and move with the troops. Coverage is close to comprehensive. NATO's countermeasures help, but the period is past when close support and interdiction missions can count on easy success. Corrective actions consequently seem imperative, because failure to achieve local air superiority "in the clutch" could cause NATO to lose land battles.[40]

Sea Control Reinforcement and resupply, now high priority projects, call for secure lines of communication from Western Europe to North America and the Middle East.

In the absence of armed escorts, allied shipping would be plagued by very heavy losses from submarine attacks. Essential avenues would have to be kept open indefinitely. Failure to do so could result in the collapse of NATO's defense, due to POL and other logistical starvation, even if the land battle stabilized. Protracted anti-submarine warfare operations would be essential before NATO could reduce losses to manageable proportions.

Controlling the entire Atlantic Basin would be a practical impossibility. Therefore, NATO practices defense in depth. During time of war, its fleets would take advantage of geographic "choke points" to help confine enemy naval forces, but Soviet submarines could circumvent that screen initially by infiltrating to patrol stations during peacetime.

Strategic Mobility To function effectively, NATO must be able to move immense amounts of men and materiel from the United

[39] U.S. Congress, Senate, *NATO and the New Soviet Threat*, pp. 15-16.

[40] *Ibid.*; Schneider and Hoeber, *Arms, Men, and Military Budgets*, pp. 174-75; Erickson, *Soviet-Warsaw Pact Force Levels*, p. 38; *Planning U.S. General Purpose Forces: The Tactical Air Forces* (Washington, D.C.: U.S. Government Printing Office, January 1977), pp. 24-27.

States to Europe on a continuing basis. The throughput capacity of ports and airfields at both ends is adequate, *provided installations in Western Europe escape early destruction.* Peacetime aerial ports would be supplemented in emergency by other military airfields suitable for transport aircraft and, if necessary, by civilian facilities (subject to political approval and the tactical situation). Benelux seaports that presently serve NATO would continue to do so in war. Either Rotterdam or Antwerp alone has sufficient capability to handle U.S. needs, but the threat of ballistic missile attacks and naval mine blockades casts a cloud over every base.

Future U.S. Force Requirements

Difficulties described above were severe when NATO planners presumed that any massive Soviet conventional assault would be preceded by lengthy preparations. Warning times appeared ample to mobilize and move CONUS-based reserves well before a shooting war erupted. Concepts which suited that situation, however, would ill-serve U.S. interests if the Soviets, as some claim, could launch a large-scale attack on moment's notice against NATO's center sector. Future U.S. force requirements therefore depend on early resolution of conflicting threat estimations and the doctrinal disputes that are now developing in Congress and the Executive Branch, including the influence of precision-guided munitions (PGM) on NATO's capabilities.[41]

NATO'S NORTHERN FLANK

NATO's far northern flank controls exits from and access to the ice-free Kola coast (see Figure 30), which houses immense Soviet submarine packs, plus more than half of all Soviet cruisers, destroyers, and ocean-going escorts.[42] If adjacent Norway fell into hostile hands, forward-based fighters and bombers could extend sea-denial capabilities far over the North Atlantic, to the wartime detriment of NATO fleets and merchant shipping. The strategic significance of upper Scandinavia to *both* sides thus is critical.

[41] Possible options are analyzed in J. V. Braddock and N. F. Wikner, *An Assessment of Soviet Forces Facing NATO—the Central Region—and Suggested NATO Initiatives* (Washington, D.C.: The BDM Corporation and the University of Miami, 1976), 85 pp.

[42] Erickson, *Soviet-Warsaw Pact Force Levels,* p. 72.

Figure 30
Map of NATO's North Flank

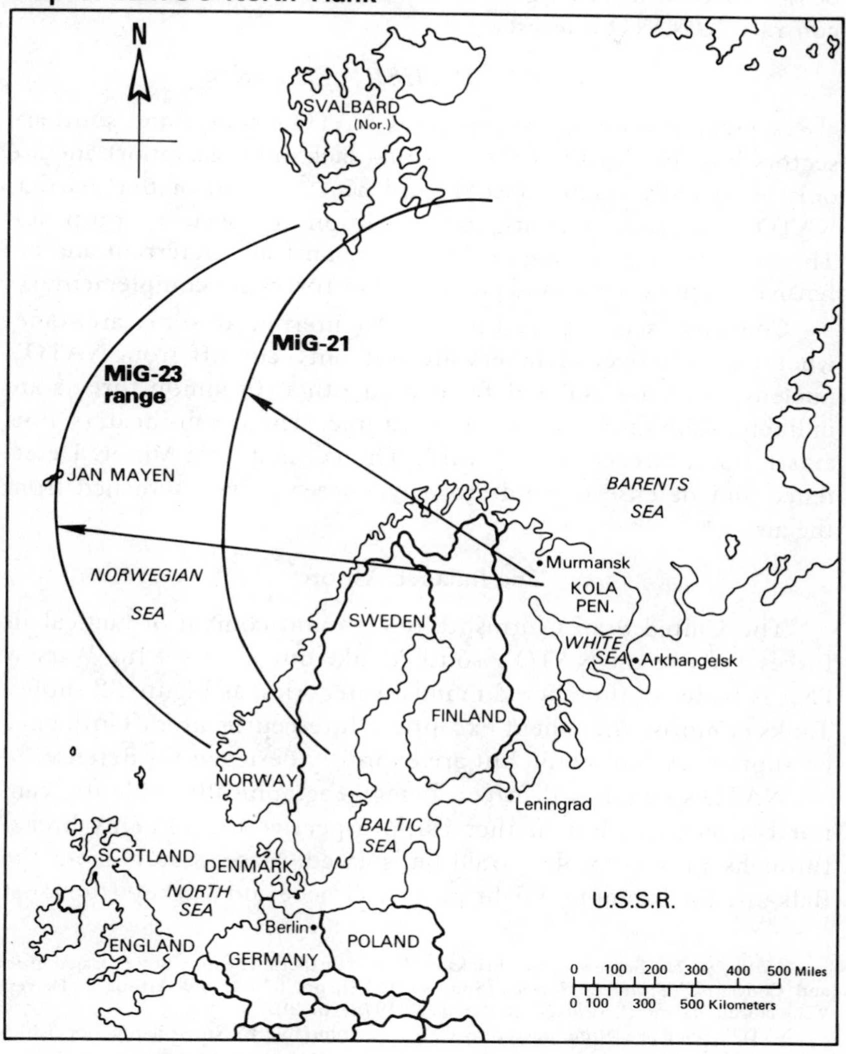

NATO, however, secures that sensitive area with a single Norwegian brigade, whose in-place opposition comprises two Soviet motorized rifle divisions and a naval infantry regiment backed by six more divisions (one being airborne) located near Leningrad. Soviet tactical air strength is substantial.[43] The implications are contrary to NATO interests.

NATO'S SOUTHERN FLANK

A great Alpine abatis separates NATO's center and southern sectors into two distinct theaters that lack mutual support and are only marginally related (see Figure 31).[44] North of that barrier, NATO comprises a contiguous coalition for security purposes. Threats against one state are threats against all. Deterrent and defensive schemes stress land power. Other forces are complementary.

Collective security measures of Mediterranean states are somewhat less cohesive. Members are not only cut off from NATO's nucleus, they are isolated from each other. Common threats are unlikely. Common fronts are infeasible. Three sub-theaters thus exist: Italy, Greece plus Turkish Thrace, and Asia Minor. Deterrence and defense depend strongly on sea power, screened from the air.

The Balance Ashore

The United States furnishes few ground combat or tactical air forces for use on NATO's south flank, but even so, the Warsaw Pact is badly outnumbered in most categories, as Figure 32 shows. Tanks comprise the salient exception. Interceptor aircraft influence air supremacy indirectly, but are expressly designed for defense.

NATO's land-based forces, being geographically separate, cannot concentrate, but neither can prospective foes. Enemy breakthroughs in any locale would be isolated. Mass assaults from the Balkans, for example, might menace Greece and Turkey (the most

[43] *Ibid.*, p. 73. See also John Erickson, "The Northern Theater: Soviet Capabilities and Concepts," *Strategic Review* (Summer 1976), pp. 67-82; and Arthur E. Dewey, "The Nordic Balance," *Strategic Review* (Fall 1976), p. 49.

NATO's poor peacetime posture is directly attributable to Norwegian policy, which permits no allied forces in the country, except in response to emergencies.

[44] NATO's south flank includes Italy, Greece (which has withdrawn from the Alliance militarily, at least for the moment), and Turkey. Opponents are primarily Bulgaria, Rumania, Hungary, and forces from southwestern U.S.S.R.

Figure 31
Map of the Mediterranean Basin

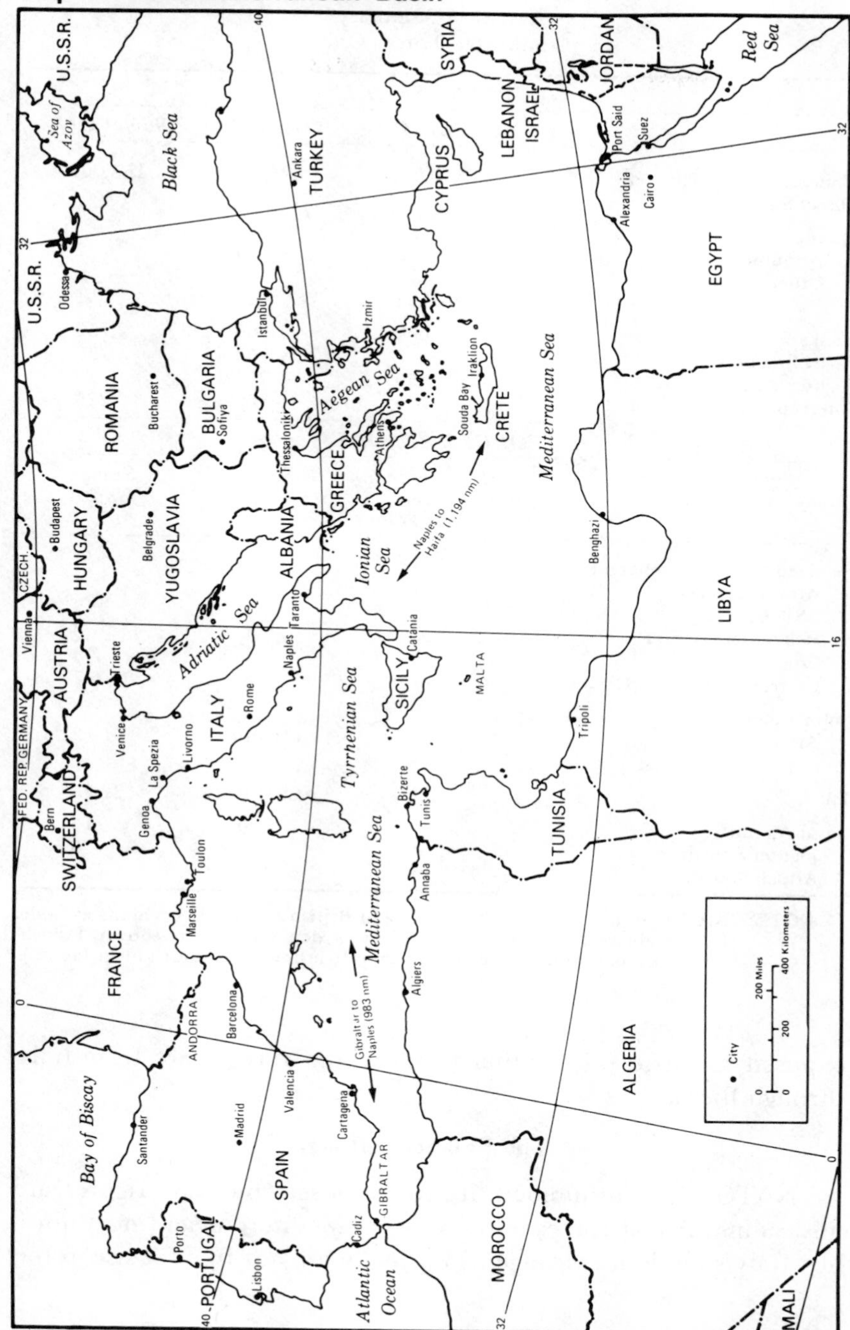

FIGURE 32 NATO'S SOUTH FLANK
Statistical Summary
(Committed Forces Only)

	FORCES ASHORE		
	NATO	*Warsaw Pact*	*Soviet Only*
Combat Troops, Plus Direct Support	540,000	395,000	155,000
Divisions			
Armored	2	5	2
Other	32	23	2
	34	28	4
Tanks	4,000	7,500	2,750
Light Bombers	0	50	50
Fighter/Attack Aircraft	450	225	75
Interceptors	275	350	225

	FORCES AFLOAT		
	Mediterranean Members of NATO	*U.S. Sixth Fleet*	*Soviet Union*
Selected Surface Combatants			
Attack Carriers	0	2	1
ASW Carriers	0	0	2
Helicopter Carriers	0	1	0
Cruisers	1	1	2-3
Destroyers/Frigates/Corvettes	66	15	6-7
Submarines			
Attack	32	Classified	11-12
Total	99	Classified	22-25
Naval Aircraft	0	200	15
Fighter Squadrons	0	4	0
Attack Squadrons	0	6	1

NOTE: NATO ground forces include U.S. and British units. Air strengths exclude U.S. dual-based squadrons. Normal naval deployments are shown. French forces are omitted. So are NATO aircraft in inventory, but not in tactical units.

exposed countries), but other states would stay secure from Italy through Iberia.

The Balance Afloat

NATO also outnumbers its rivals at sea, but raw figures are misleading. Allied forces in the western Mediterranean (most notably, Italy's modern contingent) normally are not free for use in the eastern basin, where the Soviets have significant surge capabilities, if the Turkish Straits stay open.[45]

Maneuver room is minimal in the Mediterranean, but Soviet submarines still are difficult to detect in those shallow waters, where thermal layers and many merchantmen confuse ASW devices by distorting sounds. Anti-ship cruise missiles also inject serious uncertainties into strategic equations. Peacetime contacts with Sixth Fleet are so close that U.S. reactions to sneak attacks might be measured in seconds.

Soviet logistic lines from the Black Sea are short but, being controlled by NATO at present, constrained. Moscow maintains no formal base rights in the Mediterranean, merely a presence. However, underway replenishment procedures are improving. Selected anchorages not only simplify resupply, but overlook every choke point from Suez to Gibraltar. Overall Soviet opportunities to compete with Sixth Fleet in the eastern Mediterranean are consequently impressive, especially if conflict were short.

The western Mediterranean is a much different matter. Soviet abilities to conduct combat operations in that area against numerically superior NATO are strictly limited, for short wars as well as long ones.[46]

Connections with Center Sector

Soviet breakthroughs along NATO's south flank would cause psychological shock waves to buffet the Atlantic Alliance, but the center sector could still stand.

Greek and Turkish armed forces defend a discrete region, nothing more. Reducing freedom of action for Soviet reserves in south Russia is their only direct connection with plans and operations in Western Europe. Airfields, NADGE installations,[47] and most com-

[45] Moscow massed 95 ships south of Turkey during the Arab-Israeli outburst of 1973 (Sixth Fleet totalled 60, including three attack carriers), plus 30 in the Indian Ocean, a spectacular feat for a force devoted to coastal defense in the recent past.

[46] For further background, see U.S. Congress, Senate, Report of Senator Gary Hart to the Committee on Armed Services, *U.S. Naval Forces in Europe,* 95th Congress, 1st Session (Washington, D.C.: U.S. Government Printing Office, 1977), 9 pp.

[47] A full discussion of U.S. bases is contained in U.S. Congress, House, Report prepared for the Subcommittee on Europe and the Middle East of the Committee on International Relations by the Congressional Research Service, United States Military Installations and Objectives in the Mediterranean (Washington, D.C.: March 27, 1977), 95 pp.

NADGE stands for NATO Air Defense Ground Environment, designed to provide early warning to air attack and direct interceptor actions.

munications sites are only significant locally. Aegean ports improve Sixth Fleet's posture in the eastern Mediterranean, but are not crucial beyond that basin.

If war ensued with the Warsaw Pact, far distant France would be free from fear of waterborne invasion if NATO held the Sicilian narrows. Italy would still be intact, subject to incursions only by airborne/amphibious assaults across the Adriatic (specialized Soviet sealift being in short supply), or along difficult axes in northern Yugoslavia. Assuming the Italian outlier fell, aggressors still would have to breach the Alpine obstacle before they could overrun NATO's heartland. All told, therefore, the Mediterranean seems an unlikely avenue for turning NATO's south flank, as so often alleged.[48]

[48] For further background, see T. R. Milton, "NATO's Troubled Southern Flank," *Strategic Review* (Fall 1975), p. 31.

NET ASSESSMENT APPRAISAL

Comparing NATO and Warsaw Pact capabilities is far more difficult than assessing the balance of U.S. and Soviet strategic forces. As Collins points out, there are sharply different force structures even in a given region on a given side. Equipment types differ as radically as their relative capabilities. There is little standardization of training, tactics, and support structures even within the Soviet dominated Warsaw Pact.

There are different ways that both sides can go to war, and each war could involve different levels of escalation. While it is common to talk about a general conventional war in the Center Region, there are different ways even this scenario could occur, and more ways that the forces involved can be counted.

Thus, the NATO "numbers game" is extremely complex. As a result, every expert tends to have his own view and force count.

These are almost impossible to refute, or even fully understand, without extensive access to the classified information.

The most that can be done to add to Collins's analysis, therefore, is to explain in more detail some of the major differences between experts regarding the NATO and Warsaw Pact balance, and to show some broad force trends which have recently become available in unclassified literature. I am particularly indebted to Dr. Phillip A. Karber for his help in providing much of this information.

DIFFERENCES IN U.S. AND SOVIET DEPENDENCE ON ALLIANCE WARFARE

Much of the discussion about the balance in the NATO Center Region tends to focus on basic numbers of men, divisions, tanks, and aircraft. The I.I.S.S. Military Balance provides some useful supplementary material summarized in Chart One.

DIFFERENCES IN DEPENDENCE ON ALLIANCE WARFARE

These I.I.S.S. figures closely approximate official U.S. data, and reflect the familiar Warsaw Pact advantage in men, tanks, and aircraft described by Collins. They also, however, reflect a basic strategic difference between the NATO and Warsaw Pact Alliances. NATO is an alliance of independent states. Only 23% of the manpower, 37% of the tanks, and 20% of the combat aircraft in the forward NATO countries of the Center Region are U.S. In contrast, 47% of the total manpower, 50% of the tanks, and 43% of the Warsaw Pact aircraft are Soviet. The United States is thus committed to all the advantages and disadvantages of true Alliance warfare while the U.S.S.R. is not. The U.S. must be a partner to survive. The U.S.S.R. can probably fight even after substantial defections by its Warsaw Pact Allies.

This essential strategic difference should allow the United States to safely maintain much smaller land and tactical air forces in proportion to its economic and manpower capabilities than the

CHART ONE

Summary I.I.S.S. Data on the NATO and Warsaw Pact Balance

A. Mainland Deployed Active Forces in the Center Region[1]

NATO	Manpower		Equipment	
	Ground	Air	Tanks	Aircraft
United States	193	35	2,000	335
Britain	58	9	575	145
Canada	3	2	30	50
Belgium	62	19	300	145
Germany	341	110	3,000	509
Netherlands	75	18	500	160
France	732	193	6,405	1,344
	50	—	325	—
Total	782	193	6,730	1,344

Warsaw Pact	Manpower		Equipment	
	Ground	Air	Tanks	Aircraft
Soviet Union	475	60	9,250	1,300
Czechoslovakia	135	46	2,500	550
East Germany	105	36	1,550	375
Poland	220	62	2,900	850
Total	935	204	16,200	3,075

[1] *Includes only French Forces in the FRG, no NATO forces in Denmark, France, and the UK, and no Warsaw Pact forces in Hungary.*

CHART ONE (Continued)

Summary I.I.S.S. Data on the NATO and Warsaw Pact Balance

B. Total NATO and Warsaw Pact Forces

	Northern and Central Europe[a]			Southern Europe[b]		
Ground Forces Available in Peacetime (division equivalents)[a]	**NATO**	**Warsaw Pact**	**CTR (of which USSR)**	**NATO**	**Warsaw Pact**	**(of which USSR)**
Armoured	10	32	22	4	6	2
Infantry, mechanized and airborne	17	38	23	33	27	9
Combat and Direct Support Troops Available (000)	630	945	640	560	390	145
Main Battle Tanks in Operational Service in Peacetime[b]	7,000	19,000	11,000	4,000	7,500	2,750
Tactical Aircraft in Operational Service						
Light bombers	150	125	125	–	50	50
Fighter/ground-attack	1,500	1,350	925	625	325	125
Interceptors	400	2,050	900	200	1,000	425
Reconnaissance	300	550	350	125	200	150

Adapted from: THE MILITARY BALANCE, *1977–1978, I.I.S.S., London*

[b]1976–1978 edition

Soviet Union. Moreover, the U.S. advantage in Europe should combine with a similar advantage in Asia. The U.S. is not threatened by China; the Soviet Union is.

Alliance warfare, however, imposes risks. The size of NATO forces disguises a lack of standardization and integration in virtually every meaningful aspect of military planning. There still are no common NATO tactics, few common items of military equipment, radically different force structures on the same front, and major differences in training and command and control. Even where standardization seems to exist, it is often an image rather than a reality. Most NATO ammunition, for example, differs significantly in precise performance capability. Guns calibrated on U.S. artillery shells and firing tables will not be accurate for many items of Allied artillery ammunition. The NATO "standard" tank gun has not led to a NATO standard tank round, and ironically, British and German tank ammunition is far superior in performance and reliability to U.S. made rounds.

In balance, however, the Warsaw Pact also shows much the same increasing divergence in force mix and equipment types. The accelerating modernization of Soviet forces during the last decade have steadily degraded the interoperability of the equipment held in the Warsaw Pact, and has systematically reduced commonality in Pact division structures and tactical air capabilities.

THE NATO DEPLOYMENT PROBLEM

These standardization problems are exacerbated by major problems in NATO deployments. They are the ultimate result of the occupation agreements after World War II, when the Center Region of NATO was divided into a British sector in Northern Germany, a U.S. sector in the south, and a small French sector in the southwest. As NATO grew and British strength declined, Dutch and Belgian forces were given responsibility for defense of the North German Plain—which they originally were to man with five to six divisions each, rather than the few brigades each side has today. West German rearmament then led to F.R.G. forces filling the gap as British, Belgian, and Dutch forces failed to approach their original NATO force requirements.

Finally, in 1966, France withdrew from the military planning activities of the Alliance and forced NATO to abandon most of its lines of communication (LOCs) through France. This left the United States defending less critical sectors of the F.R.G. front, and would leave the U.S. isolated from its new lines of communication through Belgium and the Netherlands if a Soviet attack were to by-pass U.S. forces and sweep through the North German Plains.

These ground force deployment problems were accompanied by similar problems in air and tactical nuclear forces. NATO national air forces adopted significantly different tactics and command and control structures. The attempt to create a NATO surface-to-air missile "belt" of Nike and Hawk forces failed to cover the entire Center Region effectively, and the attempt to create a NATO Air Defense and Ground Environment (NADGE) became a political football that then Secretary of Defense McNamara eventually came to treat with ill-concealed contempt. NADGE resulted in a system that was obsolete before it was built.

These problems were further compounded by NATO's "forward defense" strategy. This committed NATO to a rigid defense of Germany at the border or as far forward as possible. This in turn meant that NATO could not create the reserves and flexible deployments that would allow it to re-deploy forces across individual Allied national corps sectors, since it could not sacrifice space for improvements in deployment. Further, a political myth in NATO planning that all national corps sectors were equal, regardless of major imbalances in the firepower deployed per kilometer of front, meant that NATO could never objectively deal with its maldeployment problem.

In contrast, the Soviet Union systematically built up a ground and air force deployment structure which covers the entire central front. Soviet units are no closer to the front than NATO units—in fact, they average distances of 125 KM versus about 100 KM for NATO—but they are far better located and concentrated for movement. They are uploaded and much more movement ready, and they practice rapid movement, and dispersal into breakthrough or maneuver attack mode. NATO movements are far more complex, compete for the same route, and tend to rigidly lock NATO forces into their corps zone wartime deployment positions. NATO forces have negligible training in reinforcing each other across national corps zones, or making such movements under attack conditions. The current deployments on both sides are shown in Chart Two—Part A.

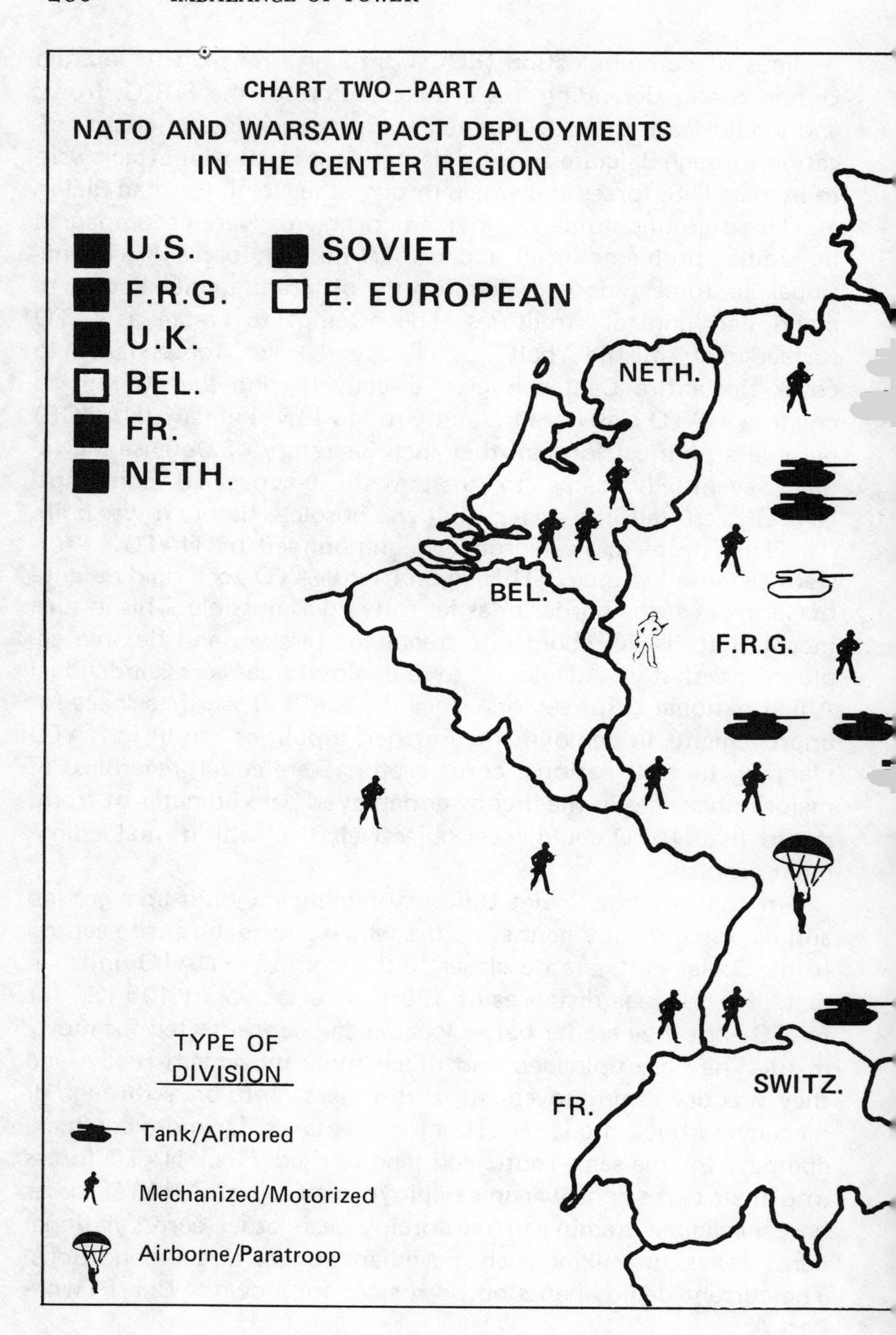

CHART TWO—PART A
NATO AND WARSAW PACT DEPLOYMENTS
IN THE CENTER REGION
U.S.
F.R.G.
U.K.
BEL.
FR.
NETH.
SOVIET
E. EUROPEAN
NETH.
BEL.
F.R.G.
SWITZ.
FR.
TYPE OF
DIVISION
Tank/Armored
Mechanized/Motorized
Airborne/Paratroop

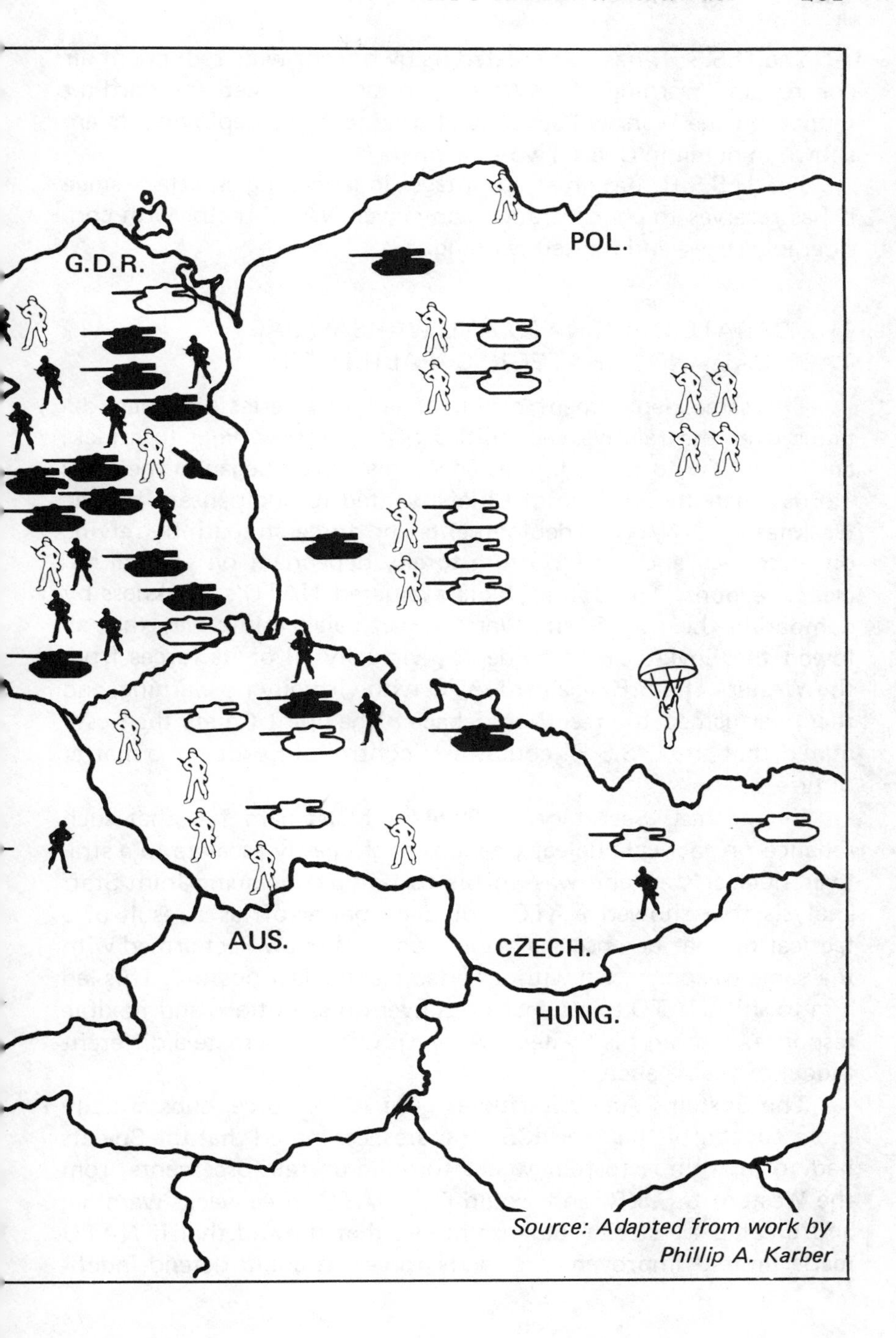

Source: Adapted from work by Phillip A. Karber

The U.S.S.R. has also created its own front-wide redundant air control and warning (AC&W) system, and enforced standard air tactics on the Warsaw Pact. Soviet ground force deployments are shown in detail in Chart Two—Part B.

The U.S.S.R. has great advantages in launching an attack since it has reserves to concentrate on any given NATO sector with considerable speed and limited warning.

THE DEBATE OVER NATO AND WARSAW PACT TACTICAL AND STRATEGIC CAPABILITIES

The same deployment problems led to a series of expert debates over the relative capabilities of the Warsaw Pact to attack, and of NATO to defend. In essence, this debate began in the early 1960s when the U.S. Joint Chiefs wanted to compensate for the weaknesses in NATO's deployments and force strength by relying on a strategy and force posture totally dependent on tactical nuclear weapons. The Joint Chiefs evaluated NATO's weakness by comparing the NATO and Warsaw Pact balance in a way that allowed the Soviet Union to deploy virtually all of its forces from the Western U.S.S.R. against NATO with virtually no warning, and then war gamed the resulting debacle. The Joint Chiefs then estimated that the U.S.S.R. could seize control of Europe in a matter of weeks.

In contrast, Secretary of Defense McNamara felt that such reliance on tactical nuclear weapons might easily escalate to a strategic nuclear war, and was unable to find a convincing Joint Staff analysis that showed NATO would be better off as a result of a tactical nuclear defense against a stronger Warsaw Pact armed with the same weapons than without a tactical nuclear posture. This led him to shift NATO to reliance on conventional options and flexible response, and led his Systems Analysis Office to create a different model of the balance.

The Systems Analysis studies gave Allied forces substantially more capability than the JCS estimates; estimated that the Soviets had to take three to four weeks to build-up reinforcements from the Western U.S.S.R. and would give NATO three weeks warning; and created a "static" defense model that showed that if NATO made limited improvements in its forces, it could defend indefi-

CHART TWO—PART B

Order of Battle GSFG

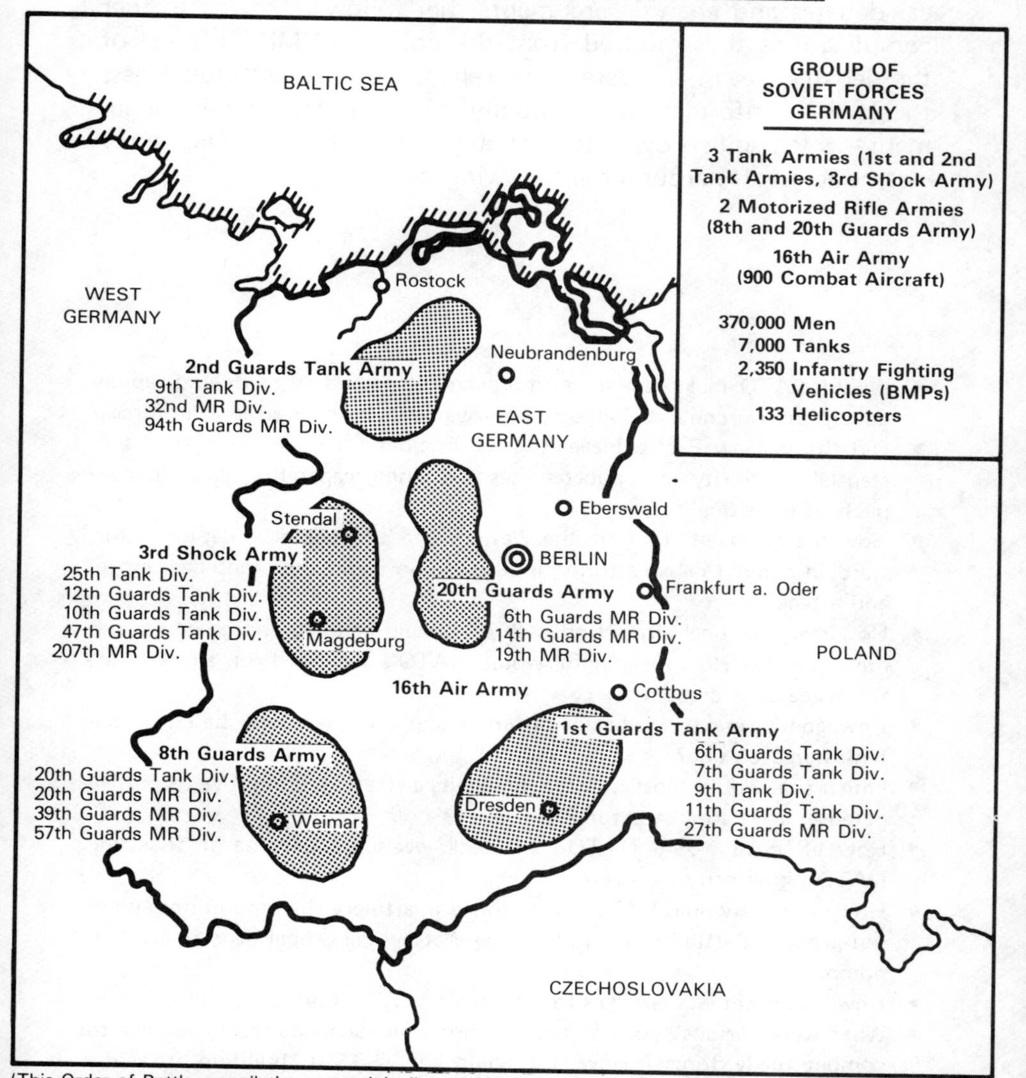

(This Order of Battle compilation was originally prepared for a study day of the Royal Corps of Transport, Scottish Command, and has been currently updated.)

Source: John Erickson, "Trends in the Soviet Combined Arms Concept," STRATEGIC REVIEW, *Vol. V, Winter, 1977, p. 8*

nitely because of a historical need by the attacker to develop a 2:1 or greater overall superiority to suppress a strong defense.

There is no way to summarize each of the decade-long statistical debates and analytic arguments that followed, or the further complications that resulted from the politics of MBFR. Most of the debates are, in any case, comprehensible only with full access to classified information. Essentially, however, this ongoing argument—which spilled over into virtually every other NATO country—came to revolve around the following issues:

- Would NATO make the force improvements necessary to defend rapidly enough to overcome the ongoing improvements in the Warsaw Pact forces?
- Did the Warsaw Pact achieve its large combat forces at the cost of substantial inferiority in support forces, sustaining capability, and other aspects of force quality?
- How many Soviet forces in the Western U.S.S.R. could be deployed forward in combat capable form; and how soon could they mobilize, move, and attack?
- How many tactical aircraft on each side could actually engage in war and how quickly? How meaningful would NATO's superiority in aircraft performance be in combat?
- How good were the radically different and unproven land based air defenses on each side?
- Which side would most quickly shelter its aircraft, and improve its passive defenses and basing structure?
- How effective would NATO's anti-tank weapons really be in offsetting NATO's inferiority in tanks?
- How effective would NATO's superiority in artillery shell munitions and in self-propelled artillery be in offsetting Warsaw Pact superiority in artillery numbers?
- How important was NATO's former superiority in armored mobility?
- What were various types of reserve forces on each side worth relative to combat ready forces? How fast could the U.S.S.R. build-up from the Western U.S.S.R. and how good was the result?
- How many war reserves and what ammunition stocks did each side have to fight with?

THE MATURING SOVIET THREAT: SURPRISE ATTACK, COMBINED ARMS, AND NEW UNCERTAINTIES

It is now a decade since the Office of the Secretary of Defense and the Joint Staff made its first real attempt to reconcile these issues as part of a National Security Council study, and one major new feature has been added to these debates since the late 1960s.

All experts agree that the Soviet Union has continued to build up its forces in the forward area every year, that it has steadily improved their quality as well as quantity, and that the U.S.S.R. has shown an outstanding improvement in the quality and flexibility of its tactical and strategic planning. Whether the U.S.S.R. has significantly outpaced NATO in its rate of force improvement is an area of debate, but some experts now believe that the improvements in Warsaw Pact forces have given it major new strategic capabilities. These views may be summarized as follows:

- The Soviet Union has now reached parity or superiority in armored mobility over NATO as a whole and has definite superiority in Armored Infantry Fighting Vehicle quality over the U.S. Army.
- The Soviet Union is nearly equal in tank quality.
- The Soviet Union has artillery which is superior in average range and rates of fire. It is correcting its past inferiority in self-propelled weapons; it has a major advantage in mass fire because of its superiority in multiple rocket launchers (MRL); it has superiority in artillery direct fire anti-tank capability; and it is correcting its past inferiority in artillery munitions, lethality, and fuzing.
- The Soviet Union has parity in anti-tank weapons launchers, and offsets its present inferiority in ATGM guidance technology and other performance characteristics with superiority in armored fighting vehicles which provide effective mobility and protection in the face of artillery suppression. New types of Soviet ATGM may correct the present performance problems in Soviet guidance and missile technology.
- The Soviets can acquire parity in tactical aircraft performance in the early 1980s, and can now achieve decisive advantages over NATO air forces in surprise air attack.
- The Soviets are superior in air defense weapons numbers and quality. This can offset any Warsaw Pact inferiority in air defense fighter performance.
- The Soviets are superior in battlefield intelligence, and in command and control.

- The Soviets can counter any NATO escalation to theater nuclear warfare (TNW) with more TNW forces, better TNW forces, better tactics, better targeting, and better command and control.
- Soviet forces have corrected any past defects in training, in sustaining capability, and support forces. They can move out of caserne in combat ready form with negligible warning, and with superior logistic mobility. They can actually reach NATO defensive positions before NATO's maldeployed forces can even occupy them.
- The Soviets have demonstrated superior capabilities to mass fires for a breakthrough that could decisively suppress a NATO forward defense, and a superiority in armored maneuver capability that could envelop and paralyze NATO forces as Germany paralyzed British and French forces in 1940.
- Both sides have grave defects in their present peacetime readiness, but the U.S.S.R. can choose its moment of attack, and correct many of its defects without warning NATO.
- The major deployment and readiness problems in NATO are so great that the Warsaw Pact could successfully attack out of caserne with virtually no warning and decisively breakthrough or overrun NATO before NATO could effectively organize its defenses. Further, the Warsaw Pact has a strong incentive to do so. The force ratio shifts in favor of NATO with every day it has time to mobilize, and the cost to NATO of escalation to theater nuclear war would be far greater if the Warsaw Pact penetrated through badly organized NATO defenses, intermingled with NATO defenses, and was located on NATO territory.
- Soviet capabilities to reinforce with units from the Western U.S.S.R. have advanced to the point where the U.S.S.R. can now move its rear divisions to the NATO front and commit them to combat in a matter of days. NATO has no guarantee of warning and cannot assume it will have a week to three weeks to get ready before combat begins.

TRENDS IN THE NATO AND WARSAW PACT BALANCE

Chart Three shows the broad trends in total NATO and Warsaw Pact manpower and division numbers in the NATO Guidelines Area (NGA). The NATO Guidelines area includes all NATO forces in Belgium, the F.R.G., Luxembourg, and the Netherlands, and all Warsaw Pact forces in Czechoslovakia, the G.D.R., and Poland. It excludes all forces in Denmark, France, the U.K. and Hungary. It shows that both Warsaw Pact manpower and division numbers have increased relative to NATO during the last decade.

CHART THREE

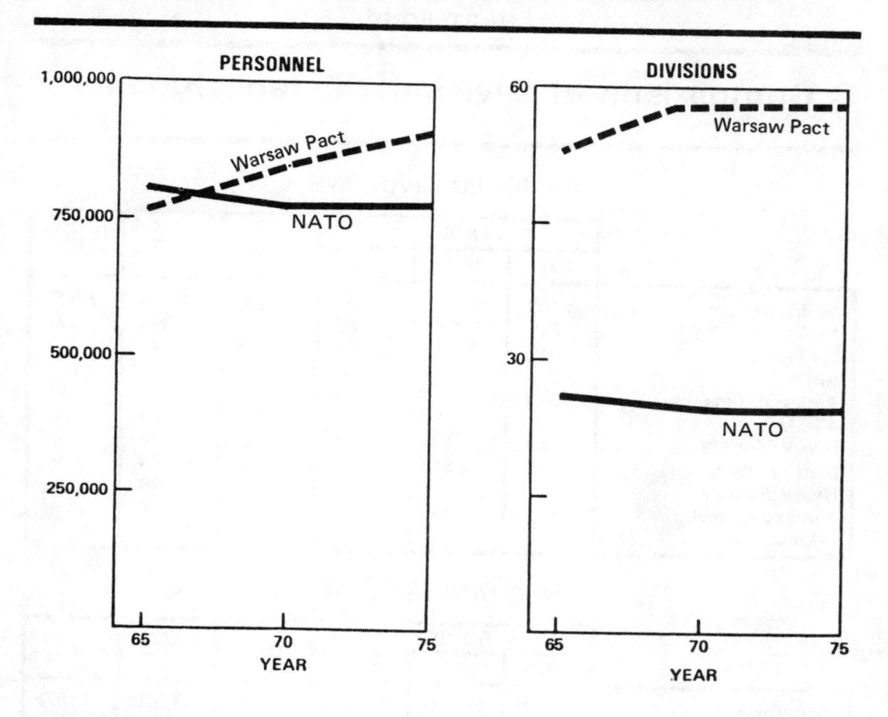

Source: Phillip A. Karber, "Evolution of the Central European Military Balance," Statement Prepared for Hearings on Western Europe: Military and Security Issues, Committee on International Relations, House of Representatives, June, 1977

Chart Four compares the changes in Soviet division structures during this period to contemporary NATO divisions.

Chart Five shows that Soviet divisions have gone from sharp inferiority in weapons numbers to near or actual parity in many areas. Many experts still argue, however, that Soviet and other Warsaw Pact divisions are inferior to NATO divisions in sustaining and combat support capability, and in their proportion of non-divisional support or "tail."

CHART FOUR

Comparison of Divisional Establishments

ARMORED DIVISIONS

	SOVIET			U.S.	FRG	UK
	1965	1970	1975			
Manpower	8,500	9,000	9,500	16,500	14,500	12,500
Tanks	316	316	325	324	300	300
Lt. Tanks	17	17	19	54	—	—
APCs	—	—	—	—	—	—
Antitank Guns	—	—		—	45	48
ATGM	9	9	105	370	29	30
Heavy Mortars	12	18	18	53	12	—
Med. Artillery	36	54	—	54	54	36
Heavy Artillery	—	—	—	12	18	—
Multiple Rocket Launchers	12	18	—	—	16	—

MECHANIZED DIVISIONS

	SOVIET			U.S.	FRG	UK
	1965	1970	1975			
Manpower	10,000	11,000	12,000	16,300	14,500	15,700
Tanks	175	188	255	216	250	162
Light Tanks	17	17	19	54	—	156
APCs	180	180	270	322	280	270
Antitank Guns	12	18	18	—	45	—
ATGM	18	36	135	426	34	48
Heavy Mortar	54	54	54	49	36	—
Med. Artillery	36	54	72-90	54	54	45
Heavy Artillery	—	—	—	12	18	—
Multiple Rocket Launchers	18	18	18	—	16	—

Source: Phillip A. Karber, op. cit.

CHART FIVE

U.S., FRG, and UK Division Strengths as a Percent of Soviet Divisions

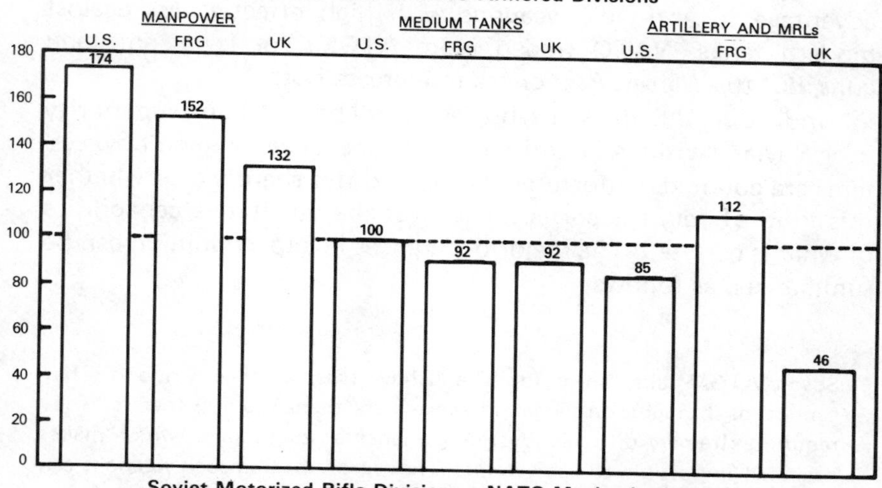

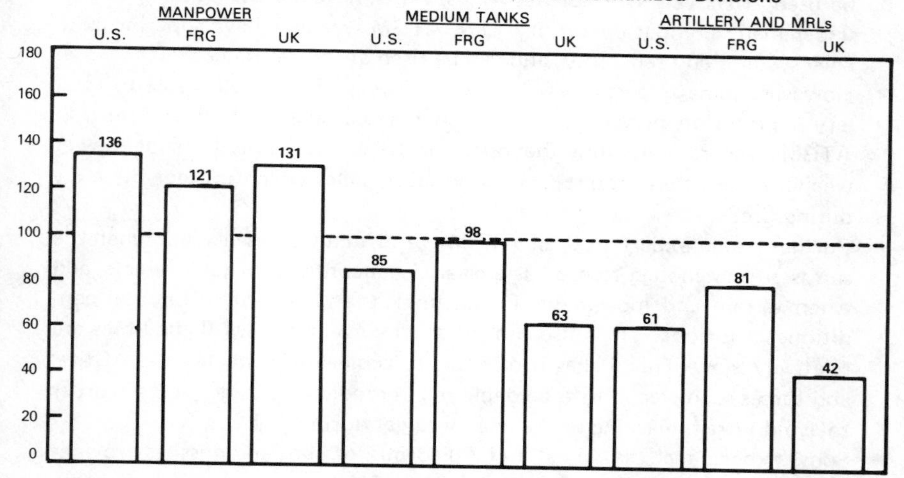

Source: Anthony H. Cordesman

THE TANK, ARMORED VEHICLE, AND ANTI-TANK WEAPONS BALANCE

The broad trends in the balance between NATO and Warsaw Pact tank, armored vehicle, and major anti-tank weapons is shown in Chart Six. Warsaw Pact strength in all categories has been increasing faster than NATO's. Further, NATO has no quantitative advantage in anti-tank weapons with high effectiveness against modern tanks. NATO traded more ATGMs for fewer anti-tank guns, but the Warsaw Pact chose to increase both.

Individual U.S. divisions, however, still have striking superiority over Soviet divisions in anti-tank weapons, depending on how systems are counted. Unfortunately, views differ sharply over whether this kind of count is correct, and over the qualitative capabilities of what is counted. The major differences in expert opinion can be summarized as follows:

- Soviet ATGM take longer to fly and have narrow range windows. They cannot be brought on target at ranges less than 300-500 meters. They require extremely difficult joystick guidance at long ranges, which makes it almost impossible to achieve hits. There is doubt that such weapons can be used effectively from Soviet AFVs, which have narrow visibility angles, under many combat conditions. U.S. ATGM weapons have superior guidance systems and reliability, and can be fired at shorter ranges.
- However, some experts feel that ATGM weapons are vulnerable to artillery suppression because they rely on manual operation under fire. U.S. ATGMs generally require the operator to expose himself while Soviet weapons are often mounted in armored vehicles which protect the crew during fire.
- Further, some experts feel all ATGM are difficult to use except when the sun is strong enough to provide a clear visual contrast for tank targets, and when terrain and movement do not mask the target—often unusual conditions in Europe. They also note that all current ATGM flight times are relatively slow. This makes it difficult to achieve reasonable rates of fire, and forces such weapons to be deployed in predictable target areas in order to avoid terrain masking of the enemy target during flight.
- Some experts feel the Soviet and Allied mix of guns and missiles provides the high rates of fire missing in U.S. forces. They note acknowledged weaknesses in the reliability, lethality, and range of the U.S. Law infantry anti-tank rocket, and cite the proven effectiveness of Soviet unguided infantry weapons. They feel the count is heavily biased in favor of NATO.

COMPARATIVE ARMORED MOBILITY

Some experts feel that the data in Chart Six also disguise major changes in Soviet capability which are still only beginning to show up in Warsaw Pact forces numbers, and weaknesses in the NATO tank mix. Both Dr. Phillip Karber and John Erickson, for example, feel that the Soviets are shifting away from their old emphasis on mass and tank superiority to maneuver, mobility, and "daring thrusts." This would be a shift in Soviet strategy from the classic breakthrough to the "indirect approach," or fluid penetration of the weaknesses in enemy defense followed by envelopment in de-

CHART SIX

Central European Balance
Armored Vehicles 1965-1975

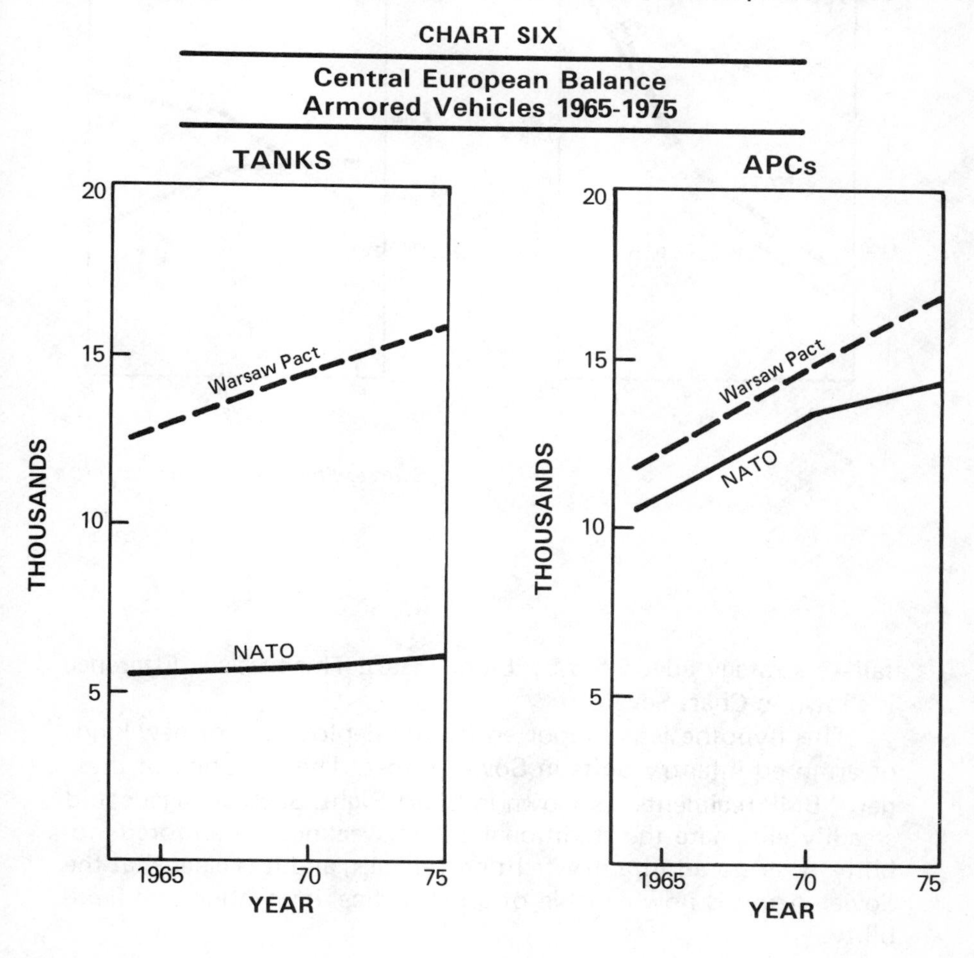

CHART SIX (Continued)

Central European Balance Armored Vehicles 1965-1975

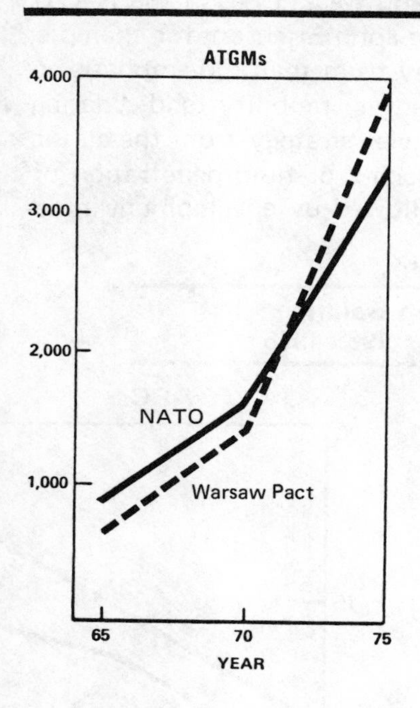

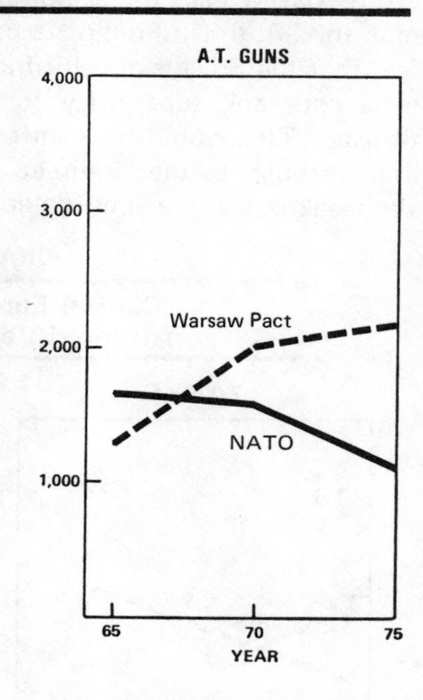

Source: Phillip A. Karber, op. cit.

tail—a strategy advocated by Liddell Hart. This tactical difference is shown in Chart Seven.

This hypothesis is supported by the deployment of new kinds of armored infantry units in Soviet forces. The evolution of these new "BMP regiments" is shown in Chart Eight. Such changes could steadily eliminate the traditional Soviet weakness in armored mobility or armored infantry fighting vehicles, and it is clear that the Soviet Army is now capable of great tactical innovation and flexibility.

CHART SEVEN

New Soviet Emphasis on Pre-Emptive Maneuver

A. 1945 to 1969: Reliance on the Classic Breakthrough

- LINEAR DEFENSE
- MASS FOR ATTACK
- CONCENTRATED FIREPOWER
- DIVISION/ARMY LEVEL

B. 1974: The Concept of "Daring Thrusts"

- GRANULAR DEFENSE
- MEETING ENGAGEMENT
- MANEUVER ON MULTIPLE AXES
- REGIMENTAL LEVEL

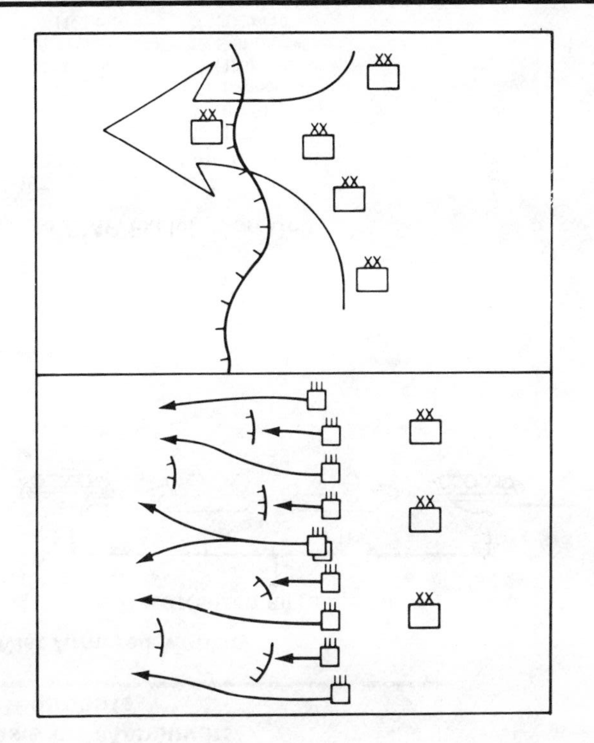

Source: Phillip A. Karber, "The Tactical Revolution in Soviet Military Doctrine," March 2, 1977. Appeared in EUROPÄISCHE WERKUNDE, *June, 1977, pp. 365–74.*

CHART EIGHT

The New Soviet Army Emphasis on Maneuver: Deployment of BMP Regiments

Karber Estimate of Changes in Soviet Armored Mobility

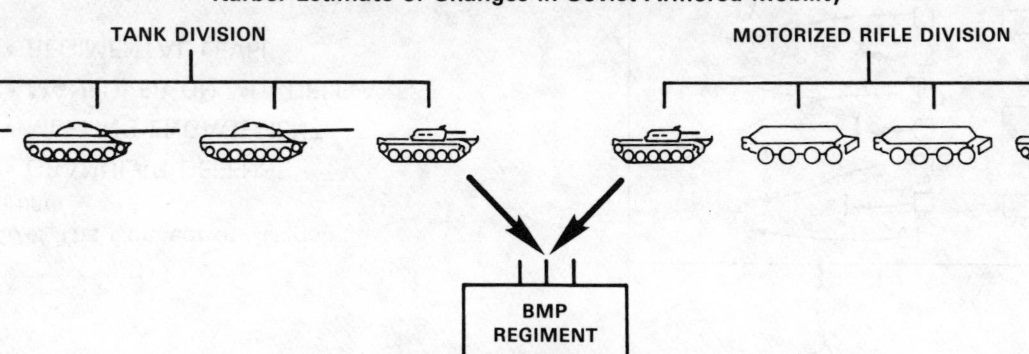

Conversion of Motorized Rifle Regiment to BMP Exploitation Unit

	1967 MRR	1972 MRR	1977 BMP Regiment
Medium Tanks	31 (T-54/55)	31 (T-62)	40 (T-72)*
Armored Personnel Carriers	60 (BTR-152)	90 (BTR-60)	105 (BMP)
ATGM (Man Portable & Vehicle Mounted)	27 (Swatter)	27 (Sagger)	132 (Sagger)
Mortar	9 (82mm)	18 (120mm)	18 (120mm)
Artillery	0	6 Towed (122mm)	18 Self-propelled (122mm)
SP Air Defense	0	4 (ZSU-23/4)	8 (4-ZSU-23/4 + 4-SH-9)**
Lift Capacity	180 Tons	270 Tons	350 Tons

*Up to 51 more tanks attached from independent tank battalion.
**Possible addition of SA-8 battery.

For additional data, see: Donnelly, C., "Military/Political Infrastructure," SOVIET WAR MACHINE, 1976 and Erickson, J., "Trends in the Soviet Combined-Arms Concept," STRATEGIC REVIEW, Winter, 1977.

Source: Phillip A. Karber, "Tactical Revolution," March 2, 1977

CHART EIGHT (Continued)

The New Soviet Army Emphasis on Maneuver: Deployment of BMP Regiments

ERICKSON ESTIMATE OF CHANGES	SOVIET MOTORIZED RIFLE REGIMENTS	
	OLD STYLE	NEW STYLE
Personnel, Officers and Men	BTR-60PB REGIMENT 2,400	BMP REGIMENT 2,300
Armored Vehicles		
AFVs (T-62)	40	40
PT-76	3	5
BMP	—	102
BTR-60PB	105	—
BRDM	34	28
Fire Support		
122-mm (self-propelled)	—	6+ +
122-mm D-30	18	—
120-mm (mortars)	18	18
ZSU-23/2	6	—
ZSU-23/4	4	4
SA-7	30	30
SA-9	4	4
RPG-7	197	267
SPG-9	6	6
BRDM (SAGGER/SWATTER)	9	9
SAGGER (manpack)	12	12
Combat Support		
Minelayers	3	3
KMT-4	9	9
Bulldozers	4	4
Trench diggers	3	3
Motorboat	1	1
MTU (bridging)	1	1
K-MM	4	4
Vehicles(excluding trailers)	560	520
Transport Capacity	270 tons	350 tons

Source: Adapted from John Erickson, "Trends in the Soviet Combined-Arms Concept," p. 6

TRENDS IN ARTILLERY AND INDIRECT FIREPOWER CAPABILITY

The overall trends in artillery strength are shown in Chart Nine. They show that the Warsaw Pact has sharply improved its overall lead over NATO since the mid-1960s, although NATO is now improving at a rate slightly faster than the Warsaw Pact. This chart may, however, disguise a temporary slow down in Warsaw Pact increases in strength while Soviet self-propelled artillery goes into large scale production.

A few experts now believe that the fact that Warsaw Pact would be an attacker would also allow it to select a given NATO Corps Sector and that it has improved to the point where it could concentrate its artillery tubes and ammunition in such numbers that it could achieve almost the same effect in suppressing NATO defenses as tactical nuclear weapons.

They also point to the fact that most NATO armies would face an agonizing dilemma trying to defend under such conditions. NATO could only attempt cohesive defense by constantly transmitting, using radios which the Soviets are well equipped and trained to "DF" or target. If NATO forces did transmit, they would then greatly improve the effectiveness of Soviet artillery. These experts also feel the reliance of some NATO armies, like the U.S., on unprotected ATGM launchers would make them highly vulnerable to artillery suppression. These views, however, are highly controversial.

There is broader agreement that the qualitative trend favors the Warsaw Pact in many areas other than artillery munitions. Most Warsaw Pact artillery is superior in range, rates of fire, and barrel life. Modern Warsaw Pact towed weapons are highly mobile and easy to deploy even when they are not self-propelled. Obsolete and underpowered artillery vehicles are disappearing from Soviet forces. The Warsaw Pact is also acquiring a wide range of modern command and control and target acquisition vehicles which may be equal in quality to those of NATO.

Further, NATO divisions have no advantage over Soviet divisions in artillery numbers. This is shown in Chart Ten.

Yet NATO could rapidly develop a major advantage over the Warsaw Pact if three current development efforts are successful:

CHART NINE

Central European Balance
Fire Support Weapons—1965-1975

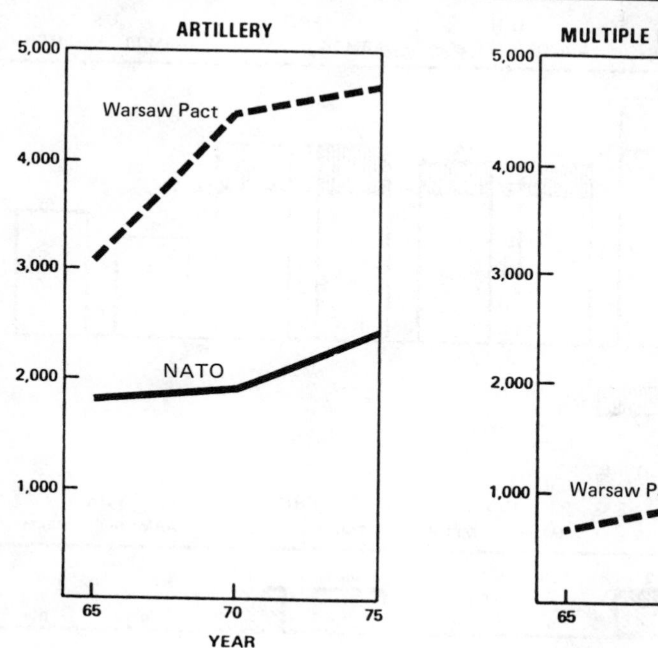

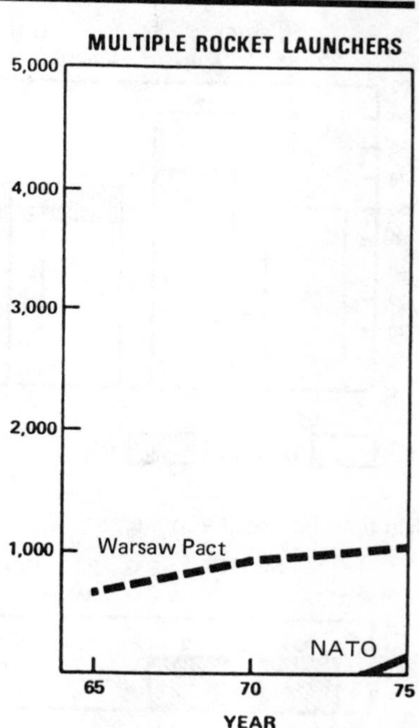

Source: Phillip A. Karber, op. cit.

- Cannon Launched Guided Projectiles could give NATO artillery massive new lethality against Soviet armor.
- Developmental U.S. anti-tank mine rounds and glide bombs would rapidly change artillery and air "fire and target" capabilities against Soviet tanks.
- NATO armies may procure multiple rocket launchers and other artillery capable of mass fires of anti-tank mine rounds.

The future success of these developments is still unclear, as is the capability of the U.S.S.R. to match them within a given period.

CHART TEN

NATO and Soviet Division Artillery Strength

A. Medium and Heavy Artillery

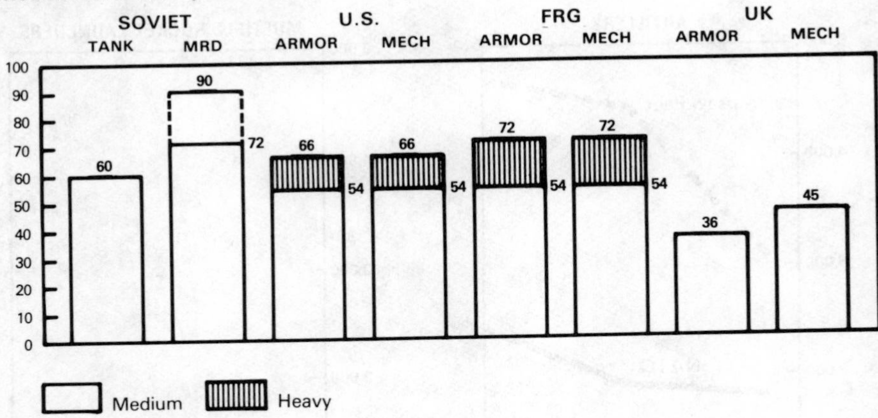

B. Multiple Rocket Launchers

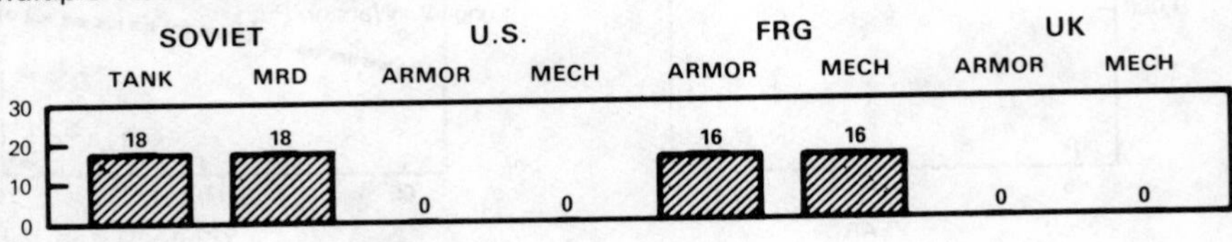

C. Heavy Mortars

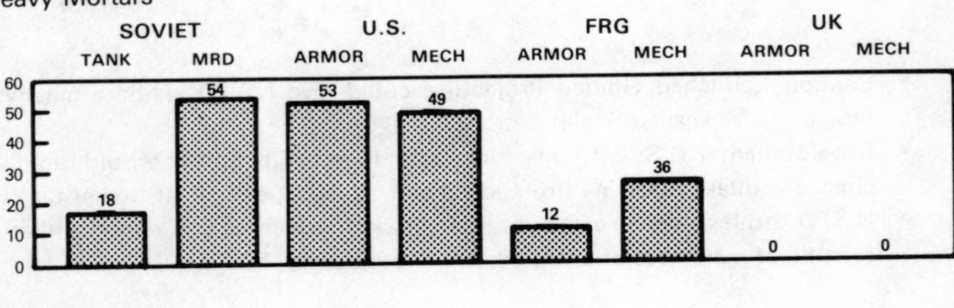

CHART TEN (Continued)

D. Total Artillery and Mortar Strength

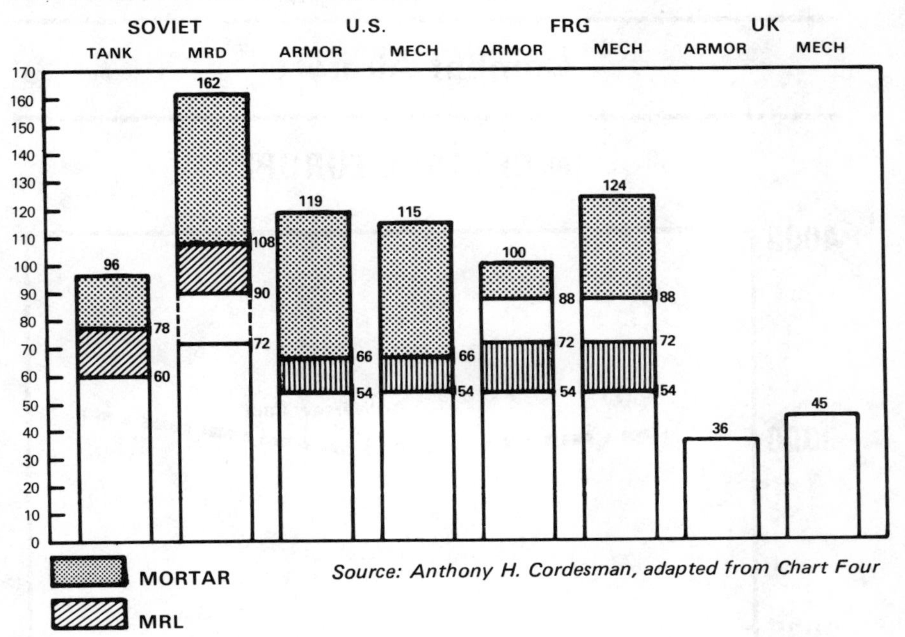

Source: Anthony H. Cordesman, adapted from Chart Four

NATO AND WARSAW PACT TACTICAL AIRCRAFT STRENGTH IN THE CENTER REGION

Chart Eleven shows the overall trends in both tactical aircraft strength and land based air defense strength in the Center Region.

It should be noted, however, that Chart Eleven potentially disguises the actual tactical air balance. It does not include 405 tactical aircraft in France, large numbers of U.S. and U.K. aircraft in the U.K., or 116 tactical aircraft in Denmark. All these NATO aircraft have the range-payload to be effective in Center Region missions. Warsaw Pact aircraft now counted in the forward area, however, lack meaningful mission capability if they must fly Lo-Lo-High or Lo-Lo-Lo attack profiles over NATO territory from their peacetime bases. Both sides also have massive reserves in the U.S. and U.S.S.R., and the U.S. forces in CONUS are generally far better trained and equipped than those of the U.S.S.R. and probably capable of much more effective strategic mobility.

CHART ELEVEN

Combat Aircraft

IN CENTRAL EUROPE

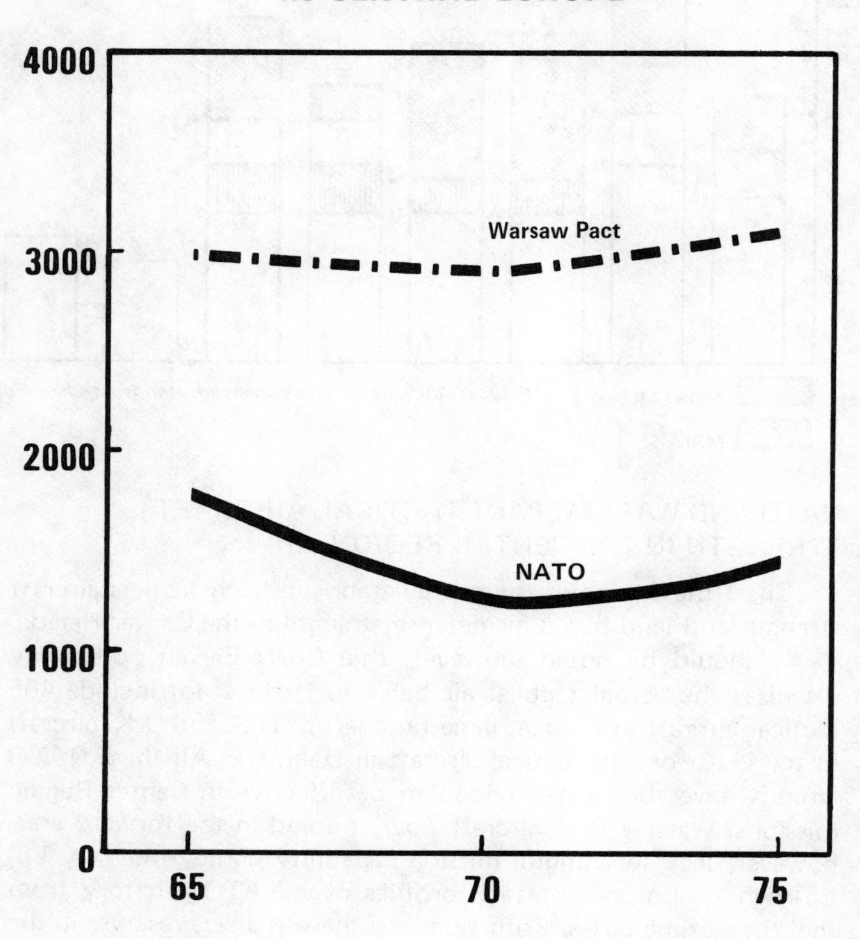

CHART ELEVEN (Continued)

Air Defense Weapons—1965-1975

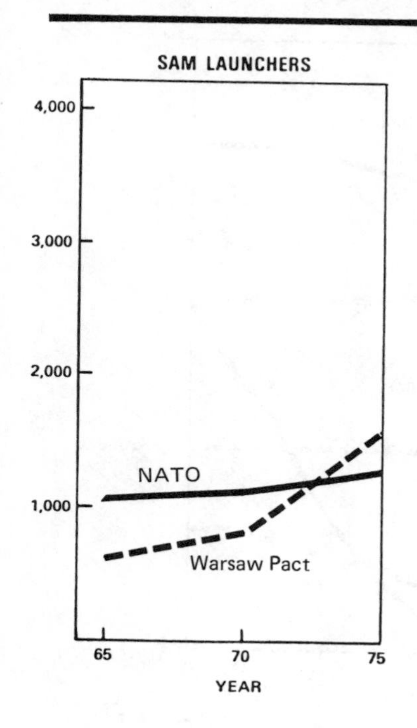

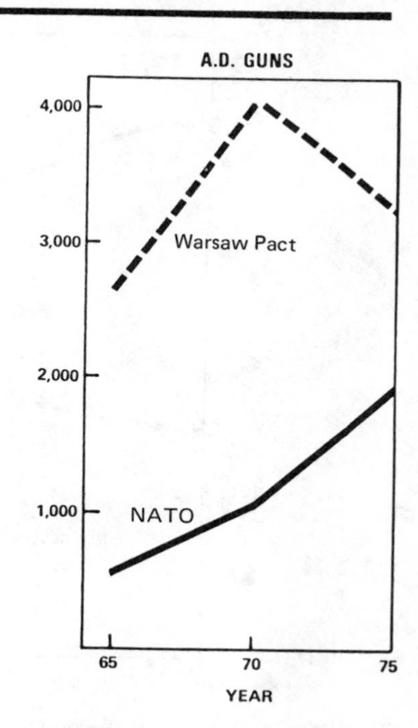

Source: Phillip A. Karber, op. cit.

Collins has already discussed the rapid changes taking place in Soviet air forces and the fact that new Soviet aircraft are coming to equal U.S. aircraft in range-payload. This modernization is compared with the rate of change in U.S. forces in Chart Twelve, and it is clear that the Soviets will soon convert virtually all of their force structure while U.S. modernization is moving much more slowly.

Experts generally agree that the U.S. A-10, F-15, and F-16 will be more effective than their Soviet counterparts when deployed. They do not agree on whether U.S. air-to-surface missiles will have much effectiveness against Warsaw Pact armor under the poor weather and visibility conditions which are common in central

CHART TWELVE

US/USSR FIGHTERS
(Central Region)

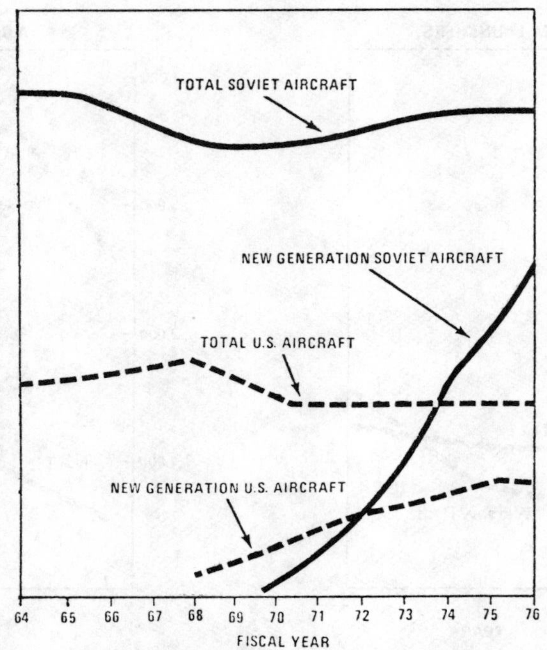

* U. S. FIGURES INCLUDE AIRCRAFT BASED IN UK: SOVIET FIGURES DO NOT INCLUDE AIRCRAFT BASED IN WESTERN MILITARY DISTRICTS

Source: DOD Annual Report, *p. 89*

Europe or over the quality of the passive defenses, shelter programs, basing systems, or AC&W and GCE/GCI systems on both sides. There are particularly sharp arguments over the future value of AWAC's, and the best way of upgrading NATO's overall air command and control systems and tactics.

Chart Thirteen shows the U.S. also has a clear advantage over the U.S.S.R. in terms of Allied aircraft quality, and this is not offset by any non-Soviet Warsaw Pact advantage in aircraft numbers.

Finally, Chart Eleven shows an overall trend in land based air defense which is consistently in favor of the Warsaw Pact, and may understate the current Soviet advantage. The decline in Warsaw

CHART THIRTEEN

Non-U.S. NATO and Non-Soviet Warsaw Pact Tactical Aircraft[7]

NATO[3]				WARSAW PACT[8]			
Country	Type	Mission	No.	No.	Type	Mission	Country
Belgium[4]	F-104G	FBA	36	80	Su-7	FGA	Czechoslovakia
	Mirage V	FBA	54	36	Mig-17	FGA	
	F-104G	AWX	36	42	Mig-21	FGA	
	Mirage V	Recce	18	250	Mig-15-21/ L-29	AD	
			144	72	Mig-21/L-29	Recce	
Canada	CF-104D		48	480			
France[1]	Mirage IIIE	FBA	140	35	Mig-17	FGA	GDR
	Mirage V	FBA	48	270	Mig-21	AD	
	Jaguar	FBA	120	115	L-29/Mig-21	Train	
	F-100D	FBA	10	12	Mig-21	Recce	
	Mirage IIIR/RD	Recce	58	432			
	Mirage IIIB/BE/C	OCU	30				
	Jaguar A/E	OCU	15				
			421				
FRG[5]	F-4F	FBA	60	6	IL-28	LBA	Poland
	F-104G	FBA	144	160	Mig-17	FGA	
	G-91	FBA	84	30	SU 7	FGA	
	T-4F	AWX	60	28	Su-20	FGA	
	TF-104G	OCU	18	100	Mig-17	AD	
	G-91T	OCU	55	340	Mig-21	AD	
	RF-4E	Recce	88	72	Mig-15/21	Recce	
			509	5	IL-28	Recce	
Netherlands[6]	F-104G	FB	36	(385)	Iskra, Mig 15/17/21, IL-28	Train	
	NF-5A/B	FB	54	741			
	F-104G	AWX	36				
	RF-104G	Recce	18				
	NF-5B	Train	20				
			164				
United Kingdom	Vulcan	MBS	50	30	Su-7	FB	Hungary (Not included in totals as does not seem to be included in Warsaw Pact plans)
	Buccaneer	LBS	56	30	Mig 17/19	FB	
	Harrier GR-3	FBA	48	116	Mig 21	AD	
	Jaguar GR-1	FBA	72	31	Mig 21, IL 29	Train	
	Lightning F-6	AWX	24	(207)			
	Phantom FG-1/FGR-2	AWX	84				
	Vulcan SR-10	NBR	10				
	Jaguar GR-1	Recce	24				
	Canberra PR-7/9	Recce	24				
	Nimrod	ECM/AEW/MR	58				
	Canberra B-6	ECM	4				
			454[2]				
TOTAL			1740	1653			

[1] Does not include 156 nuclear bombers and homeland defense fighters, or 44 Navy fighters.
[2] Does not include 90 combat "conversion" aircraft.
[3] Does not include 20 FX-35XD FBA, 40 F-100 D/F FBA, 40 F-104G AWX, and 16 RF-35XD in Denmark.
[4] 116 F-16 and 33 Alphajet on order.
[5] 10 F-4F and 175 Alphajet on order.
[6] 84 F-16 on order.
[7] Totals do not include low quality training aircraft.
[8] Does not include 19 IL-14 Recce Aircraft.

Adapted from the I.I.S.S., THE MILITARY BALANCE, *1977–1978*

Pact air defense guns seems to result from a conversion to new armored vehicles with SAMs and radar-guided AA guns which are far more effective than the unguided light AA guns they replace.

Further, Soviet land based air defense weapons are now arguably equal in quality to NATO weapons. The Soviet land based air defense force relies on a mix of different overlapping systems, unlike NATO which tends to rely on HAWK and SHORAD, but collectively the Soviet system seems equal or superior in quality to that of any given NATO Corps sector.

NATO AND WARSAW PACT BUILD-UP CAPABILITIES

Both the United States and U.S.S.R. have extensive land and air forces located in their home territories which can be used to reinforce NATO and the Warsaw Pact. In addition, most European countries in both Alliances have combat-ready units deployed away from the front, reserve combat units, extensive additional reserve manpower, and reserve aircraft.

Accordingly, the NATO and Warsaw Pact balance is highly sensitive to the assumptions made about how these forces can be deployed at the front within a given period and how effective they will be once they are deployed.

NATO BUILD-UP CAPABILITIES

An unclassified estimate of NATO build-up capabilities in the Center Region is shown in Chart Fourteen. This estimate has significant inaccuracies, but it gives a good picture of how different the reserve and build-up capabilities of NATO nations really are. In summary, each nation's forces may be described as follows:

- The Belgian forces have limited capability. They at most approximate a single Soviet Category II or III division in fighting power, and probably are so uncertain in training and readiness that they would have little value against combat ready Warsaw Pact armor.
- Canadian forces would be useful on the flanks, but are unequipped for armored warfare.
- Danish reserve units may be useful in local defense roles, but have only token capability against armor.
- French active forces in France are being reorganized and have not been properly modernized in recent years.

CHART FOURTEEN

NATO Build-Up Forces

Country	Location	Combat Ready	Reserves
1. Belgium	Belgium	1 Mech Bde 5 Recce & Mech Bns. } a/ 1 Paracommando Rgt. }	1 Mech. Bde. 1 Mot. Inf. Bde.
2. Canada	Canada	1 Combat Group b/	
3. Denmark	Denmark		5 Mech. Bns. 21 Inf. Bns. 7 Arty Bns. ? ATK sqns.
4. France	France	3 Mech. Divs. c/	2 Inf. Divs. 1 Alpine Divs. 1 Airportable Divs. 11 Armored Car Rgts. 2 Mot. Inf. Regts. 2 Parachute Bns. 10 Infantry Bns.
5. FRG	FRG	1-2 Bdes. Requiring Reserve Fill In.	6 Home Defense Bdes.
6. Netherlands	Netherlands		1 Arm. Bde. } 2 Inf. Bdes. } ? Lt. Inf. Bdes.
7. United Kingdom	United Kingdom	1 Div. of 3 Bdes. e/ 1 Para Bde. of 2 Bns. 1 Mobile Bn. Gp. 1 SAS Regt. 1 Inf. Bn. 13 Inf. Units	2 SAS Regts. 38 Inf. Bns. 11 Ulster Bns.

a/ I.I.S.S. counts as brigade equivalent
b/ I.I.S.S reports two brigades, but these do not move in published Canadian Plans
c/ 485 tanks in force
d/ Some elements of divisional organization
e/ Forces now being significantly reduced

CHART FOURTEEN (Continued)

Country	Location	Combat Ready	Reserves
8. United States	United States	**Reforger with Prepositioned Equipment** 1 Armor Div. 1 Mech. Div. 1 Armed Cav. Rgt.	**National Guard** 2 Armor Div. 1 Mech. Div. 5 Inf. Div.
		Strategic Reserve 2 Armor Div. 2 Mech. Div.[f] 3 Inf. Div. 1 Airmobile Div. 1 Airborne Div.	3 Arm. Cav. Rgts.
		Marine Corps[g] 2 Lt. Mech. Div. Equiv. with 1 air wing ea.	**Army Reserves** 12 Training Divs. 3 Training Bdes.
		U.S. Air Force 7 sqn. on call and 100 + Dual-based Aircraft At Least 400-900 rapidly available fighters.[h]	**Air National Guard** About 900 Combat Aircraft[h] **Air Force Reserve** About 200 combat aircraft[h]

[f] One allocated specifically to reinforce Europe

[g] About 192 M-60 Tanks each

[h] Estimates range as high as 1,474-3,940 build-up aircraft. The lower end of this range includes aircraft in a high state of readiness. The high end includes all reserves plus available Navy and Marine forces. See R.L. Fisher. *Defending the Central Front,* Adelphi Paper 127, I.I.S.S., London, 1976. The U.S. plans significant shelter construction for such aircraft, and is negotiating extensive co-location agreements to use allied bases.

Source: Adapted from the I.I.S.S. THE MILITARY BALANCE, *1977–1978, various publications of the Congressional Budget Office, the* INTERNATIONAL DEFENSE REVIEW, AVIATION WEEK, *and Les Aspin, "A Surprise Attack on NATO—Refocusing the Debate,"* NATO REVIEW, *August 1977*

They are, however, probably still equivalent to 1½ to 2 Soviet divisions. French reserve units are undergoing a major re-structuring, but most presently have little value except in local defense missions.

- The West German (FRG) reserves have few combat units, but they have high overall capability. They provide the fillers and support strength that can rapidly expand Germany's twelve combat ready divisions and give them sustaining power. They are almost certainly the best organized and most effective reserve units in NATO.
- The three mechanized Dutch reserve units have done surprisingly well in mobilization and training tests. They might be equivalent in fighting power to half or more of a Soviet division.
- The United Kingdom forces normally deployed in the U.K. are being reduced, some are dispersed in Ireland, many are poorly equipped, and all are of uncertain value in modern armored warfare. They could play a useful role on the flanks, but they lack the tanks, AFUS, artillery, and air defense necessary to directly defend against Soviet armor. Their organization is the result of history and underfunding, not current military purpose.
- The United States can theoretically re-deploy four combat-ready brigades, and an armored calvary regiment to unit equipment prepositioned in Europe as part of its "Reforger" force. This force should be deployable in less than two weeks but its actual readiness is uncertain, and it is unclear whether its equipment has yet been replaced after the loss of much of the equipment to Israel following the October 1973 war.

The availability of the remaining U.S. units is highly controversial. DOD studies tend to use planning factors or authorized readiness, rather than measure actual capability. DOD movement models are generally unrealistic, and countless studies during the last ten years have pointed out that the rate of U.S. Army build-up in combat power in NATO against Soviet forces is far slower than seems required by the Soviet threat. U.S. Army readiness reporting also can grossly exaggerate the actual capability of units in CONUS, and unclassified Congressional Budget Office studies indicate overall manning levels disguise significant shortages of key specialists and that active U.S. Army units are missing elements of battalion size. Many of these U.S. units also lack some of their equipment, and recent unclassified Senate studies raise serious doubts about whether the Army has the ammunition, war reserve equipment, or spare parts to maintain even its Reforger elements in intensive armored combat.

The U.S. Army force structure may, in fact, be too large for the funding it is given, and may maintain more active combat units than could be employed in combat without extensive manpower call up, and long term equipment and munition production.

Unfortunately, unclassified DOD reporting makes it virtually impossible to tell which units could fight when, or which planned improvements are likely to be real and which are this year's promises and rhetoric. The U.S. Army also has an appalling record of monitoring its own readiness, and it is unclear that it can produce an objective build-up estimate even at its own command level. The U.S. Marine forces have considerable readiness, but it is doubtful they have the weaponry for deployment against Soviet armor.

In contrast, much of the U.S. Air Force could deploy rapidly to Europe. Thirteen squadrons, or about twenty percent of the total Air Forces available to augment forces in Europe are "dual based" and the U.S. has "Rapid Reaction" squadrons which could begin movement to NATO in two to three days. In addition, most of the 38 Reserve and National Guard units could move to Europe, and 13 of these squadrons have the same equipment as active U.S.A.F. units.

The main problems the U.S. Air Force would face in rapidly deploying more than a thousand aircraft would be lack of shelters, problems in C3 and EW capability, still uncertain ability to use Allied air bases, problems in poor weather and night operations, and lack of munitions parts, and O&M capability. The Air Force would probably be base and support limited in operating in Europe, rather than be limited by readiness or number of aircraft. Attrition of aircraft in combat might, however, allow reinforcement to maintain initial M-Day strength in a conventional war.

WARSAW PACT BUILD-UP FORCES

Warsaw Pact build-up forces are more standardized than those of NATO and the U.S.S.R. can move its units in the rear by road and rail in days to weeks, while the U.S. is limited to slower and far more complex moves by sea and air. A rough estimate of total Warsaw Pact build-up capabilities for the Center Region is shown in Chart Fifteen. It should be noted, however, that this estimate may sharply exaggerate the readiness of N.S.W.P. units, and Soviet build-up capabilities from the Western U.S.S.R.

Unfortunately, virtually no analysis exists of how Soviet units in the U.S.S.R. and Warsaw Pact Category II and III reserve divi-

CHART FIFTEEN

Warsaw Pact Build-Up Forces

A. Units in Eastern Europe

Country	Category I		Category II		Category III		Total	
	Tank Divisions	MRD and Other	Tank Divisions	MRD and Other	Tank Divisions	MRD and Other	Tank Divisions	MRD and Other
A. In GDR								
Soviet	10	10	—	—	—	—	10	10
GDR	2	4	—	—	—	—	2	4
Total	12	14	—	—	—	—	12	14
B. In Czechoslovakia								
Soviet	2	3	—	—	—	—	2	3
Czech	5	3⅓	—	2	—	—	5	5⅓
Total	7	6⅓	—	2	—	—	7	8⅓
C. In Poland								
Soviet	2	—	—	—	—	—	2	—
Polish	5	6	—	4[a]	—	—	5	10
Total	7	6	—	4[a]	—	—	7	10
D. Total								
Soviet	14	13	—	—	—	—	14	13
NSWP	12	13	—	4[a]	—	—	26	19
Total	26	26	—	4	—	—	26	32
E. Soviet in Hungary[h]	(2)	(2)	—	—	—	—	(2)	(2)
F. Soviet Units in the Belorussian, Baltic, and Carpathian Military Districts								
1. *Aviation Week* Estimate	14 Tank and 7 MRD/Airborne					9	14	16
2. I.I.S.S. Estimate							13	26
3. Total Including Soviet and NSWP in E. Europe	40 Tank and 37 MRD/Other				—	9	39	58[i]
G. Total Available Soviet Build-Up Forces in Hungary and the USSR[c/g]								
1. I.I.S.S.	13	31[b]	12	36[d]	11	38[e]	36	105[f]
H. Total Build-Up Strength								
1. Soviet	27	44[b]	12	36[d]	11	38[e]	50	118
2. NSWP	12	13	—	4[g]	—	—	26	17
3. Total	39	57	12	40	11	38	76	135

NOTE: Footnotes are on following page

CHART FIFTEEN (Continued)

A. Units in Eastern Europe (Continued)

I. Soviet Aircraft: I.I.S.S. estimates 400 additional fighters available in less that 72 hours. Other unclassified estimates range from 780-3900 combat ready Soviet aircraft.

---

[a] Two are Polish MRD
[b] Two airborne
[c] Includes four Soviet Category I divisions in Hungary
[d] 32 are MRD
[e] 37 are MRD
[f] 98 are MRD
[g] I.I.S.S. estimates 80 divisions in USSR are deployable in several weeks.
[h] Not normal counted in Central Region Threat
[i] 56 MRD
[j] R. L. Fisher provides this estimate in *Defending the Central Front.* The low end assumes 40% at Soviet Tactical Aviation is for the Central Front. The high end includes all Soviet Tactical Aviation.

NOTE: Category I divisions have about 100% of their equipment and 75-100% manning; Category II divisions have most or all of their fighting vehicles and 50–75% manning; Category III divisions have most of their fighting vehicles, and 33% manning or less.

Source: Adapted from I.I.S.S. THE MILITARY BALANCE, *1977–1978, and* AVIATION WEEK, *22 August 1977*

sions compare in combat effectiveness with NATO forces. It is obvious that the Warsaw Pact build-up has far more combat elements, tanks, artillery, and fighting vehicles than NATO, but even the combat ready divisions in the Western U.S.S.R. have an average of only 75% manning, and Soviet and non-Soviet reserve units have far lower manning levels, and uncertain equipment quality and maintenance. Further, all Warsaw Pact forces are dependent upon reserve support manning for much of their sustaining and support capability, and the quality of such support forces seems likely to be quite low in many cases.

The U.S. Army is only now beginning to conduct meaningful studies of how its combat and reserve forces compare in combat effectiveness with those of Warsaw Pact combat ready and reserve forces.

In the past, U.S. analysis has tended to assume that Soviet mobilization and reserve training programs would produce relatively effective forces, and to use movement times as the main limiting factor in measuring Soviet build-up or to degrade the effectiveness of the built-up units using more or less arbitrary planning factors. Little objective study has been made of whether the Soviets have the ammunition and supplies to effectively use such units. The U.S. Army, in fact, has measured its combat readiness against itself rather than against its potential enemies or the standards set by other armies. While simple, this measurement technique effectively decouples Army readiness planning from war fighting requirements.

Russian military literature contains considerable self-criticism about the lack of effectiveness of Soviet conscripts and reserves and problems in Soviet logistics. At the same time, however, a great deal of Soviet military writing on readiness has an almost mechanical character, and does not seem sensitive to the fact the Soviet reserve system may not work. Soviet writers often seem to assume that because someone has directed the system to work, and showpiece exercises have been conducted, it will work. Anyone familiar with other Soviet planning is likely to find such a conclusion doubtful.

As for Soviet air build-up planning, it has long been limited by the quality of Soviet aircraft and training. Soviet units in the U.S.S.R.,and many Warsaw Pact air units in Eastern Europe, have lacked the skilled pilots and aircraft performance to have a great deal of probable effectiveness at the front. It is also doubtful whether such build-up aircraft can be supported with suitable munitions, be sheltered and dispersed within suitable range of NATO, be given proper AC&W, and be given proper maintenance. While the Soviets often practice showpiece forward deployments from the U.S.S.R., the effectiveness of even these showpiece movements seems doubtful, and the German and British commentators who credit the Soviets with high forward deployment capability because of such demonstrations seem somewhat credulous.

The uncertain purpose of NATO and Warsaw Pact Reserve Structures in Chart Sixteen illustrates another problem in comparing NATO and Warsaw Pact build-up capabilities. It shows very

CHART SIXTEEN

Comparative NATO Allied and Non-Soviet Army Reserve Manpower

	Ready Reserves	Total Reserves	Total Reserves	Ready Res.	
Belgium	10,000	50,000	300,000	—	Czechoslovakia
Canada	14,900	29,100	200,000	—	GDR
Denmark	45,500	123,400	500,000	—	Poland
France	52,000	400,000	130,000	—	Hungary
FRG	504,000	1,056,000			
Netherlands	145,000	145,000			
United Kingdom	110,000	167,700			
	881,400	1,971,200	1,130,000		

Source: Adapted from the I.I.S.S., THE MILITARY BALANCE, *1977–1978*

rough estimates of the total mobilizable allied reserve manpower on each side.

More of the manpower is probably committed to combat and combat support units suited for modern armored warfare in the Warsaw Pact than in NATO, and some NATO reserve forces—such as the Danish, French, and British—have been underfunded so long that their value is questionable. At the same time, however, NATO has a reserve strength of nearly two million men, and this figure includes only reserves with units and wartime assignments. Such a build-up must have a major impact on the balance, but this impact is now almost impossible to estimate for any given conflict scenario. Both sides could more than double the number of men now deployed on each side in less than two weeks without U.S. or Soviet reinforcements, but little, if any, analysis exists of how this massive increase would affect the outcome of a conflict.

ASSESSING COMPARATIVE BUILD-UP CAPABILITY

It should be apparent from the previous discussion that little is really known about the comparative build-up capabilities of each side. Chart Seventeen provides an estimate as good as any. But it does not address the critical issues involved:

- What combat power and weapons are built-up? What do the manpower and unit counts for each side mean in terms of military contingency capability?
- What does support manpower do in terms of building up combat capability? What logistics and O&M capability exists to support build-up forces? What munitions POL and war reserve equipment exists to support them?
- How many aircraft can be used and supported at the front? What shelter, dispersal, AC-W, maintenance, and supply capability exists to make these aircraft combat effective?
- What kinds of warnings are likely? How soon would NATO react to such warnings?
- What would happen if war started before build-up began? How effective are each side's build-up plans once LOC's are under attack?
- Most importantly, what is the readiness and combat effectiveness of each side's build-up forces?

At least in the U.S., attempts to analyze build-up capabilities have foundered over the inability to (a) judge the combat effectiveness of mobilized and/or Category I Soviet forces in the Western Military Districts once they were moved forward, (b) judge the effectiveness of mobilized N.S.W.P. and NATO allied forces, and (c) translate the theoretical readiness of U.S. Army forces to move to Europe into actual readiness. During the last ten years both the U.S. and U.S.S.R. have greatly improved their air mobility, the U.S. has corrected many defects in its prepositioning of equip-

CHART SEVENTEEN

Comparative NATO and Warsaw Pact Build-Up Strength

A. Mobilization and Reinforcement (divisional manpower in 10,000S)[a]

	M	+7	+14	+21	+28	+35
NATO[a]						
Belgium	2.5	2.5	3.3	3.3	3.3	3.3
Britain[c]	4.1	4.5	5.9	6.6	6.6	6.6
Canada	0.3	0.3	0.3	0.3	0.3	0.3
Denmark[d]	1.7	2.0	2.0	2.0	2.0	2.0
France[e]	3.4	8.2	9.2	9.2	9.2	9.2
Luxembourg	—	—	—	—	—	—
Netherlands	3.0	3.0	5.0	5.0	5.0	5.0
United States[f]	8.4	9.0	12.2	14.9	15.8	18.6
West Germany[g]	18.0	18.5	21.0	21.0	21.0	21.0
Total	41.4	48.0	58.9	62.3	62.3	66.0
Warsaw Pact						
Czechoslovakia	8.3	8.3	10.3	10.3	10.3	10.3
East Germany	6.0	6.0	6.0	6.0	6.0	6.0
Poland	12.4	12.4	14.4	14.4	14.4	14.4
Soviet Union	29.7	39.6	62.7	62.7	62.7	62.7
Total	56.4	66.3	93.4	93.4	93.4	93.4

B. Mobilization and Reinforcement with Degradation Factor[h]
(divisional manpower in 10,000s)

	M	+7	+14	+21	+28	+35
NATO						
Warsaw Pact						
Czechoslovakia	8.3	8.3	9.4	9.5	9.7	9.8
East Germany	6.0	6.0	6.0	6.0	6.0	6.0
Poland	12.4	12.4	13.5	13.6	13.8	13.9
Soviet Union[i]	29.7	39.6	56.5	58.0	59.6	61.0
Total	56.4	66.3	85.4	87.1	89.1	90.7

Note: Only about 75,000 men of the forces shown at M+35 are in reserve units ir peacetime. All active units are assumed to be ready.

CHART SEVENTEEN (Continued)

C. Ratio of Warsaw Pact to NATO

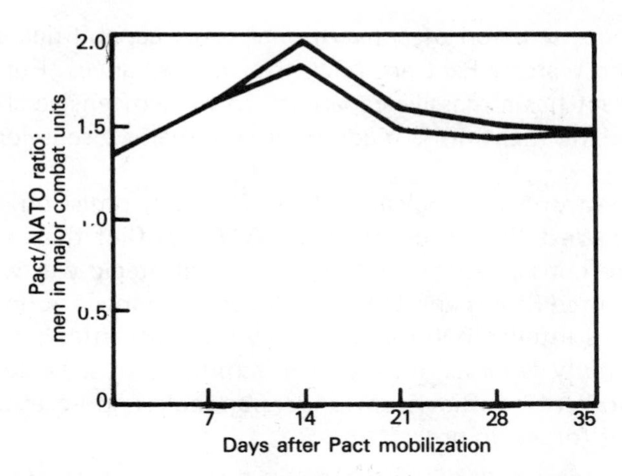

[a] NATO M-day is not the same as Pact M-day but lags behind it. Both Pact and NATO M-day strengths are of limited relevance, since both sides would have to end peacetime routine and move to combat positions.
[b] Companies, batallions and brigades for home defense (except six West German home defense groups) omitted.
[c] Active and TAVR units equivalent to five large infantry brigades arrive between M+5 and M+21. Other active and TAVR units in Britain are withheld for home defense and for Norway.
[d] Danish brigades are assumed to be rounded out by M+7. Danish regional battalions and home defense forces are not counted.
[e] French territorial forces are assumed withheld.
[f] 2⅓ division-eqivalents between M+7 and M+21, plus I/II division per day from M+5 onwards, plus a Marine division on M+30.
[g] One home defense group with assumed availability on M+7, five more by M+14.
[h] Degraded to fleet training and readiness problems.
[i] Divisions are assumed to arrive in category order (Category I first, etc.).

Source: Adapted from Robert Lucas Fisher, "Defending the Central Front: The Balance of Forces," Adelphi Paper 127, I.I.S.S., London, Autumn, 1976

ment, and the U.S.S.R. has improved its rail and road movement capability. Yet, there has been essentially no progress in getting any expert agreement as to the combat effectiveness of each side's forces once they arrive, or as to the exact movement rates each side is capable of, using real-world criteria.

COMPARING NATO AND WARSAW PACT THEATER NUCLEAR AND CHEMICAL AND BIOLOGICAL WARFARE CAPABILITIES

For all the discussion of conventional force capabilities, both NATO and the Warsaw Pact are nuclear armed alliances. Further, the Warsaw Pact has a massive capability to wage offensive chemical warfare. Little mention is made of this in many discussions of the balance.

There are strong ideological reasons for such omission. Few people in the West like to consider a NATO conflict that would have the same consequences in Europe as a strategic war would have in the United States and U.S.S.R. A general conventional war somehow seems intellectually and morally more acceptable to discuss, and certainly is easier to discuss in familiar terms. Tanks and aircraft methods of conflict seem understandable and acceptable; theater nuclear forces do not.

THE RISK OF THEATER NUCLEAR WAR

Yet, this approach to assessing the NATO balance may be highly misleading for several reasons:

- Chart Eighteen provides a rough portrayal of the theater nuclear forces on each side. There are about 6,000-7,000 U.S. nuclear weapons in Europe, less the U.S. Polaris submarines committed to NATO, and probably about 200-300 British and French weapons. The number of Warsaw Pact weapons is impossible to estimate, but both NATO and the Warsaw Pact deploy theater nuclear delivery systems in virtually every major combat unit, and much of the air strength on both sides is nuclear capable. The U.S. also has approximately 1,000 nuclear-capable naval aircraft in Europe, and NATO and the Warsaw Pact can deploy at least several hundred nuclear SLBM/SLCM. Theater nuclear capability permeates the force structures of both sides. There is no way nuclear and conventional capability can be decoupled.
- Both sides are vulnerable to a nuclear first strike under two conditions: While they are in their peacetime locations and once their forces are located and targetable in combat. While the cost of escalation to large scale theater nuclear war may be devastating, the side that strikes first gains a major and perhaps critical advantage. Both sides have given strong indication that they regard such escalation as likely or even inevitable once theaterwide conventional war is started.

CHART EIGHTEEN

NATO and Warsaw Pact Theater Nuclear Delivery Systems

Category	NATO						Warsaw Pact					
	Weapons Type	Operated By	Mach. No.	Yield	Max. Range in Miles	No. of Systems	No. of Systems	Max. Range in Miles	Yield	Mach. No.	Operated By	Weapons Type
M/IRBM	S-2	FR	–	150KT	1,875	18	20	2,685	3X25-50K	–	USSR	SS-20
							100	2,300	1MT	–	USSR	SS-5
							500	1,200	1MT	–	USSR	SS-4
							600					
SLBM/SLCM	Poseidon	U.S.	–	–	–	(c)/						
	Polaris A-3	UK	–	3X200KT	2,880	64	54	750	1-2MT		USSR	SS-N-5
	M-1	FR	–	200KT	1,550	32	27	350	1-2MT		USSR	SS-N-4
	M-2	FR	–	500KT	1,900	16	100-424(e)/	450	450KT		USSR	SS-N-3
	M-20	FR	–	1MT	3,000	16	181-506(e)/					
SRBM	Pershing	U.S.			450	108		500		–	USSR	Scaleboard
		FRG			450	72	750	185		–	USSR	SCUD B
						180		50		–	USSR	SCUD A
								185			NSWP(f)/	SCUD B
	Lance	U.S.				36	130	50			NSWP(f)/	SCUD A
		FRG				12?	880					
		Neth.				6?						
		Uk.				12?						
		It.				6?						
						72						
Strike Aircraft	F-105D	U.S. Land	2.3		2,100(b)/	350	65	3700(b)/		2.3	USSR	Backfire
	F-4C-J		2.4		2,300							
	F-111A/E		2.2-2.5		3,800		1,000	2,500		0.8	USSR Land	IL-28
	A-7D		0.9		3,400			900		1.7		SU-7
	A-4	U.S. Carrier	0.9		2,055	200		1,400		1.5		TU-22
	A-6A		0.9		3,255			1,150		2.2		MIG-21
	A-7A/B/E		0.9		3,400			1,800		2.5		MIG-23
	F-4		2.4		1,997			1,100		1.6		SU17/20
								1,800		2.3		SU-19
	Vulcan B2	UK	0.95		4,000	50						
	F-104	Allied	2.2		1,300	(a)/						
	F-4	UK&FRG	2.4		1,600	(a)/	(a)/	2,500		0.81	PO	IL-28
	Buccaneer	UK	0.95		2,000	70		900		1.7	CZ&PO	SU-7
	Mirage IV	FR	2.2		2,000	50		1,100		1.6	PO	SU-20
	Jaguar	UK&FR	1.1		1,000	192						
Battlefield(g)/	Sergeant	FRG	–	15-25KT	85	20	450	10-45(e)/	10-45KT		USSR	FROG 3-7
	Pluton	FR	–		75	24	200	10-45(e)/	10-45KT		NSWP(f)/	FROG 3-7
						86	650					
	Honest John	U.S.	–	25	25	0						
	Honest John	Allied	–	25	25	112						
	M-110 8"HOW	U.S.	–	1	10	(a)/						
	M-115 8"HOW	Allied	–	1	10	(a)/						
	M-109 155mm	U.S.	–	2	10	(a)/						

(a) Dual capable in conventional and theater nuclear missions. Number used in nuclear role unknown.
(b) Actual Mission Radius in realistic missions would be 30-40% of most of the ranges shown. Applies to *all* aircraft ranges shown.
(c) The U.S. allocates a large number of RVs to support NATO; since 1974, all U.S. strategic forces have been designed to provide extensive contingency support to NATO with considerable targeting flexibility.
(d) Number used in theater role unknown.
(e) Low end of range more probable.
(f) NSWP is Non-Soviet Warsaw Pact
(g) USSR may be deploying 203mm nuclear artillery.

Source: Adapted from the I.I.S.S. MILITARY BALANCE, 1977-1978, *various CBO publications, and unclassified working papers by Dr. Joseph Douglas of the Systems Planning Corporation*

- Although NATO emphasizes conventional options and controlled escalation, its theater nuclear force posture remains highly ambiguous, as to a lesser extent does that of the Warsaw Pact. Most of NATO's theater nuclear forces are not survivable. For complex reasons even most of its land-based missiles and artillery may be relatively easy to target. Accordingly, NATO theater nuclear forces have many of the features of a first strike posture. NATO cannot ride out a Warsaw Pact attack. It can only launch its aircraft, Pershing missiles, and many of its battlefield weapons if it strikes first. It would still have submarine and U.S. strategic forces after a Soviet attack on its theater systems but remains very vulnerable, and cannot inspire Warsaw Pact confidence in NATO restraint.

- Little unclassified evidence exists as to what the top Soviet leadership thinks or knows about theater nuclear war in Europe. It is clear, however, that fear of U.S. theater nuclear systems and West German use of such systems may cause Soviet leadership to show restraint once a conflict begins. Or they may feel that large scale war is so likely that they would accept use of nuclear weapons in a preemptive role as inevitable. Fear may cause restraint only up to a point; once war starts it may well encourage a major Soviet strike.
- The views of Warsaw Pact military planners, as distinguished from Soviet leaders, seem less ambiguous. In spite of Western discussion of Soviet interest in conventional options, there is virtually no significant evidence that Soviet military planners regard theater nuclear war as anything less than the probable outcome of a large scale conflict in Europe. Soviet writings and doctrine have been consistent on this point for more than a decade. There have been virtually no Soviet or Warsaw Pact exercises which did not make extensive use of nuclear weapons, and Soviet writings on the use of conventional forces in Europe show a grim consistency in using theater nuclear weapons as an extension of conventional firepower.
- The Warsaw Pact has put a massive effort into providing chemical-biological-radiation protection for its ground forces and some of its aircraft shelters. Calling its improved ground and air forces conventional is misleading. They are being designed at great cost for dual-capable operations in a nuclear environment.
- Soviet military operations research literature and tactical exercises do not reflect NATO's interest in carefully controlled escalation. It instead emphasizes the technical trade-offs between conventional, nuclear, and chemical weapons as interchangeable forms of firepower. The Soviets focus on how command and control, maneuver, and targeting can be made to work, and how nuclear weapons can ease the task of "conventional" forces.
- Unlike NATO, the Warsaw Pact has a major capability for chemical offensive warfare, and practices its use interchangeably with that of conventional and theater nuclear weapons. There is no evidence of moral constraints regarding the use of chemical weapons in Soviet military literature.
- It is far from clear how the Soviets view strategic parity, but they may feel it is possible to decouple theater and strategic nuclear war. The United States has firmly asserted that such decoupling is impossible, and Secretary of Defense James Schlesinger changed U.S. and NATO strategy and targeting to allow the controlled use of U.S. strategic forces in NATO wars,

but the Soviets may still feel the U.S. would not risk major strategic exchange over NATO.

- NATO's nuclear forces are co-located near NATO conventional land and air forces. A Warsaw Pact strike against NATO strike aircraft, Pershing missile units, and nuclear weapons storage sites would produce massive "bonus" damage to NATO's conventional capability even after deployment of NATO forces to their wartime positions.

None of these factors indicate a Soviet desire for theater nuclear war, but they collectively raise serious doubts about the wisdom of assuming that any large scale conflict would be conventional. The Soviets may well have a different view of the balance, and this is reflected in the effort they have put into improving theater nuclear capabilities over the last ten years.

THE EVOLVING NATURE OF SOVIET FORCES

Since the early 1960's, Soviet theater nuclear planning has moved from reliance on a few relatively low capability systems, and simplistic ideological analyses of how the Soviets might "win" a nuclear war, to a highly sophisticated use of operations research. Constant exercise testing planning is now more realistic and sophisticated at the tactical level than NATO's, and it reflects the rapidly improving nature of Warsaw Pact theater nuclear forces. For the first time, the Warsaw Pact may be acquiring enough capability relative to NATO to fight a successful theater war:

- The Soviets are now deploying the MIRV SS-20 IRBM. This weapon has potential ICBM range, and three nuclear warheads of at least 25-50 KT yield. It can be silo or mobile deployed, and can be rapidly reloaded and refired in some configurations. Most importantly it probably has the accuracy, reliability, and reaction times to allow the Soviets to make major improvements in their tactics and war plans for a first strike.
- The old SS-4 and SS-5 had poor reaction times, low accuracy and reliability, negligible re-targeting flexibility, poor reload capability, and fuel systems which took at least half an hour to an hour to bring to launch readiness as a force. They then could only be kept ready for limited periods, and probably suffered significant loss of reliability after a comparatively limited number of such alerts.
- The new SS-20 is based on current ICBM technology. It should allow launch on early warning against NATO strikes, allow far more reliable mass

strikes, provide improved ride-out capability, and allow an up to five-fold increase in the number of warheads targetable against NATO with greater lethality, targeting flexibility, and targeting capability per warhead. In contrast, the NATO Pershing missile is also improving in reaction time and accuracy, but is deployed in small numbers and has limited range. French IRBMs have limited accuracy and poor reliability, and are suitable only as countercity weapons.

- Soviet and Warsaw Pact doctrine has long stressed the use of strike aircraft, but the U.S.S.R. was forced until comparatively recently to rely on low performance aircraft with limited low altitude range and poor avionics. Deployment of the Backfire, the improved MIG-21, the MIG-23, the MIG-25, and the SU-17/20 gives the Pact the kind of aircraft necessary to support tactical strikes before or after an IRBM strike, and greatly reduces Warsaw Pact dependence on forward vulnerable airbases. NATO has long had such strike aircraft capability, but is now losing a key superiority over the Warsaw Pact.
- Aviation Week and other sources have indicated that the U.S.S.R. now has replacements for its aging and low performance FROG and SCUD missiles. The nature of these replacements is still unclear, but it seems certain that they will eliminate the grave problems in present Soviet SRBM and battlefield systems. The FROG and SCUD have poor reaction times, low reliability, poor operational accuracy, and primitive manual interface to Soviet targeting and command control system; and they are vulnerable to aircraft, HAWK, some SHORAD, and SAM-D. The U.S. Lance has similar improvements, but the Honest John rocket is inferior in range and performance to the FROG. The French Pluton is roughly equivalent to FROG F.
- Some press discussion has taken place on Soviet development of an anti-tactical Nuclear Ballistic Missile. Such development seems uncertain, but would be far simpler than an ABM, and is well within the state of the art. The SAM-D would have a substantial ATBM capability, and some existing Soviet SAMs may have limited ATBM capability.
- There is some evidence Soviets now have modern nuclear artillery capability. Primitive nuclear artillery was deployed during the Krushchev èra and then withdrawn. NATO has an extensive nuclear artillery capability.
- Soviet SLBM and SLCM systems may be improving to the point where they could supplement the Warsaw Pact's land and air forces with a significant capability against land targets. Older systems lack the accuracy, flexibility, and response times for such use. U.S. and U.K. SLBMs already have such capability. French SLBMs are unreliable and are suitable only for targeting against cities.
- Soviet command and control, and targeting systems, have evolved to the point where they should retain significant capability even after the damage imposed by high levels of theater nuclear war. Soviet communications and

command capabilities now are probably able to function without radio transmission. In contrast, NATO can now only fight a nuclear war by clearly identifying the precise location and type of many of its delivery systems in ways the U.S.S.R. can easily target.

- Soviet "conventional" forces are acquiring what may be effective CBR protection for the first time. They also practice dispersal and maneuver effectively in ways that might allow effective operation in a theater nuclear environment. NATO has virtually no CBR capability, and conducts only token TNW maneuver practice, much of which would be ineffective if high yield weapons were employed.

In short, the Soviets are gradually building up a force that could launch an effective first strike; are massively increasing the long range delivery systems and warheads they can target against NATO; and are acquiring the shorter range systems, forces, and command and control capability necessary for conflict management in a tactical nuclear war. There are other indications, however, that this means the Soviets would not launch such a war unless mutual deterrence failed for some unpredictable reason. But they have greatly improved their advantage in striking NATO first and in a massive escalation if they feel theater nuclear war cannot otherwise be avoided.

Further, for all their progress in tactics and force improvements, the Soviet military have shown a disturbing lack of flexibility in their strategy for theater war. They write as if a theater nuclear war had no political objective other than destroying the enemy and seizing control of NATO territory. They show little overt interest in terminating theater nuclear wars before they have devastating effects, in reducing collateral damage except where it is convenient for war fighting, or in controlling escalation. There are disturbing indications that militarism and ideology coincide; that Soviet military plan to fight out a "final victory" against capitalism even if it is far more costly than accepting a settlement with NATO.

NATO: THEATER NUCLEAR STRATEGY WITHOUT FORCE IMPROVEMENT

The trends in NATO have been very different. During the late 1950's, NATO viewed theater nuclear weapons as a means of offsetting the Warsaw Pact's superiority in conventional firepower. It had a near monopoly in such weapons, and the backing of massive U.S. strategic superiority.

But, NATO never developed the command and control, communications, targeting doctrine, CBR protection, or maneuver capability to actually fight a theater nuclear war. NATO strategy instead relied on a "conventional tripwire" which led to a nuclear explosion when NATO's conventional defense failed. This strategy made sense when the enemy could not shoot back, and when such a high level of deterrence was provided.

By the early 1960s, however, Soviet forces had improved to the point where the risks of NATO's "tripwire" strategy were obvious. Accordingly, Secretary McNamara shifted the U.S. and NATO towards a strategy of "flexible response" beginning with his speech at the NATO meeting in Athens in May 1962, and again in his Ann Arbor speech in June 1962.

However, NATO did not actually formally agree to the new strategy until 1966–1967, and then NATO's "flexible response" existed only on paper. Secretary McNamara left the actual targeting and war planning to the Joint Staff, and the Joint Staff did nothing. NATO had no theater nuclear "flexible response." It improved its conventional options, but its theater nuclear plans consisted of nothing more than "massively preplanned strikes" linked to the use of the U.S. SIOP.[1]

Unfortunately, this may well have given the Soviets the impression of being faced by an Alliance oriented towards a preemptive first strike with its vulnerable tactical aircraft which claimed to emphasize conventional options only to mislead Soviet planners.

The shift in strategy to flexible response also had another effect. Conventional options were given so much emphasis that the new strategy led to a virtual suspension of U.S. and NATO thinking about how to fight with theater nuclear weapons. NATO discussed theater nuclear planning at the political level, but this had little impact on the actual NATO war planning.

Further, for bureaucratic reasons, the focus of U.S. TNW planning shifted from seeking superiority to minimizing the consequences of theater nuclear war and reducing collateral damage.

[1]Lynn E. Davis, Limited Nuclear Options, Adelphi Paper 121, I.I.S.S., London, Winter 1975/76.

For nearly a decade, the few U.S. planners interested in theater nuclear systems attempted the hopeless task of reducing the yield and effect of theater nuclear weapons to the point where the Administration would accept them as a "legitimate" tool for use in "flexible response."

Ironically, therefore, "flexible response" locked NATO into reliance on nuclear strike aircraft and war plans which offered little real alternative to a massive all-out strike against Warsaw Pact targets. These plans included targets that would have entailed massive damage to cities deep in Soviet territory.

It was not until January of 1974 that anything was done to create an actual change in U.S. and NATO strategy. At this point, Secretary of Defense Schlesinger started serious planning of real limited nuclear options or LNOs. Work is still underway to restructure NATO forces, targeting plans, and command and control to make this shift away from "spasm war" effective.

Unfortunately, the rhetoric behind the shift in U.S. and NATO strategy to LNOs was so confusing that it was not apparent to many policy makers outside the Pentagon that it was intended to do little more than make "flexible response" a reality. In general terms, "LNOs" meant NATO was to move from reliance on conventional war fighting options, and deterrence of theater nuclear war by the threat of a massive tactical nuclear strike, to a strategy of controlled escalation at any required level of conflict. NATO was to deter, control, and de-escalate, and terminate Warsaw Pact conventional attacks and Soviet and Warsaw Pact military targets with a carefully chosen level of nuclear escalation.

This strategy was to be implemented by:

- Reducing the Warsaw Pact incentive to launch a preemptive theater strike against NATO strike aircraft by specifically allocating SLBM warheads for NATO theater nuclear missions.
- Seeking to design options which could combine a number of complex objectives: preventing the U.S.S.R. from obtaining a NATO objective while holding Warsaw Pact military targets hostage; convincing the Warsaw Pact that the NATO attack was limited and conflict could be terminated on mutually acceptable terms; gaining control of escalation by being able to inflict superior damage on the Warsaw Pact at any higher level of escalation; and denying the U.S.S.R. the ability to attack other objectives at low cost.

- Creating escalation barriers—uses of NATO forces where the Soviets could not use their forces at a matching level of intensity, but could not escalate without suffering more than NATO.
- Selectively using U.S. strategic forces to strike at targets in Warsaw Pact territory outside the U.S.S.R., or in highly selective ways at targets within the U.S.S.R., to deter prolongation of Soviet conventional or theater nuclear attacks on NATO.
- Striking at Warsaw Pact ground forces with weapons that could control collateral damage to NATO territory, but destroy enough Pact forces to halt a Warsaw Pact advance.
- Halting a major Warsaw Pact conventional or tactical nuclear break through NATO's forward defenses with nuclear escalation if NATO's conventional defenses decisively failed.

There is no question that the new LNO strategy was a significant step forward, and that it greatly improved the sophistication of NATO theater nuclear war planning. Unfortunately, the new strategy also had two significant defects: First, it depended on a matching Soviet response, and the Warsaw Pact "rationally" terminating conflict because this was the most favorable alternative. Second, much of the rhetoric behind this sophisticated game of "chicken" assumed U.S. superiority in strategic forces, and NATO superiority in theater nuclear forces. It assumed that the U.S. and NATO would be superior at virtually every level of nuclear escalation, and could create situations where any feasible Warsaw Pact option for sustaining conflict or escalation would result in more damage to the Warsaw Pact than NATO.

The planning behind the new strategy did not really examine whether NATO had or could achieve such superiority, what would happen if the Warsaw Pact applied the same strategic doctrine to NATO with superior forces, or what would happen if the Warsaw Pact initiated war at a high level of theater nuclear conflict and destroyed much of NATO's theater nuclear strength in a first strike.

Much of the material released on the new strategy by the Department of Defense has an almost theological cast in which the Soviets are not allowed to play by the same rules with potentially superior forces. Further, NATO has not matched its improvements in tactics, command and control, and targeting under the new strategy with improvements in its nuclear forces. The limited improvements in Pershing, NATO strike aircraft, and deployment of Lance have not done much to reduce the vulnerability of most NATO forces.

NATO must now rely on the British Polaris and U.S. Poseidon for survivable systems, and this might make it possible for the Soviets to "de-couple" a nuclear strike on NATO land and air forces since Britain or the United States would then have to ecalate to what are normally viewed as strategic forces.

Many NATO improvements are also so sensitive to warning and wartime deployment that they may increase the military benefits to the Soviets of a first strike on NATO before it deploys.

The NATO emphasis on reducing collateral damage under the new LNO strategy also has inherent major problems. First, the U.S.S.R. has shown little overt interest in similar reductions, and there can be no such thing as unilateral theater nuclear restraint. Second, most U.S. weapons can only be used in ways which reduce the collateral damage effects of individual weapons by restricting what can be targeted. Some studies indicate that larger numbers of weapons are then required, and that the cumulative effect may be more damaging to civil populations than fewer and higher yield strikes. Ironically, the result may be that NATO would create as much collateral damage as the Warsaw Pact. Enhanced radiation weapons may alter this situation, but this is not clear from available unclassified studies.

THE CURRENT BALANCE OF THEATER NUCLEAR AND CHEMICAL WARFARE IN EUROPE

It is unclear whether the present uncertainties in the theater nuclear balance create a significant risk. Deterrence seems highly secure, and the Schlesinger doctrine should create serious doubt in the minds of Soviet planners about the nature of a U.S. and NATO response even if the Warsaw Pact could execute a successful preemptive attack.

Further, even the most successful Warsaw Pact nuclear attack would still have devastating results. It could involve hundreds to thousands of nuclear weapons, and while Europe would certainly survive such a conflict, it could kill hundreds of thousands to tens of millions on both sides. Such a war might be "won," but it is scarcely likely to offer the Soviets an attractive option as long as other courses of action do not threaten their vital interests.

The problem, in view of the history of the twentieth century, is what should happen if for any irrational reason war should occur. Some aspects of Soviet military writing and planning are particularly frightening in this regard because they are so rigidly tactical in strategic aim. They make no distinction between conventional forces and theater nuclear chemical forces, and concentrate single-mindedly on destroying NATO forces and seizing control of Europe, rather than flexibility and control.

Hopefully, Soviet political leaders plan differently, and have the command and control capability to enforce this view on their generals. Yet, it is difficult to imagine how the Soviet leaders now interpret the Schlesinger strategy in view of the past contradiction between the McNamara strategy and NATO war planning that lasted for more than a decade, and in view of the still vulnerable "first strike" oriented posture of many of NATO's theater nuclear forces.

[Chapter 7]

Wrap-up

THE PRESENT BALANCE IN PERSPECTIVE

Quantitative changes in U.S. and Soviet armed forces since 1970 favor the Soviet Union, with scattered exceptions. U.S. *qualitative* leads, less pronounced than in the past, cannot completely compensate. The full significance of such trends is beyond the scope of this unclassified study, but a few findings stand out.

First and foremost, essential equivalence with the Soviets across the conflict spectrum is neither necessary nor desirable. Opposing *capabilities* would be quite different in many cases if *forces* were identical in size and structure, because circumstances and strategies are dissimilar. Possible Soviet preemptive employment of ICBMs and/or anti-ship cruise missiles, for example, would pose counterforce perils far out of proportion to reciprocal U.S. actions constrained by second-strike concepts.

Procuring unnecessary or unusable assets simply for the sake of achieving apparent parity (a strong argument supporting some U.S. systems) seems a poor way to influence perceptions among enemies, allies, or uncommitted countries. Practical power would pay greater dividends.

It is clear, however, that current trends curtail U.S. freedom of action. The upshot impinges increasingly on American abilities to deter attacks against the United States, defend this country if deterrence fails, and safeguard associates whose security is closely linked with our own.

In the strategic nuclear field, Mutual Assured Destruction seems less mutual than it was in 1970. U.S. conventional capabilities have also faded, in comparison with those of the Soviet Union. U.S. land, sea, and air forces alike consequently would be hard pressed to support NATO plans at existing levels and cope concurrently with large-scale contingencies, including those caused or sustained by the Kremlin.

THE PROBLEM OF U.S. PRIORITIES

A second set of asymmetries, between complementary U.S. systems, bears directly on abilities to maintain a satisfactory military balance with the Soviet Union. Airlift, for example, is adequate to commit forces quickly, but sealift is insufficient to sustain them in crucial contingencies. Marines are ample, but lack amphibious lift.

Congress and the Executive Branch, with focus on forces and funds, clash annually over expensive programs, each considered essentially in isolation, and each with a life of its own. Interrelationships with enemy systems and each other commonly get short shrift, except for matching counts with Moscow. Political expediency and technological excellence rather than real requirements too often are the tests. Misplaced priorities consequently stress inessentials in many important cases while slighting critical sectors.

Problems will prevail as long as U.S. decision-makers bank on bigger budgets to cure defense ills without reference to better strategy. More money will ensure substantial improvements only if connected with force sufficiency factors that match meaningful U.S. ends with measured means in ways that minimize risks and reinforce weak spots.[1]

[1] For force sufficiency factors, see U.S. Congress, Senate, *The United States/Soviet Military Balance,* pp. 41, 47-54.

[Appendix A]

Nicknames for Selected Weapons Systems

UNITED STATES

United States Armed Service Designation				Nickname
Missiles				
	Strategic			
		ICBM		
			LGM-30F/G	Minuteman II/III
			LGM-25C	Titan II
		SLBM		
			UGM-27C	Polaris A-3
			UGM-73A	Poseidon C-3
			UGM-93A	Trident C-4
	Air-to-Surface			
			AGM-12	Bullpup
			AGM-45	Shrike
			AGM-65	Maverick
			AGM-69A	SRAM
			AGM-78	Standard
			AGM-84	Harpoon
			AGM-86A	ALCM
	Surface-to-Air			
			CIM-10B	BOMARC
			MIM-14B	Nike-Hercules
			MIM-23A	Hawk
			XMIM-104	SAM-D

Category	Subcategory	Designation	Nickname
	Shipborne		
		AIM-54L	Phoenix
		RIM-8G	Talos
		RIM-24B	Tartar
		RIM-2F	Terrier
	Air-to-Air		
		AIR-2	Genie
		AIM-4	Falcon
		AIM-7	Sparrow
		AIM-9	Sidewinder
Aircraft			
	Bombers		
		B-52	Stratofortress
		FB-111	(None)
	Fighter/Attack		
		F-4	Phantom
		F-14	Tomcat
		F-15	Eagle
		F-16	(None)
		F-100	Super Sabre
		F-101	Voodoo
		F-102	Delta Dagger
		F-105	Thunderchief
		F-106	Delta Dart
		F-111	(None)
	Naval Aircraft		
		F-8	Crusader
		A-4	Skyhawk
		A-6	Intruder
		A-7	Corsair
		AV-8	Harrier
		P-3	Orion
		S-2	Tracker
		S-3	Viking
	Helicopters		
		UH-1	Iroquois
		CH-34	Choctaw
		CH-47	Chinook
		CH-54	Flying Crane
		SH-3	Sea King
		CH-46	Sea Knight
		CH-53	Sea Stallion
	Airlift		
		C-5	Galaxy
		C-7	Caribou
		C-97	Stratofreighter
		C-119	Flying Boxcar
		C-123	Provider
		C-124	Globemaster

	C-130	Hercules
	C-141	Starlifter
	KC-135	Stratotanker
Artillery		
	M-101	105mm Howitzer
	M-102	105mm Howitzer
	M-107	175mm Gun
	M-109	155mm Howitzer
	M-110	8-inch Howitzer
	M-114	155mm Howitzer
Anti-Tank		
	M-47	Dragon
	BGM-71A	TOW
Armor		
	M-48	(None)
	M-60	(None)
	M-551	Sheridan
	M-113	(None)
	LVTP-7	(None)

SOVIET UNION

	Numerical Designation	Nickname
Missiles		
Strategic		
ICBM		
	SS-7	Saddler
	SS-8	Sasin
	SS-9	Scarp
	SS-11	Sego
	SS-13	Savage
	SS-17	(None)
	SS-18	(None)
	SS-19	(None)
IRBM		
	SS-5	Skean
	SS-20	(None)
MRBM		
	SS-4	Sandal
SLBM		
	SS-N-4	Sark
	SS-N-5	Serb
	SS-N-6	Sawfly
	SS-N-8	Sasin
	SS-N-X17	(None)
	SS-N-X18	(None)

	Air-to-Surface		
		AS-2	Kipper
		AS-3	Kangaroo
		AS-4	Kitchen
		AS-5	Kelt
		AS-6	(None)
		AS-7	Kerry
	Surface-to-Air		
		SA-1	Guild
		SA-2	Guideline
		SA-3	Goa
		SA-4	Ganef
		SA-5	Gammon
		ABM	Galosh
	Shipborne		
		SA-N-1	Goa
		SA-N-3	(None)
		SA-N-4	(None)
	Air-to-Air		
		AA-1	Alkali
		AA-2	Atoll
		AA-3	Anab
		AA-5	Ash
		AA-6	Acrid
	Tactical Shipborne		
		SS-N-2	Styx
		SS-N-3	Shaddock
		SS-N-9	(None)
		SS-N-10	(None)
		SS-N-11	(None)
		SS-N-13	(None)
		SS-N-14	(None)
Aircraft			
	Bombers		
		TU-16	Badger
		TU-22	Blinder
		TU-95	Bear
		TU-	Backfire
		M-4	Bison
		IL-28	Beagle
	Fighter/Attack		
		MIG-17	Fresco
		MIG-19	Farmer
		MIG-21	Fishbed
		MIG-23	Flogger
		MIG-25	Foxbat
		SU-7	Fitter-A
		SU-17	Fitter-C
		SU-9	Fishpot

Category	Designation	Name
Recon/Intercept		
	SU-15	Flagon
	TU-28	Fiddler
	YAK-25	Mangrove
	YAK-28	Firebar (Brewer)
Naval Aircraft		
	BE-6	Madge
	BE-12	Mail
	IL-38	May
	YAK-36	Forger (VTOL)
Helicopter (Naval)		
	KA-25	Hormone
	MI-4	Hound
	MI-24	Hind-A
Airlift		
	AN-12	Cub
	AN-22	Cock
	IL-76	Candid
Artillery		
	M-55	203.2mm
	S-23	180mm
	M-43	152mm
	M-46	130mm
	M-1938	122mm
	M-1955	100mm
	M-1975	152mm(SP)
	ZSU-23-4	23mm (AA)
Anti-Tank		
	AT-2	Swatter
	AT-3	Sagger
Armor		
Tanks		
	T-55	(None)
	T-62	(None)
	T-72	(None)
APC/AFV		
	BTR-50P	(None)
	BTR-60P	(None)
	BTR-152	(None)
	BMP-76PB	(None)

Note: Naval ships are identified by class throughout the study.

[Appendix B]

Abbreviations

AAA	Anti-aircraft armament
ABM	Anti-ballistic missile
ACR	Armored cavalry regiment
ADCOM	Air Defense Command
AFCENT	Allied Forces Central Europe
AFV	Armored fighting vehicle
ALBM	Air-launched ballistic missile
ALCM	Air-launched cruise missile
APC	Armored personnel carrier
ARNG	Army National Guard
ARPA	Advanced Research Projects Agency
ASM	Air-to-surface missile
ASW	Anti-submarine warfare
AT	Anti-tank
CBR	Chemical, biological and radiological
CD	Civil Defense
CEP	Circular errors probable; Circle of equal probability
CGS	Command and general support
CIA	Central Intelligence Agency
CONUS	Continental United States
CRAF	Civil Reserve Air Fleet
DCSLOG	Deputy Chief of Staff for Logistics
DIA	Defense Intelligence Agency
DMC	Defense Manpower Commission
DOD	Department of Defense

ECCM	Electronic counter-counter-measures
ECM	Electronic counter-measures
EUCOM	U.S. European Command
EUSC	Effective U.S. controlled (fleet)
GNP	Gross National Product
HE	High Explosive
ICBM	Intercontinental ballistic missile
IR	Infrared
IRBM	Intermediate-range ballistic missile
K	Warhead lethality factor
KT	Kiloton
LAMPS	Light airborne multi-purpose system
LAW	Light anti-tank weapon
LOC	Line of communication
MAC	Military Airlift Command
MAD	Magnetic anomoly detector
MARAD	Maritime Administration
MaRV	Maneuverable reentry vehicle
MICV	Mechanized infantry combat vehicle
MIRV	Multiple independently-targetable reentry vehicle
mm	Millimeter
MRBM	Medium-range ballistic missile
MRV	Multiple reentry vehicle
MSC	Military Sealift Command
MT	Megaton
NADGE	NATO air defense ground environment
NATO	North Atlantic Treaty Organization
NCA	National command authorities
NCO	Non-commissioned officer
NDRF	National Defense Reserve Fleet
Nuke	Nuclear
OASD (Compt)	Office of the Assistant Secretary of Defense, Comptroller
OASD (M & RA)	Office of the Assistant Secretary of Defense for Manpower and Reserve Affairs
PACAF	Pacific Armed Forces
PGM	Precision-guided munitions
PH	Passive homing
POMCUS	U.S. pre-positioned stocks
PONAST	Post-nuclear attack study
psi	Pounds per square inch

RDT&E	Research, Development, Test and Evaluation
R&D	Research and development
REDCON	Readiness condition
RRF	Ready Reserve Force
SAC	Strategic Air Command
SALT	Strategic Arms Limitations Talks
SAM	Surface-to-air missile
SEATO	Southeast Asia Treaty Organization
SLBM	Submarine-launched ballistic missile
SLCM	Sea-launched cruise missile
SRAM	Short-range attack missile
SRF	Strategic rocket forces
TAC	Tactical Air Command
UE	Unit equipment
USAF	United States Air Force
USAFE	United States Armed Forces Europe
USMC	United States Marine Corps
V/STOL	Vertical/short takeoff and landing
VTOL	Vertical takeoff and landing